Merchants of Medicines

Merchants of Medicines

The Commerce and Coercion of Health in Britain's Long Eighteenth Century

ZACHARY DORNER

The University of Chicago Press
Chicago and London

PUBLICATION OF THIS BOOK HAS BEEN AIDED BY A GRANT FROM THE BEVINGTON FUND.

The University of Chicago Press, Chicago 60637
The University of Chicago Press, Ltd., London

Published 2020
Printed in the United States of America

29 28 27 26 25 24 23 22 21 20   1 2 3 4 5

ISBN-13: 978-0-226-70680-1 (cloth)
ISBN-13: 978-0-226-70694-8 (e-book)
DOI: https://doi.org/10.7208/chicago/9780226706948.001.0001

Library of Congress Cataloging-in-Publication Data

Names: Dorner, Zack, author.
Title: Merchants of medicines : the commerce and coercion of health in Britain's long eighteenth century / Zachary Dorner.
Description: Chicago : University of Chicago Press, 2020. | Includes bibliographical references and index.
Identifiers: LCCN 2019044527 | ISBN 9780226706801 (cloth) | ISBN 9780226706948 (ebook)
Subjects: LCSH: Pharmaceutical industry—Great Britain—History. | Great Britain—Commerce—History.
Classification: LCC RS67.G8 D675 2020 | DDC 338.4/7615100941—dc23
LC record available at https://lccn.loc.gov/2019044527

♾ This paper meets the requirements of ANSI/NISO Z39.48-1992 (Permanence of Paper).

CONTENTS

FIGURES AND TABLES

Introduction

Consider the medicine chest. Over the long eighteenth century, these wooden cases filled with glass containers appeared in country estates, apothecary shops, sugar plantations, merchant vessels, military hospitals, timber camps, and trading posts. The wood, often from common species but sometimes more valuable ones, reflected the material wealth of the chest's owner and his connection to trade circuits that afforded access to Caribbean mahogany or New England pine. The glass vials, sometimes hundreds in a single chest, were by this time often of standard manufacture, rather than varying in size and shape, clearness, and color as they once had been. The vials' contents came from across the globe and passed through many hands: opium and China root from Asia, Peruvian bark and guaiacum from the Americas, gum arabic from West Africa, and a variety of plants, salts, and chemicals from Europe. These items, sometimes sweet, other times bitter, and often aromatic, were imbued with the hopes of those who prepared, sold, and ingested them. They could comfort a child bedridden with fever, clean the battlefield wound of a soldier, or coerce laborers to work in a field or aboard a ship.[1] Medicines were purchased with the profits of overseas empire, traveled according to its itineraries, and supported its goals. They offered a solution to the nearly ubiquitous labor shortages of the armed forces, trading companies, and plantations, while informing the categories of difference that organized such institutions. Medicines became a consideration of state power and an object of long-distance trade during the long eighteenth century, similar in some ways to other commodities of political-economic importance such as gunpowder or wood.[2] For those who could take advantage of the trade, medicines offered the prospect of power and wealth in a globalizing, but still highly local, world. For others, they were part of the mechanisms of enslavement. Emergent healthcare practices lumped mobile individuals—captives, slaves, soldiers, seamen, servants—into groups conducive to bulk treatment in a range of locations. This approach to long-distance medicine rested on principles of abstraction, commensurability, and interchangeability that drew on credit instruments and the slave trade alongside long-standing modes of trust and corporeal knowledge. A medieval world of personalized treatment seemed to be giving way to a modern world of universal subjects and mass markets at a time of rapid expansion of British state power abroad. This was no coincidence.

If the medicine chest is one artifact of this history, the ledger of a medicine manufacturer is another. These account books represented the health of a business, recording credits on one side and debts on the other. They offer paper traces of relationships, aspirations, and trust, while illustrating who could owe someone a debt and who could not. The records of Plough Court pharmacy, one of the most prolific manufacturers and exporters of medicines in eighteenth-century London, reside today inside the polished glass and steel of GlaxoSmithKline's West London headquarters: a far cry, at least sensorially, from the open flames and noxious fumes of the early modern pharmacy.[3] Though the trade in medicines has been studied less than other trades, its records reframe the story of medicines in the early modern world as a commercial one, exemplary of certain commodities' contributions to the day-to-day expressions of power.[4] Some historians have treated the eighteenth century as a wilderness waiting to be tamed by the medical reforms of a later age; it remains in the scholarship an era amid long-distance commerce and empire, but simultaneously of neither.[5] Recent work has begun to identify the historical connections between commerce, health, and power in broad perspectives.[6] In considering medicines as commodities, this book makes clear that over the span of several decades in the eighteenth century, manufactured medicines began to circulate at unprecedented scope and scale, reshaping healthcare practices and expectations for many. Business records, however, reveal only part of the story. Who appears in this archive does not represent the range of healers or patients who used medicines, nor does it fully capture the diversity of people's relationships with medicine. This book broadens the business archive to trace the infrastructures of healthcare that emanated from Europe and extended across the Atlantic and Indian Oceans.[7] Rather than examining what arrived in metropolitan locations as a way to discuss the codependence of science and commerce in an imperial frame, it identifies the commercial drivers of an emergent long-distance medicine in the commodities flowing out of such places.[8] In so doing, the book describes how material forces transformed both practices and ideas related to nature and human nature through the eighteenth-century medicine trade.

Though historians have tended to associate industrialization with the cotton mill, similar processes were occurring in the laboratory. The eighteenth-century term *laboratory*, like *workshop* or *factory*, encompassed the breadth of manufacturing, commercial, and intellectual work relating to medicines occurring inside.[9] Laboratories were hot, busy, and dangerous places where raw materials were converted into medicinal commodities. In general, early modern remedies divided into two categories: drugs and medicines—two words now often used interchangeably but that speak to the historical difference between raw materials and refined products, as well as two distinct but interrelated trades. Eighteenth-century merchants, healers, and tax collectors understood the category of *drug* to include any animal, vegetable, or mineral substance that could be used as a medicinal ingredient, whereas *medicine* referred to the manufactured goods prepared from those ingredients—similar, in many ways, to today's *pharmaceutical*.[10] One could ingest a drug directly (as a "simple"), but one could also, through a range of physical and chemical processes, make drugs into medicines with uses, values, or effects different from those of their constitutive parts. Britain's emerging industrial sector relied on imported, multifunctional drugs since many of them also served as dyes and catalysts in textile or metallurgical processes. In Latin, for example, the verb *medico* meant to medicate and to dye. Importation of drugs from Asia and the Americas to Europe rose at least twenty-five-fold in the seventeenth century, setting up the increasing production of medicines in the eighteenth.[11] As manufactured goods, medicines formed part of a consumer society in which everyday goods were more readily available to a wider range of people.[12]

Men and women across the early modern world sought to ward off discomfort and death, prompting an iterative approach to health rooted in desperation. Medicines embodied a range of hopes and fears. They offered palliation or even recovery from bodily harm or illness, as well as a way to control fertility, for instance. Ultimately, modern observers cannot recreate how people responded to medicines in the past even if some can be said to have transhistorical effects: ipecac presumably induced vomiting in the eighteenth century just as it does today.[13] The causes and consequences of a system for delivering these medicines, by contrast, occupy the chapters that follow. In an environment rife with sickness and pain, whether in London, Jamaica, or Madras, people ingested manufactured medicines for a range of physiological, political, and cultural reasons. They looked to one thing, then another and another, searching for answers to questions of health that remained complicated and deeply personal, influenced by humoral, emotional, economic, political, and spiritual interpretations of the body.[14] Given their material forms and intellectual contexts, medicines also offered convenient answers to questions of manpower commonly faced by plantation owners, ship captains, naval surgeons, and merchants. The expectation that certain medicines would offer answers to such questions at a transoceanic scale illustrates the convergence of capitalism, empire, and medicine.[15]

During a moment of profound change in the ways many people saw, understood, and moved through the world, manufactured medicines became global products. The British medicine trade at once sprung from these changes and helped produce them. To understand these items and the consequences of their circulation, this book applies the new histories of capitalism and empire to the history of science and medicine. As is well established, the period from the late seventeenth to the early nineteenth century saw the maturation of the British fiscal-military state, colonialism in South Asia and the Americas, the plantation complex, and a decentralized Atlantic trading system. It also witnessed the opening of epistemic possibilities: new credit instruments, categories of bodily difference, and recognition of human potential to bend nature to its ends.[16] Medicines braid these strands in a broad frame. The book follows medicines from their manufacture in Britain along trade routes to the edges of empire to reconsider what medicines were, what they did, and what they meant at those various stages. Rather than proceed purely chronologically, chapters take an episodic approach organized around different themes and geographies to connect chemists in London, enslaved laborers in the Caribbean, trading company employees in India, merchants in Philadelphia, and settlers in the New England woods. Because of the potential they represented, medicines held political-economic value for the institutions of early modern capitalism and empire, and thereby joined the lives—and the deaths—of seemingly disparate people. They also embodied concepts of state power, commercial integration, connectivity, bodily knowledge, authenticity, and trust that motivate the book's arguments.

At the book's core lies an observation about exchange. As more Britons struggled to survive overseas away from their usual healthways, more medicines than ever before were exported to help them do so. On its own, such an observation means little and explains even less. Recognizing that most of these medicines went to regimes reliant on a spectrum of unfree labor, however, brings into sharper relief the implications of this trade for the codependence of medicine, plantation agriculture, and military fiscalism on a global scale. The book argues that the British empire, particularly its commercial form, fundamentally reshaped medicine in the eighteenth century. As certain manufactured medicines became imperial commodities, they inaugurated a new kind of mass healthcare in markets comprised of slaves, soldiers, and trading company employees, which had significant consequences for how people conceived of their bodies, received care, and were inscribed with categories of difference. From an institutional standpoint, this scale of healthcare encouraged the kinds of abstraction constitutive of empire and capitalism. Seemingly portable, predictable, and commensurable medicines spread wildly because of what they promised for maintaining health, wealth, and authority in a world of increasing connectivity, deepening power disparities, and nearly unlimited labor needs.

The globalization of manufactured medicines owed to the long-distance trade, resource extraction, and enslavement formalized by European imperialism as well as financial developments in Europe. Medicines flowed through transoceanic channels irrespective of imperial boundaries during the eighteenth century, helping integrate the Atlantic world. In London, apothecaries, chemists, and druggists began to act like merchants to manufacture and distribute medicines at a scale that marked medicines as bulk, mobile commodities, offering efficacy across the geographies linked by the infrastructure of overseas empire. Planters, bankers, slave traders, trading company employees, and military officers supported the production of such medicines in London and soon in other Atlantic port cities as well. These merchants of medicines distributed another option within the iterative structure of healthcare that offered certain advantages within the British empire. Many people continued to make medicines, but most could not access the material benefits derived from distributing products to the plantation complex, armed forces, and trading companies. The trade owed much of its structure and expanding volume to patterns of unfree labor but just as critically reinforced the power dynamics crucial to maintaining them through the expectations that emerged from using medicines in those contexts.

As they transitioned from local to global commodities, manufactured medicines represented a shift from treating the person to treating the disease by more standardized means. This kind of bodily abstraction aligned with an imperial system based on commodifying people and nature. In such a system, considering the constituents of large patient groups as mostly interchangeable proved economically and logistically expedient given the sheer numbers of patients and the limited ability to treat them all. An embrace of this way of thinking by merchants of medicines and their customers drew on European intellectual trends to support the treatment of disease with bulk medicines rather than by individualized constitutional ones as had often been the practice. Such an approach to healthcare offered the possibility of treating anybody, whether free or unfree, old or young, poor or rich, suffering from a certain ailment with a similar pill or tincture from a laboratory in London. It also suggested a universal body across patients, a concept that confronted and was eventually foreclosed by emergent racial categories. By that time, however, medicines had become so entrenched in the day-to-day operations of imperial institutions and so central to global flows of people, goods, and credit that the erasure of one intellectual underpinning had little practical effect on the arrangement as a whole. A reliance on medicines remained into the next century despite credit crises, failures, frauds, and revolutions, as did the expectation that such commodities would bring health and wealth to some but not all.

In the early decades of the eighteenth century, the long history of using manufactured remedies departed from patterns that had reigned for centuries. Rather than adapting treatment to a patient's particular circumstances, some practitioners began to take a more standardized approach to treating illness. This turn away from particularity built on what some historians have recognized as a seventeenth-century medical revolution that departed from the study established by classical theorists. Patterns of medical spending across Britain tipped toward purchasing medicines expected to do something curative, though many forms of advice-based and home care persisted alongside the commercializing market for medical goods and services.[17] Such changes in the metropolitan marketplace also reflected the challenges of long-distance medicine in an age of empire and commerce: distant patients, long voyages, unfamiliar environments, shortages of practitioners, and impersonal interactions. While the commercialization of European medicine helped transform a relatively small-scale and parochial medicine trade into one with global scope, bulk scale, and state support, so too did the experience of long-distance trade and global mobility contribute to the transformation of the medical marketplace.

Eighteenth-century Britain was hardly the first time or place that medicines compounded from multiple ingredients were sold in bulk. The terms *pharmacy* and *chemist* trace to the end of the sixteenth century, and the antecedents of early modern pharmacy date to much earlier. People applied a variety of substances to their bodies on account of perceived therapeutic properties, although many would seem unfamiliar to a twenty-first-century patient.[18] The seventeenth century saw the beginning of significant shifts in the availability, composition, and marketing of care. The consumption of imported drugs became more widespread and was no longer limited to the English elite following an expansion of overseas trade and a proliferation of vendors. Popular print carried descriptions of remedies for sale, while more porous occupational boundaries encouraged diversification in the medicine trade. Potential patients could hope to purchase medicines or medical services in a commercial setting to supplement familial and folk care.[19] Many of the plants that proved most useful to European medical practitioners appeared in print from the early 1600s. Likewise, advances in chemical medicines emerged in the first half of the seventeenth century when Paracelsian and Van Helmontian chemistry spurred the adoption of new mineral, metallic, and chemical medicines, as well as the introduction of laboratory-produced ingredients. Chemical medicines gained popularity after 1670, a harbinger of how manufacturing would pervade European chemistry in the eighteenth century.[20]

The eighteenth century brought new attitudes toward medicines, as it did for many consumer goods on a global scale. People gained acquisitive aspirations as they worked more and had additional income to spend on nonessential items as a result. Healthcare itself was increasingly understood as a commodity to be bought and sold.[21] Several parliamentary decisions during the period from 1704 to 1730 demonstrated that the medicine trade's old boundaries no longer held. By midcentury, apothecaries, druggists, and chemists in London faced only nominal oversight, in contrast to elsewhere on the continent, and many of the occupational distinctions between medical practitioners faded.[22] Practitioners of all sorts began to traffic in medicines, contributing to a flourishing trade after 1700. A commercializing medicine trade both at home and abroad provided experiential impetus for several strains of thought that proved advantageous to an expanding trade. A gradual shift in conceptualizing disease—from an internal imbalance toward an entity attacking the body from outside—helped erode long-standing place-based and physiological requirements for treatment in favor of a more widespread take-this-for-that approach. Pharmacopoeias continued to expand with the influx of colonial produce, so patients could choose from a larger array of products while wholesale drug prices fell. Though some, such as the poor, still lacked regular access, recourse to medical care by adults of certain means had become nearly universal in cities across Europe by the end of the century according to some metrics. Commercial, charitable, and family care continued to coexist long after this time. Because of the iterative nature of much healthcare, a turn toward commercial medicine did not necessary imply a rejection of magical or spiritual healing practices either. Nevertheless, patterns in the consumption of medicines mirrored those of other goods as more and more people could afford them, forming part of a broader uptick in household consumerism. In short, the sick had begun to rely more heavily on commercial sources of medicine by 1800, though they still constituted only one part of the panoply of European medical practice.[23]

While the influence of commercial values on early modern medical practice is becoming better understood, the larger global picture of this interaction remains less so.[24] Medicinal plants had connected Asia, the Americas, and Europe for centuries via long-distance trade routes. In the early modern period, the search for new, valuable drugs—another form of resource extraction—motivated the political economy of the emergent European empires. The rapidly expanding scope and scale of manufactured medicine exports presented an additional layer to this project in the long eighteenth century. From one perspective, the British empire was organized to remove resources from colonial landscapes, the labor requirements of which placed significant monetary and political value on the maintenance of health. Medicines therefore appeared crucial in an empire based on extensive labor and land control.[25] As merchants of medicines turned their attention overseas to augment their local and regional businesses, they linked the growth of the medicine trade to the extension of British fiscal-military power on a global scale. Many labor-intensive institutions responsible for entrenching British territorial authority overseas—namely plantations, trading company settlements, and navy bases—became lucrative markets for medicines manufactured in London and contributed to a reconceptualization of the potential of bodies, medicines, and disease in the process.[26]

The British medicine trade operated within a similar legal regime as other manufacturing trades, but it also found large markets overseas owing to the slightly different status of medicines in the eyes of the mercantilist state. The eighteenth-century British state lacked many of the features of a strong state, but such constraints made it effective at exercising what power it did have.[27] State involvement increasingly extended into the realms of health and medicine through contracts and trade policy since England had little formal regulation of medicines, despite a long history of such activity on the continent. Amid a general wave of customs reform early in the century, officials removed the duty on exporting manufactured medicines and made it simpler to import several key ingredients in the 1720s.[28] Medicine exports generally followed some trends in overseas trade, while bucking or outpacing others. The volume of English trade grew at a steady rate over the eighteenth century. The period from 1745 until 1763, in particular, is seen as the most accelerated one for British overseas trade between the 1680s and 1780s. The Atlantic colonies emerged as an important early industrial export market for metal goods and woolens due to rising populations and purchasing power. Manufactured medicines comprised a small but expanding share of these kinds of exports.[29] Under these circumstances, medicine exports from London rose exponentially from the early 1700s until 1775, continuing even when British-manufactured exports generally struggled after 1765 due to falling European demand and prohibitive duties. By 1770, the medicine trade stretched across the Atlantic and Indian Oceans and was the fastest growing of any of the trades in British manufactures.[30] The volume of exports to slave societies by the mid-eighteenth century suggests an earlier turn toward European remedies as one option to maintain the health of enslaved workforces before the policies of self-sustaining reproduction of the 1780s, for example.[31] A particular region of Europe was certainly not on a foreordained path toward medical primacy when compared to the medical systems across Asia, Africa, and the Americas.[32] Nevertheless, British medicines gained traction in colonial locations during this period buttressed by military might and local hierarchies, which they in turn bolstered.

Like the medicines intended to allay them, illness and pain were daily experiences that reflected and reified socioeconomic distinctions across the British empire. While death came for all regardless of status or wealth, underfed slaves certainly were more vulnerable to some ailments than others in this system. The ubiquity of death and dying among colonial populations heightened the stakes of efforts to maintain health, which in turn meant that merchants, planters, and practitioners ordered medicines in advance of illness, permanently weaving them into the fabric of Atlantic trade. Though life expectancies in London improved during the second half of the eighteenth century, such was not the case in the Caribbean, where neither rank nor skin could protect against mosquitoes and the diseases they carried. Medical practitioners there worked urgently to combat the staggering mortality that greeted newcomers, planters and African captives both. The brutality, demographic turnover, malnutrition, overwork, warfare, and anxiety of the eighteenth-century Caribbean fueled the medicine trade to that region and informed larger concerns about manpower.[33] Though the sugar plantation presents a starker example of the lethal conditions of overseas empire, the British West Indies were not so different from New England or South Asia in terms of medical concerns.

The creation of colonial demand for manufactured medicines cannot be attributed to trends in real income or fashion among a broad middling population without considering the impact of overseas labor exploitation. In the medicine trade, enslavement provided markets for the mass consumption of manufactured commodities rather than raw materials, which is the configuration typically identified as contributing to Britain's industrial growth in the second half of the eighteenth century. Efforts to keep people alive were also crucial to the capacity of states and trading companies to project power. Planning a commercial or military venture therefore required consideration of medical supplies that could often seem like a determinant of victory or defeat. The various links of the commodity chain, encompassing production, distribution, and consumption, all shaped the material and discursive characteristics of the medicines that arrived overseas.[34] The challenges of raising sufficient capital to manufacture medicines at scale and sell them at a distance based solely on promises to pay, for example, pushed exporters to make their medicines seem more portable and impersonal. At the receiving end, manufactured medicines struck a chord with plantation owners, bureaucrats, and surgeons, who sought treatments for the populations they oversaw. While some could expect to choose their medicines, many unfree migrants were excluded from commercialized healthcare or forcibly subjected to it.[35]

The term *unfree migrant* offers a means to broadly compare people who comprised substantial overseas markets for medicines, such as African captives and impressed seamen, while also accounting for the differing circumstances of their mobility. It encompasses a spectrum of capture, impressment, enslavement, and migration experiences to underscore several similarities in healthcare from the perspective of the medicine trade. In Europe, the practitioner-patient relationship shaped diagnosis and treatment, so what happened when a patient was neither given a voice nor option to consent, as on a plantation or in a military hospital, offers another important site for the negotiation and contestation of medical practice. While such a generalizing term threatens to obscure the violence of and adaptations to some mobility, *unfree migrant* conveys the centrality of long-distance movement to the histories of capitalism, medicine, and empire, as opposed to other categories that could have been used, such as *consumer* or *subject*. The idea, in part, borrows from the term "involuntary consumers," initially used to scrutinize servants' dress, but also invokes a broader range of consumption on a transoceanic scale while giving more analytical attention to the consequences of long-distance movement itself.[36] Not everyone was mobile in the early modern world, nor wanted to be, so movement and the resultant separation from systems of healthcare stand out among the conditions making a kind of bulk, portable medicine appear increasingly necessary and convenient compared to other approaches in this moment.

Manufactured medicines offered a convenient solution to the omnipresent challenge of manpower by the logic that certain treatments could work on anybody irrespective of external characteristics or internal complexion. Yet an effective universal remedy would require treating all people afflicted by a particular disease alike, which was a rather radical proposition at the time, not least because it presumed an unchanging, ontological view of disease. While medicines designed to be used in such a way arrived in large quantities for populations of unfree migrants overseas, something similar can be said for hospital patients and the poor in England as well.[37] The focus here, however, remains the long-distance distribution of medicines. As this book shows, medicines played a material role in abstracting groups of people, not only as part of an epistemic project, as others have described, but as a physical reminder of that abstraction.[38] In other words, manpower concerns introduced manufactured medicines to issues of political economy, such as improvement, balance of trade, and white supremacy, that shaped the routines of the British empire. Medicines offered another opportunity for some to link enslaved men and women's embodied value to commercial pursuits in ways that eventually threatened recognition of their humanity.[39] The medicine trade also contributed to the process whereby learned, professional men attacked the expertise of women, the poor, and non-Europeans in efforts to remove medical care from the folk, domestic, and female spheres that had long defined it. Examples of women healers in Britain or black doctors in the Caribbean underscore that a variety of practitioners continued to ply their trades even as manufactured medicines and their infrastructures crowded other healthways.[40]

Medicines could be many different things across the Anglo-American world: a cause for hope, an intoxicant, a store of value, or a punishment. They offered the prospect of health and wealth, while also underwriting the gendered and racialized regimes at the core of globalizing, commercial empires. The physical and discursive violence inherent to their production, distribution, and consumption—and necessary for their transformation into global commodities—is often obscured by their association with healing. Long-distance exchange in conjunction with local conditions over the eighteenth century provoked a reconceptualization of medicines conducive to their bulk trade while still retaining hopes for broad efficacy. Tensions between health and violence, timeliness and trust, quality and profit, self-sufficiency and expediency shaped these objects and their attendant discursive frameworks. That state power, first British then American, asserted a particular thread of European corporate medical practice as normative while obscuring other systems relied on an expanding medicine trade during the long eighteenth century.[41] Medicines illustrate the complicated legacy of many global commodities and serve as a reminder of the dependencies between power, health, and violence that can often seem imperceptible or have been made to appear that way.

Imperial institutions enabled certain kinds of medicines to become global commodities at a particular moment in the eighteenth century, a development that shaped expectations about all sorts of things—bodies, diseases, profits, goods—in local and global contexts.[42] The material forms and uses of medicines arising from the exigencies of this imperial system of healthcare influenced the formulation of the common therapeutic body, including what features were seen to be unique and how medicine would target the disease as opposed to the individual.[43] Likewise, capitalism, another manner of seeing, valuing, knowing, and doing, required extensive infrastructure designed to alter people's understanding of themselves, their neighbors, and the world around them. Medicines both reflected and contributed to the slippages between discursive and material, economic and scientific, and public and private making that ideology possible and profitable at an unprecedented scale in the long eighteenth century. Medicines, in other words, stand at the center of the story about the changing relationships between nature, people, and objects. This book's story begins in London with the laboratory, the ledger, and the medicine chest.

ONE

Toward an Industry

Before a medicine could become an imperial commodity, it first had to be made. In eighteenth-century London, druggists, chemists, and apothecaries facilitated the bulk production of medicines by acting like merchants and manufacturers. These merchants of medicines united production and distribution under a single roof and at greater scale than their predecessors. This trend mirrored the ongoing shift away from domestic production across the British trades.[1] It also transformed the manufacture of medicines and fused their trade to the imperial system. To consolidate their businesses within a varied marketplace, merchants of medicines turned to recent accounting and organizational techniques while continuing to rely on long-standing ones of secrecy, personal credit, and patronage. From some perspectives, a more competitive medical marketplace bred an individualism that contrasted with markets held together by personal relationships. Yet competition also encouraged collaboration at a time when the general decline of regulation permitted many people to enter the medicine trade.[2] Centuries of medicinal practices, whether of wise women and men, of the rural and poor, or of non-Europeans, were in some cases marginalized and in others borrowed as the British medicine trade reached a scope and scale unlike ever before. Whether such an arrangement would hold in the colonies would have to be seen.

Selling medicines had been characterized as different from other trades, and those who did so were often maligned by contemporaries who failed to grasp that the medicine trade was changing. The author of *The London Tradesman* (London, 1747), for example, chided the druggist to "confine himself to the Sale of Drugs only" and avoid making medicines, so "his Want of Understanding can be of no Damage to any but himself."[3] Despite such advice, many druggists, chemists, and apothecaries had already begun to outgrow the traditional boundaries of their occupations. According to the customary categorization, apothecaries could make medicines but not prescribe them or provide care. Physicians gave advice, though usually did not touch patients or make medicines, whereas surgeons handled patients. Meanwhile, chemists prepared chemical remedies, and druggists served as middlemen for botanical ingredients. By the early eighteenth century, however, these boundaries meant little in an expanding medicine trade.[4] Thomas Corbyn (1711–91), an apothecary, chemist, or druggist depending on who asked, established himself as one of London's most successful merchants of medicines. He built a bulk, long-distance trade using credit, partnerships, and personal ties to endure the currents of the mercantile economy and the particularities of a medical marketplace marked by weakening regulation, shifting state policy, and rampant competition. His business more closely resembled the nearby mercantile houses than the apothecary shops of medieval Europe.

Many followed the example Corbyn set, though few were as successful. Nevertheless, the London medical marketplace soon featured a cohort of well-connected partnerships that offered entry into a medicine trade no longer restricted by its nominal gatekeepers, the Society of Apothecaries and College of Physicians.[5] From 1730 to 1790, people selling medicines proliferated across Britain, with particular concentration in London. According to some accounts, more than 150 chemists and druggists worked in the city by 1748 and supplied more than 700 shops in London alone.[6] A meeting of 200 apothecaries in 1794 estimated that competition from chemists and druggists selling medicines cost each of them nearly £200 in annual revenue, a significant sum at the time.[7] The use of credit, especially in the form of partnership agreements, enabled certain people to capitalize a larger scale of production and distribution best represented by their laboratories. These were dynamic places where matter could be transformed into different, valuable things, requiring significant labor and raw materials, as well as credit or capital to pay for them. Laborers bent over hot coals, breathed toxic fumes, and carried heavy loads; other employees recorded recipes, quantities, and costs in notebooks; still others packaged medicines for shipment. A common culture of folk, home-based, religious, and magical remedies coexisted with those produced in such spaces. Women and folk healers continued to sell plants to the very apothecaries and physicians who often inveighed against them. Female apothecaries, herbalists, midwives, and herb gatherers formed another facet—a more hidden, though no less important one—of this emergent industry.[8] Around these partnerships, then, developed a web of artisans, bankers, and manufacturers who provided necessary credit, expertise, goods, or services and benefited from their expanding scale. Some partnerships failed, but others thrived at the increasingly lucrative intersection of banking, healthcare, manufacturing, and state power.

Regulation, Expansion, Consolidation

Oversight of London's medical marketplace had nominally rested in the hands of two livery companies—the College of Physicians from the early sixteenth century and the Society of Apothecaries since the second decade of the seventeenth—in place of the types of judicial authority that developed elsewhere in Europe. By the eighteenth century, while versions of medical police antagonized unconventional healers on the continent, physicians and apothecaries contested their claims to regulate who could make and sell medicines in London in response to an influx of practitioners comprised largely of manufacturing druggists and chemists.[9] Druggists and chemists can be identified going back to Restoration England, but it was only in the early eighteenth century that apothecaries began to complain about competition from them. Surgeons, apothecaries, and other literate practitioners working in population centers blended therapeutic approaches and accessed trade networks, which challenged the status of the small cadre of licensed physicians and posed an ideological threat to conventional understandings of healthcare.[10]

For more than a century, officers of the College of Physicians sought to control the medical activities of practitioners outside its membership in London and within 7 miles of the city, pointing to an act of Parliament in 1523 that gave statutory authority to their institution's charter. The college could use its regulatory powers quite actively in London when the Crown supported it (as under Charles I).[11] Censors inspected the shops of apothecaries, druggists, and chemists across London; products that did not meet their standards were destroyed and the proprietor fined. By the end of the seventeenth century, these inspections often provoked resentment, but they rarely resulted in major penalties nor did they entirely stop counterfeit or bad medicines from circulating.[12] The college's power to search was effectively defunct for most of the second half of the century, apart from a short-lived revival in 1687–88, which, even then, was never more than partially implemented. The verdict of the Rose case in 1704 signaled the college's inability to reassert its legal regulatory powers at the start of the eighteenth century and affirmed a more expansive view of what constituted the practice of medicine (*physic*) in London.[13]

While often antagonistic to the physicians, members of the Society of Apothecaries similarly sought to constrain vendors outside the group's membership and also proved unsuccessful. The licensed apothecaries, those "*Best-side Banditti of* Black-Fryers" as their critics called them, represented the interests of licensed apothecaries who sought to increase their business at the expense of the college's authority. Though they gained legal recognition of their right to prescribe medicines by the verdict in the Rose case, apothecaries were often regarded as inferior by physicians due to their emphasis on empirical medicines and experience instead of humanistic learning. Thanks to a Crown monopoly granted in 1620 that they alone could compound and sell medicines in London and its suburbs, the apothecaries were able to inspect the shops of their society's members.[14] But the society's regulatory powers could not entice all those who sold, prepared, or traded medicines to join, and membership as a proportion of total apothecaries in London declined during the early decades of the eighteenth century. The situation seemed dire enough that, in December 1746, some members feared the company would soon cease to exist unless the charter was strengthened to compel all people who made and sold medicines within the city limits to join the society. A parliamentary bill to that end, however, promptly failed, in part due to opposition from the physicians.[15] Other attempts to renew their power to "prevent the making and vending of counterfeit, false, and pernicious medicines," such as a parliamentary petition from 1748 to extend the "Power and Authority to search for and destroy all deceitful, stale, corrupt, unmedicinable, and pernicious Drugs and Medicines" beyond their own membership, also failed.[16] Such attempts to regulate medical practice reflected contemporary fears that people who had not been examined by the society, of which there seemed to be more every day, could use adulterated or false ingredients to undersell licensed apothecaries in a marketplace unrestricted by institutional rules.

Not only did the eighteenth century see the influx of new entrants to the medicine trade; it also saw the accumulation of property by British apothecaries, chemists, and druggists. Fire insurance offers one way to chart the growth in the medicine trade, even though not all such vendors had the resources or desire to purchase a policy. Fire insurance took many forms in the decades after 1700 but was soon a permanent feature of the British commercial landscape. Fire remained of great concern, especially among those who manufactured medicines in wooden buildings with combustible ingredients and large heat sources, often in densely settled urban areas. The policies, therefore, enumerated the assets insured against a potential conflagration, including buildings, instruments, and merchandise.[17] The number of apothecaries, chemists, and druggists purchasing fire insurance policies rose more than 60 percent from the 1730s to the 1790s, with the most rapid expansion of participation occurring after 1760. Trade reports, medical licenses, subscriptions, and apprenticeships corroborate an opening of the medicine trade. Not only did more people identify as these occupations in cities and towns across Britain, but also more of them had money to spend on insurance. In general, the average value of assets insured against fire by people trading medicines grew by more than 30 percent during this period. Druggists and chemists in London accounted for much of these gains (the average value of their insured assets increased by nearly 60 percent); in the countryside, apothecaries added insured assets at a faster rate than druggists and chemists because they were more likely to provide a range of medical goods and services. While the bulk manufacturers in London appear to have been better capitalized with greater production capacity, rural apothecaries greatly increased their property as provincial spending on healthcare caught up to urban medical consumption. Rising wealth, combined with an increase in the number of policyholders, suggests that entrants into the medicine trade had sufficient access to cash or credit to establish larger businesses than had previously been commonplace.[18] Unlike in other industries, these large enterprises did not crowd out smaller ones.[19] Those with the most assets, usually druggists and chemists operating in partnerships, proliferated in urban areas, while people with more modest means also began to purchase insurance policies. Many quickly vanished from the policy registers, while others who could not afford to purchase insurance never appeared in them at all. Nevertheless, many found niches in the trade, including, for instance, William Radley, a chemist who for years kept a shop at the Queens Head near Grays Inn Gate and was insured for a fraction of the value of Thomas Corbyn's nearby shop and laboratory.[20]

Mirroring the ongoing erosion of strict occupational roles, there was less of a division between so-called popular, learned, and empirical healthcare in England than seems apparent from contemporary texts. Practitioners of all stripes borrowed from various sources all the time, while tenets of do-it-yourself medicine endured as well.[21] Meanwhile, a flood of material on self-dosing informed the sick of the range of medicines available in the marketplace. These works encouraged customers to buy medicines that they could now expect to solve their ailments and ease their pains as part of an iterative health routine. Potential customers could peruse the medicinal qualities of foreign plants in print, while gardens sprang up across the country where residents could grow some of the plants they read about.[22] In *The Friendly Physician* (1773), Francis Spilsbury, a chemist, provided instructions to assemble portable medicine chests and popular recipes for that purpose.[23] Patterns of state service and periodic warfare also fostered a thriving genre of naval and imperial medicine texts by the mid-eighteenth century.[24] Medical discourse would incorporate the ideas of merchants of medicines, military men, and planters on the practice and trade of medicines in the British colonies to profound effect. As the ranks of printers swelled alongside literacy rates, merchants of medicines could advertise through broadsides, handbills, and newspapers, supplementing the oral means of earlier centuries.[25] Soon, printed advertisements and instructions for using many medicines circulated throughout Britain. Tracts warned against cheap counterfeits—"spurious sorts of a very inferiour quality [that were] offered for Sale in almost every Town"—as those who claimed to sell the "true" or "genuine" products sought to publicly expose and distinguish themselves from hawkers of imitations.[26] Regardless of whether these concerns stemmed from an interest in public health or personal profit, they reflected how quickly the British medicine trade had become crowded and connected. These trends can best be illustrated at street level.

Building a Business

Thomas Corbyn's figurative journey began with a literal one. In September 1728, Corbyn arrived in London after a journey of several days from the West Midlands to begin an eight-year apprenticeship to Joseph Clutton, an apothecary known for keeping a "good shop" and for the breadth of his scientific interests. Clutton published on medicine and pharmacy, experimented with chemicals, collected botanical specimens, and constructed *materia medica* cabinets. One cabinet he made in 1729, now in the Oglander Collection at Oxford University, consists of 1,032 specimens and cost more than £21 (nearly £3,000 today). Coming from a Quaker family of modest means in Worcester, Corbyn would have arrived with the literacy and numeracy required of an apothecary's apprentice. He must have impressed since Clutton employed Corbyn to stay on at the shop after the apprenticeship concluded in 1736. Once Corbyn demonstrated his knowledge of medicinal plants and preparing, handling, and dispensing medicines in 1743, he received his freedom of the Society of Apothecaries to work for himself and became a partner in Clutton's business. Corbyn did not rely on the society for medicines but instead built his own capacity at a time of growing demand and unprecedented commercial opportunities across Britain, especially in the medicine trade.[27]

Corbyn soon turned the Clutton family business, established in the 1650s, from small-scale medicinal preparation to bulk manufacturing and long-distance trade. From their premises at 300 Holborn, Corbyn and

Clutton maintained a steady trade in chemical and Galenic medicines to a variety of retail and wholesale customers, including county hospitals. Meanwhile, Corbyn agreed to "occupy [his] own money in trade anywhere except in England" and so started exporting medicines to and importing drugs from the Atlantic colonies.[28] Corbyn could hone his strategies for long-distance trade without the imminent threat of failure while he still had access to the partnership's capital stock, despite the high costs of money and a relative lack of collateral. But this arrangement did not last long. Later in 1743, Joseph Clutton died, and his share of the partnership transferred to Mary Clutton, his widow. Corbyn transacted business in partnership with Mary until 1747, but by 1745, "several differences and disputes" had arisen between them over Corbyn's growing authority within the company.[29] As members of the Society of Friends, they were required to take up these disputes within the jurisdiction of the Quaker Meeting. No official resolution was recorded, but Mary left the partnership, while her son, Morris Clutton, and Corbyn formally entered into partnership in 1747 using £2,200 loaned from Mary to form the new capital stock.[30] In the past, sellers of medicines had not lent much, and then usually only small sums within tightly knit networks, but Corbyn's financial relationships operated on a more ambitious scale. While in partnership with Morris, Corbyn borrowed money widely to provide liquidity as he waited for returns from his overseas trade.[31]

Even as an expanding pool of metropolitan credit made purchasing medicines easier, druggists, chemists, and apothecaries struggled to raise sufficient capital to make them. They depended on a variety of credit instruments to build capital stocks, buy out partners, weather downturns in trade, and facilitate bulk sales. Participants in the medicine trade turned to the stocks, transferrable credit, and partnerships to mitigate the myriad challenges of eighteenth-century trade. The bankruptcy rate in London at this time settled around one in thirty and was even higher for those involved in distribution, particularly transatlantic trade. Failure, whether insolvency or bankruptcy, occurred regularly due to the prevalence of risk taking and competition. Merchants experienced bankruptcies more than any other occupation in the eighteenth century. After 1700, speculators and investors, buyers and sellers could access transferrable credit backed by new institutions (including the Bank of England), a national debt, a secondary market in that debt, and a host of joint-stock companies.[32] The widespread acceptance of new credit forms in everyday transactions enabled Corbyn to hold increasingly large amounts of debt, which in turn enabled him to provide a wider range of goods and services. Yet in many cases, seemingly old conceptions of cooperation, kinship, trust, and honesty never disappeared from the credit relationships that arose.

The exact source of Corbyn's early profits remains unclear, but he conducted a high volume of overseas trade during these years. Correspondence from abroad was often copied into bound letter books due to its significance, which has led many of these records to survive, although far fewer than originally existed. By contrast, local or regional sales could be noted in more ephemeral forms that often did not endure. Thus, in Corbyn's case, a comparatively robust documentation of his long-distance trade appears in the archive, despite what appears to be similar growth in the volume of his local and regional trades. From 1742 to 1747, the value of medicines Corbyn exported exceeded his earnings from domestic trade in partnership. The partnership's trade overall had increased under Corbyn's direction. During Corbyn's eight years in partnership with Morris Clutton, their business grew by as much as 350 percent and turned a profit of at least £2,480.[33] Censors from the College of Physicians visiting Corbyn and Morris's shop found their products, including preparations of rhubarb and Peruvian bark, "very good" and noted a brisk wholesale trade in 1748 and again in 1751.[34] Peruvian bark, from the cinchona tree of the northern Andes Mountains, contained alkaloids (quinine) that could kill the malarial plasmodia (*falciparum* and *vivax*) in human red blood cells, though it was not understood at the time in such terms. The bark was expensive, tasted bitter, and often induced nausea, but it gained prominence in colonial therapeutic regimes as an astringent and antiseptic for patients with intermittent or continued fevers.[35] Corbyn also informed Elijah Collins in Boston, Massachusetts, that he and Morris Clutton "return[ed] near 5000 Sterling per Annum wholesale in drugs & preparations besides what I export which is fully on my own account."[36] Gradually, Corbyn and Clutton abandoned any semblance of medical practice to focus on bulk sales, paying special attention to trade with the North American mainland and the Caribbean. Overseas trade constituted almost a quarter of the partnership's receipts, though much of its profits continued to come from the country wholesale trade. An emphasis on overseas trade represented a risky departure from the typical strategy of many apothecaries, chemists, and druggists who conducted mainly local retail and some wholesale business.[37]

When Morris Clutton died suddenly, in 1754, Corbyn assumed the partnership's debts and once again faced financial uncertainty. During their partnership, the capital stock had swelled to £14,000 from an initial investment of almost £4,000, in line with the capitalization of many of London's respectable merchant houses at the time. Business had been so good, in fact, that Corbyn claimed to control "the greatest part of the Custom [in Barbados]" for medicines. Yet this success also meant Corbyn required more capital to buy out the heirs of his deceased partner, which he did not readily have as a result of his overseas trade. Not only did Corbyn need to settle Clutton's share of the business, but he also needed to continue trading on his own account with a diminished capital stock to avoid insolvency.[38] He faced what was a common problem at the time: how to raise the funds to maintain control of a business after the death or departure of a partner. People usually turned to personal connections in such situations.[39] Corbyn, likewise, appealed to the medical and Quaker communities to borrow more than £6,000 to satisfy Morris Clutton's heirs in 1755. Loans came from others who manufactured medicines for export and from a variety of merchants whom Corbyn would have known through the Quaker Meeting or with whom he transacted business. Corbyn borrowed so much in 1755 that he was still repaying the apothecary Thomas Talwin seven years later. Despite some delays, Clutton's estate received its share in 1756, and Corbyn became the sole owner of the pharmacy.[40] He could now make medicines for domestic and overseas trade fully on his own account for the first time, nearly thirty years after his arrival in London. Like other merchants trading overseas, he continued to borrow on bond as his business expanded.[41]

Borrowing and lending remained common practice for Corbyn, especially once he was able to borrow on bond. Besides his fellow Quakers, Corbyn's creditors consisted of merchants and tradesmen in fields related to the medical marketplace. Ties of religion and kinship often provided reassurances for debt since personal credit had not fully given way to impersonal contractual obligations. Corbyn received multiple bonds from Silvanus and Timothy Bevan, large-scale druggists and chemists, who were also members of the Quaker community. The brewers, merchants, and apothecaries from whom Corbyn borrowed would also have been accustomed to trading in the new credit economy.[42] Like bankers or brewers in the eighteenth century, merchants of medicines could be active managers of capital. Corbyn lent small sums (relative to what he borrowed) to men in similar fields, such as William Arch Deacon, a male midwife, and Daniel Vandewall and Daniel Squibb, both surgeons.[43] From 1752 to 1762, John Hunt, a prominent tobacco merchant and fellow Quaker, borrowed more than £4,500 from Corbyn.[44] Occupational and religious affiliations offered opportunities for borrowing in a medicine trade that still relied on mutuality and custom despite the introduction of more impersonal or codified credit instruments.

As the loans to Hunt suggest, Corbyn's transatlantic trade had blossomed by the time of Morris Clutton's death in 1754.[45] British medicines were used alongside local drugs in many North American port cities. In New York, Peter Renaudet received medicines from Corbyn that he likely distributed to patients nearby and ship captains going abroad.[46] Corbyn regularly received returns from around the Atlantic world, sometimes at more favorable rates than the current exchange rates to minimize losses from local currency inflation in the colonies.[47] It had taken Corbyn some time, however, to build up this correspondence and to develop sufficient trust to ensure that his contacts would transact trade to his benefit. In the early 1740s, he did not yet have much experience with shipping and lacked credit.[48] Every parcel, therefore, required remittance before the next could be sent. In 1742 and 1743, Corbyn established contacts along the Atlantic coast in Newport, Boston, Philadelphia, New York, Virginia, and Maryland. He harbored particularly high hopes for sales in Philadelphia because his products were already selling in Boston and Newport at nearly fivefold markup. Although still short on credit a year later, Corbyn offered his goods on credit to colonial medical practitioners, enticing customers with additional time to pay.[49]

Throughout his career, Corbyn concerned himself with the strategies of export merchants alongside his attention to matters of pharmacy. He subscribed to publications relating to trade and wrote an article in the *Morning Chronicle & London Advertiser* entitled "Advantage of insurers or underwriters to merchants and traders exporting goods."[50] Corbyn was hardly unique in these interests among merchants of medicines, particularly since fluency in foreign currencies was a requirement for anyone wishing to trade overseas during the early modern period. Often, merchants of medicines relied on supply chains across continental Europe for raw materials to avoid the high duties on direct imports from Asia. Lacey Primatt, for instance, copied the current exchange rates for a variety of Spanish, Portuguese, Venetian, and French coins into his partnership's record book in the 1750s. The same held true for those engaged in transatlantic trade. Anyone trading in the Atlantic world needed to recognize, compare, and convert old, foreign, or worn coins.[51]

Corbyn's credit improved as he gained experience trading overseas, which enabled his long-distance enterprise to expand as well. Soon he had contacts in Connecticut and was receiving regular returns from Pennsylvania. Corbyn's medicines sold for smaller markup in Philadelphia than in Boston, but also on shorter credit so he could see returns sooner. Forgoing some potential profit in the future for more timely credit is a decision many transatlantic merchants would have considered at this time.[52] In 1746, Corbyn established new connections to Virginia, Maryland, and Pennsylvania, yet struggled for a time to find contacts in the British West Indies. He wrote to Barbados, Antigua, St. Christopher, and Jamaica for lists of "common &

saleable goods" and pondered whether to accept returns directly from the Caribbean or to send them first to Philadelphia, New York, or Boston.[53] Corbyn preferred to deal with general merchants in the colonies who did not have particular experience in the medicine trade so that he could use his expertise to influence the terms of the transactions and rely on their expertise to distribute the goods quickly, in bulk, and at good margins.[54] Selling on commission to merchants who could achieve higher markups upon resale was the priority, but, if that option was unavailable, Corbyn sold directly to physicians for lower prices to at least break even on the shipments before the medicines spoiled. By the end of 1747, he exported garments sewn by his Quaker relations in addition to medicines, evidence of his sturdier foothold in the Atlantic world.[55]

Credit and long-distance trade are only two examples of how apothecaries, chemists, and druggists were adopting the language and practices of the mercantile house. The advent of the mercantile partnership in the eighteenth century underwrote the aspirations held by men like Corbyn. As a result of the legal restrictions on joint-stock incorporation, most manufacturers relied on partnership agreements to pool resources and share risks. They also offered greater borrowing power and therefore the likelihood of more cashflow and liquidity.[56] Increasingly, those who made medicines in the pharmacies springing up across London did so using this method of association. The number of partnerships in the British medicine trade quadrupled during the eighteenth century. In 1790, 12 percent of apothecaries, chemists, and druggists who owned fire insurance worked in partnerships (up from only 3 percent in 1730); among them were the largest London companies making medicines for bulk export. In London, the percentage of merchants of medicines in partnerships was even higher, more than 20 percent.[57] Such an arrangement could bring together investors from different arenas of local or overseas trade and enable them to diversify their pursuits. Such was the case for Alexander Johnston, a medical practitioner and sugar planter from Scotland, who worked as a chemist and druggist for almost a decade in London before he met Alexander Grant, another Scotsman with ties to the Caribbean, around 1740. From Johnston's shop at the sign of the Golden Steed in Magpie Alley near Fleet Street, the two Scots marketed medicines, sugar, and plantation supplies. They worked together in the commission merchandising business until 1753, when they divided their assets and went their separate ways. Johnston continued as a merchant of medicines, supplying the East India Company in some years, and Grant acted as a sugar merchant, slave trader, and navy supplier.[58]

The Plough Court pharmacy, one of the biggest and best known of such partnerships in London alongside Corbyn & Company, embodied these trends in the medicine trade. Its founder, Silvanus Bevan (1691–1765), began his career in 1708 as an apprentice to Thomas Mayleigh (d. 1732), a Quaker druggist and merchant with extensive investments in shipping and the slave trade. Bevan gained his freedom from the Society of Apothecaries after seven years and began in 1715 to lease premises, with large basements and a laboratory, off Lombard Street at 2 Plough Court. The same year, he married Elizabeth Quare, daughter of the royal clockmaker, Daniel Quare. With his younger brother, Timothy, Silvanus carried on extensive domestic trade to practitioners and hospitals alongside overseas trade to the Atlantic colonies and continental Europe, returning thousands of pounds sterling yearly. In 1725, Silvanus was named a Fellow of the Royal Society, which offered an opportunity to extend his commercial relationships further beyond the Quaker community.[59] Soon, the Plough Court pharmacy supplied medicines to the Royal Navy and East India Company on annual contracts. The brothers received a commendation from the East India Company's Committee of Shipping in 1758: "that we had served the Company several years with great Reputation," recounted Timothy several years later.[60] Contracts to supply the armed forces and trading companies would become only more significant to those who wished to prosper in the medicine trade.

Silvanus's death, in 1765, marked a turning point for Plough Court. Timothy managed the business alone for a time before bringing on his son, Joseph Gurney Bevan (1753–1814), as a partner. After gaining control of the partnership in 1782, Joseph continued the pattern of overseas trade established by his forebears but strengthened familial and religious ties with East Anglian textile exporters and London banks via his brother, a founder of Barclay's Bank. Under Joseph's management, Plough Court's exports and cash flow reached new heights. Yet retail sales stayed relatively static during Joseph's tenure because he emphasized the overseas trade, which occupied the majority of his attention and the pharmacy's output.[61] He sold medicines to bankers and, in turn, trusted the Lombard Street Partnership and Gurney's Bank to help him cycle capital, accept bills of exchange and provincial banknotes, and access short-term credit. As private banks increasingly underwrote the medicine trade, manufactured medicines and colonial drugs entered the cabinets of the ascendant mercantile and financial elite, such as John and David Barclay, John Porker, Silvanus Bevan (the younger), and John Tritton, all of whom numbered among Plough Court's regular customers. Blood, credit, and health tied the pharmacy to the country's leading private banks and commercial houses.[62] As a result, merchants of medicines could send and receive payments across Britain and beyond.

Figure 1.1 Exterior of the Plough Court pharmacy, ca. 1800. Chapman-Huston and Cripps, *Through a City Archway*. Image courtesy Wellcome Collection.

While London continued to feature the greatest concentration of large partnerships, some merchants of medicines, like William Cookworthy (1705–80), resisted the city's pull. Cookworthy carved out a niche as a manufacturing chemist and druggist serving military contracts and a regional market in southwest England. He received instruction from Silvanus Bevan, a fellow Quaker and merchant of medicines, but did not sign formal apprenticeship papers. The Cookworthy family was said to have been impoverished by the South Sea Bubble, only to be rescued when Bevan gave the fifteen-year-old William training in the medicine trade. After six years of working alongside Bevan in London, Cookworthy returned to Devonshire in 1726 to set up shop on Notte Street, Plymouth, with Bevan's financial support. Cookworthy used an understanding of chemistry to devise more than 180 of his own recipes for medicines he supplied to apothecaries and surgeons throughout the counties of Devon, Cornwall, and Somerset. He also gained renown for his fascination with manufacturing porcelain. In 1755, Cookworthy obtained a contract to supply medicines for the hospital ship *Rupert*, which was docked at Plymouth. The Society of Apothecaries in London, which ordinarily provided medicines for the Royal Navy, lodged a complaint with the Sick and Hurt Board; but in 1778, Cookworthy was still supplying the navy to the tune of almost £700 annually.[63]

Because partnerships offered many advantages, Corbyn belonged to one for most of his career, aside from the brief period after Morris Clutton's death. By 1762, he had already entered into partnership with John Brown and Nicholas Marshall, never again to return to business on his own account. This arrangement underwrote a larger capital stock, which in turn funded a larger workforce and more equipment. The pharmacy grew significantly during the 1760s compared to when Corbyn assumed control in 1754. Yearly profits had more than doubled since the 1740s, and they continued to rise. In 1766, for example, the partnership made a profit of £600, and in 1767, the total value of the company's effects amounted to almost £13,000. For comparison, the average annual nominal earnings in Britain were only £15 in 1766. Earlier in the century, most London businessmen earned less than £750 per year, with many taking in less than £250. Those who provided medical services usually saw even smaller incomes.[64] The ten different partnerships that operated as Corbyn & Company over the next thirty years attest to the challenges of sustaining a bulk export business. The partners employed forty-seven different workers from 1762 to 1770, and, after 1770, they paid wages to more than eighty. Thomas Jones earned £10 annually with a twenty shilling raise "if [he] behaves well," the partners promised. George Simpson likewise saw £2 added to his yearly wage of £10 "in consideration of his diligence."[65] By contrast, the entire chemical industry in the United Provinces employed about one thousand people by 1800, mainly in private laboratories that employed between one and ten individuals.[66] Corbyn & Company was operating on an altogether different scale. In 1770, it generated almost as much profit (more than £2,000) in a single year as it did during the entire eight-year partnership between Corbyn and Morris Clutton decades earlier.[67]

Keeping the pharmacy in partnership also ensured the retention of institutional expertise. Brown and Marshall, for example, were already employees of Corbyn's before they became partners in the business. Elevating Brown and Marshall to partners not only replenished Corbyn's capital, but it also kept in place key sources of manufacturing know-how at a time when exports were growing rapidly. If Corbyn had not done so, perhaps these longtime employees would have left his pharmacy for another or started their own where they could have had a greater share of the returns. The agreements they signed contained the equivalent of a noncompete clause: partners could retire after giving six months' notice, but upon retiring, "shall not afterwards follow the Business of a Chymist, etc."[68] Finding new partners among current employees kept valuable skills within the company, limited competition, and ensured that work continued smoothly after a partner departed. By the same measure, a partner's unexpected death could threaten the capital stock as well as the accumulated experience, thereby endangering the entire venture. To mitigate such risks, partnership agreements necessitated more accurate and regular accounting practices for keeping track of the value of the business. By 1766, for example, employees at Corbyn & Company were taking increasingly detailed inventories of the company's warehouses.[69]

As their businesses achieved longevity, some merchants of medicines began to provide financial services to faraway customers. Without a widespread banking system, merchants were often entrusted with collecting and holding drafts, the means of settling accounts at a distance. Corbyn, for example, acted as a *de facto* banker for his long-distance clients by purchasing and discounting bills. He thereby came to accept an array of remittances, which sometimes occasioned him to invest in the transatlantic sugar and provisioning trades. Nor was providing such services limited to the largest partnerships. William Jones, a chemist and druggist in Great Russell Street, provided a variety of banking services and bought and sold India Bonds for his customers, domestic and overseas, who were often small investors or individuals with fixed incomes from annuities, such as widows, spinsters, and country clergy.[70] The birth of a securities market allowed artisans, manufacturers, merchants, and tradespeople, as well as married and unmarried women, to access new investment opportunities and thereby hold a greater economic stake in the government and trading companies.[71] As apothecaries, chemists, and druggists began to borrow, lend, and trade like mercantile houses, they also began to deploy labor and keep records like manufacturers.

Working in the City

Conducting business like a merchant would not have accomplished much without the ability to make sufficient quantities of quality medicines. To that end, Thomas Corbyn and his partners operated two laboratories, a shop, a counting house, and multiple warehouses in the neighborhoods of Holborn and Clerkenwell, just outside the old London walls. Medicine production capacity grew alongside British industrial output and per capita consumption of manufactures in the eighteenth century as part of a trend toward cheaper goods and substitutes for exotic imports.[72] The increasing concentration of labor and scale of operations at Corbyn & Company presented hazardous working conditions from chemical vapors and byproducts. Yet labor, by Corbyn's tally, accounted for only a small percentage of the final cost of medicines. Costlier were the raw materials and packaging—the bottles, labels, wax, paper, and wood that represented more than half the prime cost of a batch of Bateman's Drops, for instance. Corbyn and his partners spent more than £9,000 on raw materials alone in 1770.[73] Since labor costs stayed low compared to the costs of other materials, profit margins could be high for merchants of medicines who reached a certain scale of manufacturing. In other words, higher profits depended on lower costs of materials, mostly by buying at a discount, or on selling in bulk.

At Corbyn's retail shop on High Holborn under the sign of the bell and dragon, customers perused an impressive array of remedies. Such retail spaces were purposefully kept clean, fragrant, and well organized to attract passersby. Upon entering, potential customers could marvel at the exotic items in labeled glass containers or decorated ceramic jars representing the span of European empire.[74] Displayed in the shop window were aloes and guaiacum from the Americas. From Asia, there was gamboge in the drug room and myrrh in the shop's central glass case. Corbyn also stocked Virginia snakeroot, camphor, sassafras, and multiple types of rhubarb, some for wholesale and some for processing into medicines.[75] Drugs imported from the Americas had increased in prevalence among London shops between 1701 and 1722, peaking in the early 1750s as many gained popularity in treatments for syphilis.[76] As patterns of domestic medical consumerism shifted toward paying for goods and services, sales of elixirs, tinctures, unguents, and other medicinal commodities increased, as did the variety of such products. Finished products were kept in the closet under the stairs or in one of the many storerooms before being brought into the main shop. One side of the shop contained spirits, pills, and extracts, while a glass case in the middle of the floor displayed salts and oils. Corbyn reserved two cases solely for chemical oils and plasters. While the shop was tidy to evince a sense of trustworthiness, Corbyn's home and storerooms on the premises were a mess. Plant specimens, medicines, and packaging choked the hallways. Corbyn also maintained a small laboratory in Holborn that was separate from the shop so customers would not have to confront the pungent,

Figure 1.2 Corbyn & Company emblem, 1749. Corbyn's bell and dragon appeared on the wooden sign above his shop at 300 Holborn as well as on receipts from retail sales as seen here. Detail from bill for medicines purchased by J. Hodgkin, 12 Oct. 1749, MS 5436/12, Corbyn & Co. MSS. Image courtesy Wellcome Collection.

hot, and dangerous realities of medicine manufacture as they considered making a purchase.[77]

In the eighteenth century, laboratories became more prevalent, especially in urban areas, which improved manufacturing capacity while increasing the threat of fire. Laboratories were typically enumerated on insurance policies to note the risk they posed.[78] Alchemical techniques formerly applied to the transmutation of metals found use in the production of compound medicines in these spaces.[79] They housed machinery for grinding, pounding, and sifting drugs, as well as high-pressure boilers and furnaces for distillation and drying. The process at the heart of making many medicines, distillation, remained an intricate one. Ingredients first required washing, drying, and some form of breaking or grinding before infusion concentrated their oils (and thus their flavors and aromas). Only after such laborious preparation could the actual distilling begin. Open fires were common beneath hundred-gallon stills, evaporating pans, condensers, copper boilers, and stoves. If temperatures went unmanaged, ingredients could burn, ruining a preparation; even worse, stills could boil over or even explode. Manufacturing medicines with this equipment required significant input of energy, water, raw materials, and labor, especially before merchants of medicines began to mechanize these processes using steam.[80] Under these conditions, men earned a wage stoking fires, stirring vats, filtering solutions, and carting barrels for as little as a month or for years. Some, such as Richard Ridley, stayed with the company for more than a decade.[81] What brought these men to work in the laboratories of the medicine trade is not known, but it can be assumed that many of them had come to the city within the broader labor migrations to London. Settlement laws in 1662 and 1697 had normalized mobility for working men and encouraged relocation to areas with a demand for labor, in effect creating a new group of rootless wage earners by the start of the eighteenth century. The partners of Corbyn & Company could take advantage of the availability of the economically precarious working poor in London, despite relatively high labor costs relative to most of Europe.[82]

Corbyn's laboratories surpassed in size and capacity those of past apothecaries and many of his peers. His principal laboratory and warehouse at Coldbath Fields in Clerkenwell provided an ideal space for manufacturing medicines removed from the city's bustle. It offered less risk of fire damage or pollution and provided ample storage for the large quantities of botanicals, chemicals, and containers required to produce medicines in bulk. The upstairs storeroom held hundreds of pounds of plant matter, and in the back shed lay nearly one thousand empty glass bottles for packaging tinctures and other preparations. Proper bottling preserved medicines, whereas faulty seals would render them rotten and useless. Some ingredients were kept closer at hand near the distillation equipment. There was also a vault for chemical oils, a cellar for spirit of hartshorn, and a special room for herbs.[83] Nearby at 66 Aldersgate Street, the partners Lacey Primatt and John Maud, druggists and chemists, kept a laboratory, millhouse, and multiple warehouses. By 1794, they added spaces for other production techniques, notably a subliming room (for vaporizing and then condensing liquids), hot room, and drying room.[84] William Jones operated a laboratory in Covent Garden with a sixty-gallon still and tub for making medicated waters.[85] While wealthier apothecaries, chemists, and druggists usually had the largest-scale laboratories, many people continued to make medicines for smaller-scale distribution and consumption in other commercial and household settings. Apparatuses in kitchens could produce a variety of medicines even without the kinds of equipment owned by Corbyn or Primatt and Maud.[86] Many apothecaries, druggists, and chemists used their own equipment, whereas wholesalers from Cornwall to Scotland purchased remedies from middlemen or directly from those with laboratories.[87]

Without precise recordkeeping, however, one might fail to generate the quality medicines and profits one anticipated, especially if working in bulk. When the partners of Corbyn & Company manufactured something, they first calculated associated expenses and expected profits. Take, for instance, the production of flower of benzoin (benzoic acid), a topical antiseptic used to open internal obstructions, promote mucus secretion, and cleanse the chest. First, gum benzoin, a brittle, fragrant resin from the benjamin tree (*Styrax benzoin*) of Sumatra and Java, had to be purchased at an East India Company auction. The gum was then milled and moved to one of their laboratories where it would be pressed by two men for fifty days or resold to other druggists and chemists in London. After accounting for expenses, including the gum, wages, beer for the laborers, transportation, and machine wear and tear, Corbyn could calculate the "prime cost," encompassing all of the expenses going into the medicine, of the resultant hard-pressed flower of benzoin and then determine what to price it per ounce in order to see a profit based on that cost. Careful accounting proved necessary because such processes typically generated hundreds of pounds of products and could cost more than £50 per batch. Adding another layer of variability, slightly different medicines with different values could be made from a similar collection of materials with the proper manipulations. While flower of benzoin was typically applied topically, oil of benzoin made from the same resin was identified by John Quincy as a powerful internal medicine, even as a "Specific against the Stone and Gravel in the *Kidneys* and *Bladder*."[88] Corbyn could, for example, make an extract of Virginia snakeroot or produce a tincture from it instead and then ask another price. These kinds of manipulations do not encompass the other purposeful and accidental variations in production to lower costs or mislead customers. Corbyn also made varying qualities of medicines depending on where he intended to sell them: one grade to be bottled for retail, another to be sold wholesale, and a third provided to ships—each with its own price. Other merchants of medicines followed suit in their own laboratories.[89]

They relied on recipes that they treated as industrial secrets. Corbyn's recipes do not bear the names of patrons, as was common in domestic recipe books, suggesting that these commercial recipes were governed by a different set of rules from others. His manufacturing books remain imposing volumes comprising hundreds of pages with successive recipes for similar products in multiple hands, suggestive of many contributions during Corbyn's career and after his death. They are written in shorthand with shortcuts or steps abridged requiring a level of particular know-how to decipher, though not encrypted like some early modern secret texts.[90] This style of writing may have been for convenience in the laboratory or to limit knowledge transfer. The British chemical industry already suffered from espionage by continental rivals, like the French, in the mid-eighteenth century.[91] Merchants of medicines engaged in such schemes with their local rivals as well. Neither patents nor decorum prevented Corbyn from imitating his competitors' medicines. Corbyn reproduced Bateman's Drops, Dr. Pett's Oil, Dr. James's Fever Pills, Daffy's Elixir, and Lockyer's Pills—all popular patent medicines but what he called "quack medicines"—in his Holborn laboratory where he kept a "box of samples of quack medicines" for experimentation. These he sold alongside medicines based on recipes culled from popular printed works, including the *London* and *Edinburgh Dispensatories*, or that he devised himself.[92]

In theory, medicine recipes were supposed to follow certain standards of quality and composition. But in reality, they varied, sometimes significantly, between pharmacies or preparations. In spite of the ongoing commercialization of the medical marketplace, medicines across the healthcare spectrum remained personalized. Customers did not yet demand completely homogenous products and often still preferred individually tailored advice or prescriptions. This variability suited manufacturers who could alter the quality or composition of their products to fit their and their customers' needs, though these did not always align. Widely available printed recipe books, such as the College of Physicians' *Pharmacopoeia* and *Dispensatory*, represented an important step toward standardizing composition, or at least the perception of it, since remedies often differed in the quality and quantity of ingredients despite what was recorded in a recipe. No single text or authority could guarantee the composition of a particular medicine. Recipe collections borrowed from many sources, adding remedies "more efficacious" to those from the more well-known, or formal, medical texts.[93] Since raw material costs, rather than labor or processing expenses, largely determined the prime cost of medicines, merchants of medicines had reason to manipulate ingredients. For instance, Corbyn produced two different types of vitriolated tartar (a sulfate of potassium used as a cathartic on its own or as a common chemical ingredient in other medicines): one from the work of Herman Boerhaave (1668–1738), the Dutch botanist, chemist, and physician, and another similar to recipes in the *Edinburgh New Dispensatory* (1786) and the earlier *New Dispensatory* (1753). Corbyn's second recipe follows "a very elegant" way of making the salt in the *New Dispensatory*, suggesting he did not employ the "almost useless" substitution often used by "the wholesale dealers in medicine," as noted by the author William Lewis. Even so, Corbyn's version of the recipe is recorded much more vaguely than those in the texts, leaving room for adaptation during the production process. Lewis sought to update earlier recipes based on his forays into medical practice and experimental chemistry. After being elected a Fellow of the Royal Society in 1745, he moved to Kingston, Surrey, where he established a laboratory for experimenting with medicines. Lewis also lent his expertise to ironworks in northern England and the Midlands, again emphasizing the growing connections between manufacturing sectors in Britain. Corbyn meanwhile recorded multiple recipes for the same medicine as his preparations changed over time, sometimes quite suddenly due to the availability of supplies or the fickleness of laboratory equipment. His notes show both a carefulness but also a flexibility to accommodate commercial realities. Corbyn pondered what to value oil and salt of ambergris in the summer of 1765 to best balance his accounting irrespective of the items' compositions or purity.[94] Standardization and consistency ran up against not only routine variation but also profit margins.

Concerns about the costs of materials fostered another form of collaboration in this developing industry, political lobbying, which was already commonplace for other groups of merchants. The protestations of druggists, chemists, and apothecaries against certain fiscal policies were thus not unique at the time, but they demonstrate that merchants of medicines deliberated on broader political-economic issues even after medicine exports had been released from much of their tax burden in the 1720s.[95] Two petitions to the treasury in 1769 about the duties on drugs show apothecaries, druggists, and chemists opposing a customs system that seemed to threaten their livelihood. Among the thirty men who signed the first memorial to the Lords Commissioners of the Office of the Lord High Treasurer on February 9, 1769, were prominent merchants of medicines: Timothy Bevan, manufacturing chemist; Hugh Hughes, wholesaler; Lacey Primatt, druggist, chemist, and exporter; and Corbyn's partner John Brown. Of these signatories, nine—Andrew Grote, Timothy Bevan, Ninian Ballantine, George Hayter, Robert Kingscote, George Webster, Richard Speed, John Hopkins, and John

Figure 1.3 A chemical laboratory in mid-eighteenth-century London. This view of the author William Lewis's laboratory shows some of the apparatuses used to produce medicines in similar spaces by this time. Detail of William Lewis's laboratory in Kingston, Surrey, in Lewis, *Commercium Philosophico-Technicum*.

Brown—authored a second memorial on February 15, 1769, with a more specific proposal to reduce the rates, duties, and drawbacks on a number of drugs. These petitions appeared within a deluge of extraparliamentary lobbying beginning in the 1760s from merchants and manufacturers.[96]

As merchants of medicines organized outside the jurisdiction of London's livery companies, members of the Society of Apothecaries turned to similar tactics in an effort to reclaim a share of the trade. The apothecaries had used a joint-stock subscription since 1676 to fund chemical manufacturing at their laboratory in Blackfriars near the Thames. What was originally designed as a voluntary association to make medicines for members soon grew into one of the largest, and most highly regarded, laboratories in London totaling more than 1,500 square feet complete with a counting house as well as the "open fires and furnaces" and "operations requiring intense heat" one could expect of these spaces.[97] From the laboratory issued chemical and Galenic medicines for institutions and wholesalers, as well as for members who "depend[ed] on the goodnesse" of those medicines for their own businesses. The laboratory also represented an advantage for apothecaries in the medical marketplace the society had lost its ability to regulate.[98] Stockholders received dividends from the laboratory stock in the 1720s; and by the 1740s, they allowed nonmembers, including merchants, trading companies, and hospitals, to buy medicines directly from the laboratory.[99] Committees for Buying and Inspection (whose members served for decades in some cases) oversaw the purchase of ingredients and ensured preparations met standards of composition and quality.[100] The laboratory's operators kept account books of drugs coming in and medicines going out to calculate yearly profits and manage inventory, while the Laboratory Stock

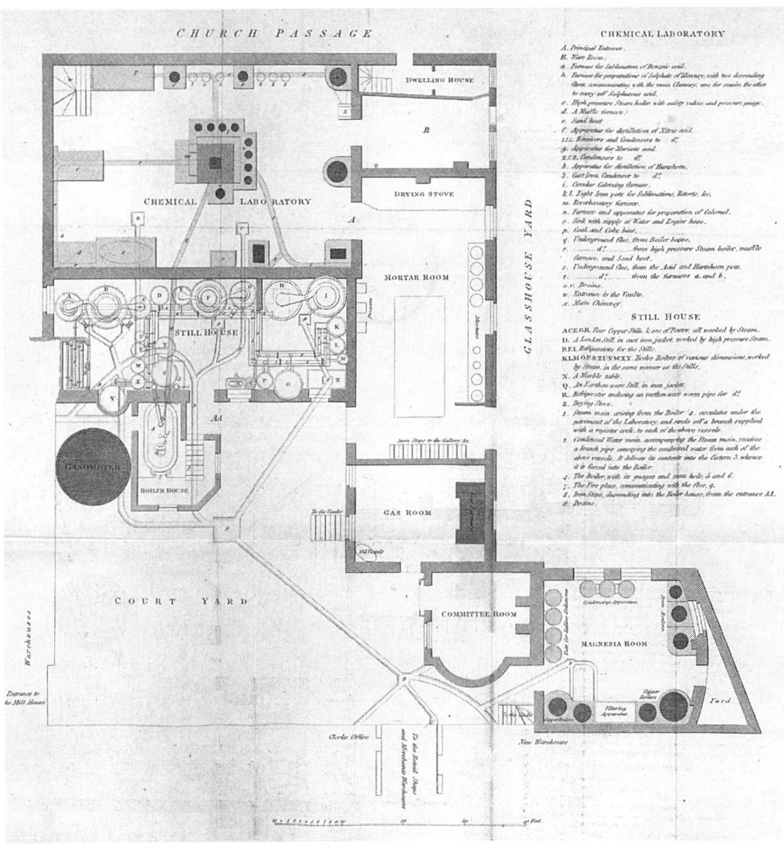

Figure 1.4 Plan of the laboratory at Apothecaries' Hall, 1823. By the early nineteenth century, the Society of Apothecaries boasted a chemical laboratory, still house, and other specialized rooms for manufacturing medicines for various institutional and individual customers from their hall near Blackfriars along the Thames in London. *Origin, Progress and Present State*. From the Library of the Royal College of Surgeons of England.

Committee adjusted prices by comparing medicines prepared at their laboratory with those manufactured by "ye most eminent Chymists" in London.[101] Working at a large scale enabled the apothecaries to give discounts to those who purchased in bulk. Subscribers could buy spirit of ammonia (smelling salts) for six pence per pound. By comparison, customers who bought more than 500 pounds of the same could have it for only three and a half pence per pound.[102] That products from the apothecaries' laboratory were often more expensive than those made elsewhere concerned committee members but was reasonable, in their opinion, given the laboratory's quality controls. Nevertheless, accusations of profiteering dogged the society, prompting new policies in 1745 and 1749 intended to bring the prices offered outsiders closer to the lower ones offered members.[103]

The Society's laboratory offered an opportunity to purchase quality medicines at reasonable prices. Yet many refused it. Those with access to laboratories, big or small, could develop their own products for sale. Thomas Corbyn, Timothy Bevan, and William Jones, for example, did not appear on the Laboratory Stock's subscriber rolls in the 1750s or 1760s, even though they were counted among the society's membership and had held nominal positions within its leadership. Instead, they competed with it. John Beaumont, for example, subscribed to the stock from 1728 until 1757, but stopped after he entered into partnership with Corbyn and thereby gained access to equipment of his own. The Laboratory Stock Committee further discouraged rival pharmacies from subscribing by preventing the wholesaling of medicines bought from the apothecaries' laboratory in 1767 (though how this was enforced remains unclear). At the other end of the spectrum, apothecaries struggling to maintain a small shop subscribed to the Laboratory Stock to line their shelves and receive the dividend.[104]

No merchant of medicines, regardless of their capital stock, laboratory, or expertise, however, could alone produce all the chemicals, botanicals, equipment, or packaging required for bulk trade. As a result, routines of exchange developed among the pharmacies clustered in the neighborhoods near Holborn and Cornhill (see fig. 1.5). They relied on one another to acquire goods from outside their own supply chains or that were someone else's specialty. Corbyn & Company, for example, had become "famous" for its ammoniac gum, an acrid stimulant with a variety of medicinal applications; so, Joseph Gurney Bevan resold gum from Corbyn's pharmacy instead of making his own.[105] Bevan stocked his laboratory at Plough Court with goods from the drug merchants Vezey, Bradney, and Roebuck; George & George Webster; and Hayter & Strong.[106] At the same time, people involved in the glass and textile industries purchased substrates and catalysts

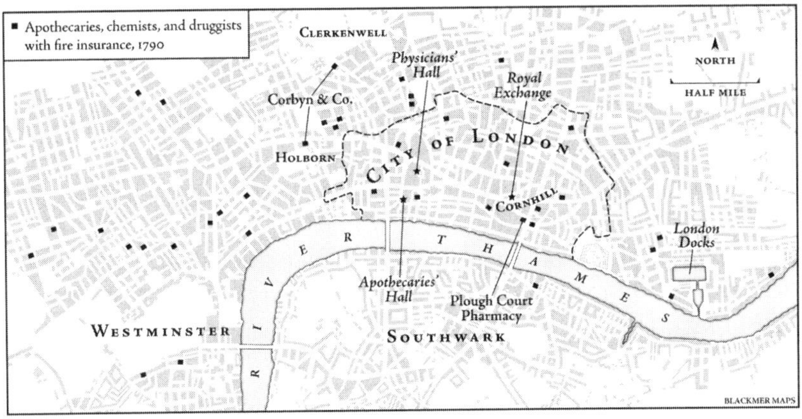

Figure 1.5 The London medicine trade in 1790. Participants in the London medicine trade had proliferated to the extent that they appear regularly in the fire insurance registers. They clustered near several neighborhoods, particularly Cornhill by the Exchange, Holborn, and Clerkenwell. Many people who described themselves as apothecaries, chemists, or druggists purchased fire insurance given their buildings' timber frames and the danger of fire from their laboratories. Still, many more likely did not purchase insurance, so this map represents only a fraction of people engaged in such activities. Several other locations relevant to the medicine trade have been added for context. MS 11936/30–33, 130–35, 364–75, 382–92, Sun Insurance. This map is modeled on G. Jones, *London* (1815). Lionel Pincus and Princess Firyal Map Division, New York Public Library. New York Public Library Digital Collections.

from Plough Court because many of the chemicals used in medicine preparation also contributed to a variety of industrial processes. Manufacturing medicines in bulk meant that Corbyn or Bevan would often have extra chemicals or plants available to sell and could acquire more.[107] Much of the trade among these businesses occurred on twelve months' credit, similar to the overseas trade; but, in contrast to that trade, most of these debts were repaid on time.[108] In the 1760s, physicians, apothecaries, druggists, chemists, merchants, and hospitals across Britain owed Corbyn money. Corbyn and his partners knew their customers, so their debt books served only as a reminder of balances and typically lacked detailed personal information. Much retail purchasing occurred on shop credit, but it remained unavailable to those without the proper wealth, status, or connections to the vendor. As crowded as the rolls of Corbyn's debtors had become by the 1770s, still more people must have been prevented from joining them.[109]

Proceeds from the medicine trade dispersed to a range of tradespeople beyond those who sold medicines. Joseph Gurney Bevan kept accounts with chemists, coopers, distillers, tallow-chandlers, scale makers, pewterers, and chest makers, just to name a few.[110] These tradespeople, skilled in their crafts, proved essential to Plough Court's day-to-day operations. A busy laboratory, shop, or warehouse required barrels and chests to store roots and leaves; scales for weighing ingredients; acids, alcohols, oils, and other solvents; and stoneware jars, glass vials, and pewter containers for the resultant tinctures, elixirs, and extracts. A constant supply of barrels, bottles, and boxes was also required for long-distance trade. Glass, in particular, remained an essential, and expensive, item for packing small quantities or individual doses of remedies for shipment. Bottles manufactured in Britain were subject to a duty based on the color of their glass, which could raise the costs of exporting certain medicines if they needed to be preserved in particular bottles.[111] Those who made medicines in bulk in eighteenth-century London also relied on a steady flow of plants imported by the English and Dutch East India Companies on the one hand and more local plants provided by gardeners, herbists, herb women, seedsmen, and sassafras cutters on the other.[112] By the eighteenth century's close, London's principal pharmacies depended on and encouraged a wide network of suppliers and artisans who all profited from the bulk sale of medicines.

Conclusion

Thomas Corbyn's and Joseph Gurney Bevan's capacity to trade on their own accounts or in partnerships with their own agents and access to credit set them apart from many of their peers on the continent.[113] They established connections to other merchants of medicines, as well as to tradespeople, bankers, manufacturers, and merchants across Europe and the Atlantic world. These ties marked the acceptance of merchants of medicines as a fixture of London's commercial life, while enabling the bulk production and wider distribution of medicines in a changing marketplace where apothecaries, chemists, and druggists gained greater access to credit, held more property, and behaved like merchants. In other words, a medicine manufacturing industry took shape in eighteenth-century Britain built on characteristics of interdependence, long-distance trade, and bulk production. This process heightened competition, but also encouraged instances of borrowing whereby seemingly older forms of secrecy, personal credit, and artisanal expertise remained integral to the new trading arrangements. While Corbyn, Bevan, and their neighbors enriched their partnerships and contributed to the general expansion of the medicine trade, they also grappled with the challenges of long-distance trade, which comprised a significant portion of their business.

In London's complex and unregulated medical marketplace of the later seventeenth and eighteenth centuries, almost anything could be had for a price. Abroad, however, medical scarcity ruled, with differing degrees of fidelity to European norms across the empires of the Atlantic world. Medicines made in London went to places that looked quite different in terms of expectations of medical practice, mortality rates, and accessibility of care. While medicines from Corbyn & Company or Plough Court sailed aboard ships bound for the British West Indies, London's medical marketplace could not be so easily transferred. How Spanish and French authorities organized medicine in their Atlantic colonies represented other strategies for reproducing European models of healthcare overseas. Spanish medical practice, perhaps the most regulated of any in Europe during the early modern period, could be more closely replicated abroad via state and church institutions. New France also saw colonists install European-style constraints on medical practice in Quebec.[114] Britain's relative lack of regulation coupled with scarcity abroad made its medical marketplace more difficult to export, but fostered a transatlantic system where medicines would travel widely.

Britain, by many metrics, remained an outlier in the history of European medicine. Despite similar conditions of mortality, sickness, and poverty elsewhere in Europe, a particular kind of medicine manufacturing and medical marketplace emerged in Britain at this moment. The loose regulation of medicines largely stemmed from the conditions of trade, rather than from concerns about their composition or application, as in other places. Nevertheless, composition and application would shift due to the rhythms of transatlantic trade with consequences for the intended consumers of those medicines in the colonies. As the pharmacies in London turned to overseas trade, merchants of medicines sought a remedy for the problem of distance in order to access emerging transatlantic markets offered by the armed forces and the plantation complex.

TWO

Distance's Remedies

By joining production and distribution, merchants of medicines could reap significant returns from long-distance trade. Yet Thomas Corbyn complained that he saw little profit from overseas trade early in his career, while nearly thirty years later Joseph Gurney Bevan lamented the cost of delays when he wrote that "Punctuality of Payment is the life of business."[1] Merchants of medicines grappled with how to effectively trade across an ocean awash with goods during the eighteenth century. Some gradually established routines for the bulk distribution of manufactured medicines around the Atlantic world. Often a complicated task, exporting can, nevertheless, be reduced to three general steps. First, goods had to be packaged, shipped, and then arrive in a far-flung port after a protracted voyage; second, they had to be sold, ideally for more than it cost to produce and distribute them; and third, the proceeds of those sales in some form had to reach the initial sender. Each stage presented a range of obstacles considering the times and distances involved. Many, such as storms, defied human intervention, though others could perhaps be mitigated with a little planning and good fortune. When so much of early modern trade rested on a promise, what steps could be taken to improve the odds that repayment would arrive in a timely, usable manner? Some answers to this question proved technological or relied on commercial diversification, while others took an epistemological tack. Even at a quickly expanding scale, the medicine trade continued to depend on long-standing forms of obligation alongside newer credit instruments. The late seventeenth and early eighteenth centuries had witnessed the birth of public credit in Britain, which neither guaranteed transactions nor obviated their social embeddedness, especially when it came to the long-distance variety.[2]

Risk and uncertainty in overseas trade stemmed from the age-old obstacle of distance, which broke down into two related problems for most merchants. The first was the scarcity of information. Success as a trader required access to current news in order to evaluate when or where to send goods and whom to trust. Merchants generally valued speed and privacy in their communication, favoring face-to-face interaction over printed forms. Yet face-to-face communication was often arduous and sometimes impossible, which required depending on someone one did not know or had never met. Correspondence could enable a transaction to occur as planned, but letters alone could do little to prevent fraud. It required experience to find trustworthy trading partners or reliable informants, so many traders preferred to remain secretive with commercial information in the absence of networks assembled from co-religionists, kin, or professional organizations.[3] The second problem was the collection of debts. With debt ubiquitous, yet money typically in short supply, the task of moving value from one place to another, often with different laws and customs, occupied much attention in the Atlantic world. Metal and paper currencies rarely held a consistent value, while credit remained based on trust and thereby capricious as well. One frequently had to consider whether a coin was worth the value stamped on it or whether the person who proffered it was worth trusting.[4] As a result of these risks, merchants devised methods of obtaining returns from transatlantic trade that revolved around commodities other than silver.

In the long-distance medicine trade, exporters diversified their enterprises to see returns by borrowing from the engines of Atlantic commerce: the slave and sugar trades. In 1728, for example, Silvanus Bevan's mentor, the apothecary Thomas Mayleigh (1671–1732), purchased with proceeds from his medicine exports a three-fourths share of the *Mary's* voyage to Jamaica that likely transported captives from the West African coast.[5] Mayleigh was hardly the only one to diversify in such a way. As the plantation complex permeated Atlantic commerce, so too did its strategies influence the long-distance medicine trade. Over time, sugar factors, commission agents, mortgages, and slaving voyages appeared in the account books of London's pharmacies as they acquired more capital and credit. So too did their partners profit from the demand for medicines on plantations and in the armed forces. The transatlantic slave trade had long been a site of knowledge production, but it also shaped healthcare by spurring the distribution of medicines around the Atlantic world.[6] British medicines were exchanged for colonial produce or local currencies and appeared on slave ships, in frontier settlements, among military encampments, and on shop shelves from Boston to Bridgetown.

Attempts to overcome the challenges of distance also reshaped perceptions of medicines in an increasingly disparate marketplace. The requirements of partnership and bulk production in London began to standardize expectations of how medicines were made and what they could do, if not standardizing their actual compositions. Exportation advanced such expectations as long-distance trade rendered medicines' value in terms of the possibilities they embodied. When details about price and quality, for instance, were omitted from correspondence, few indicators remained available to evaluate the composition or test the efficacy of an item received from far away. They would seem unknowable, but also more uniform and predicable in the absence of certain information—in other words, commensurable. This commensurability did not arise from material processes, tests, or proofs as has been described for other early modern things, but from their essential unknowability and opaqueness. Proof could come from taste, yet often proved elusive.[7] Merchants of medicines selectively drew distinctions between their products as they integrated them into transatlantic commodity flows. The same approaches to information and credit that enabled Thomas Corbyn or Joseph Gurney Bevan to export medicines in bulk to a variety of contacts also prompted an abstraction of medicines that simplified their iterative use. Remedies for distance led to complications, intended and not, as the medicine trade grew during the eighteenth century.

An Expanding Trade

The changing landscape of medicine production in London enabled dramatic growth in medicine exports following the efforts of apothecaries, chemists, and druggists to establish overseas connections and the concurrent expansion of the British armed forces and plantation complex. Thomas Corbyn's construction of transatlantic ties illustrates the takeoff in medicine exports during the 1730s and 1740s from the perspective of a particular trader. Joseph Gurney Bevan's experiences three decades later reflect the work required to maintain such ties during the booms and busts of war and peace. Exporters also had to contend with changing state policy toward medicines that in some cases encouraged long-distance trade and in others threatened its viability. As the medicine trade experienced significant growth for most of the eighteenth century, some manufactured medicines became fixtures of the Atlantic trading system, alongside the other trades from which its participants borrowed. A picture of this moment emerges from customs records and mercantile correspondence. The growth of medicine exports beginning in the 1730s would have unfolded differently, however, without the trade policies of the century's first decades.

Since the Navigation Act of 1651, government officials promoted domestic industry and a positive balance of trade by gradually reducing and then eliminating many of the export duties on goods manufactured in Britain. This trend culminated in 1721 when the recently appointed First Lord of the Treasury and Chancellor of the Exchequer, Robert Walpole (1676–1745), abolished the remaining export duties on manufactured goods and lessened import duties on many raw materials. Medicine exports fell under this change in tax code since many medicines were officially classified as manufactures, "apothecary ware" in the customs ledgers.[8] Before Walpole's customs reforms, steep duties frustrated those who attempted to export medicines in bulk. Afterwards, they faced fewer charges, though the reforms did not immediately quiet all their criticisms of trade policy. In 1712, the apothecary Joseph Cruttenden (born ca. 1660) mocked "our wise Parliament" for a 20 percent tax he thought would drive up the cost of drug exports to the North American colonies. Cruttenden tried to evade the tax but lamented several years later that "custome ruines all" in his trade.[9] Although Walpole did not invent protectionism in Britain, he began to shift the goal of taxation from generating revenue to promoting manufacturing. The reduction of many import duties on raw materials and export duties on the finished products followed. Walpole described the exportation of manufactures as a matter of national importance, saying in a speech to Parliament that "nothing would more conduce to the obtaining so publick a Good, than to make the Exportation of our own Manufactures, and the Importation of the Commodities used in the manufacturing of them, as practicable and easy as may be."[10] Despite this rhetoric, import duties remained high on some drugs used often in making medicines, reaching nearly 25 percent in many cases. Petitions condemning these decisions circulated within a matter of months, presaging the arguments made decades later.[11] On the one hand, merchants of medicines benefited from the elimination of export duties on medicines; on the other, they still faced considerable charges on raw materials.

In the decades after Walpole's reforms, medicine exports began a half-century-long ascent that would transform the production, distribution, consumption, and understanding of medicines on a global scale. By 1745, overseas destinations, mainly the North American mainland, the Caribbean, and India, received the largest part of medicine exports that now averaged nearly 200,000 pounds annually (see table 2.1, fig. 3.3). Exports to those destinations had surpassed exports to continental Europe in the past twenty years, more than doubling over that span. Such a pattern contrasted with the

Table 2.1 Average weight of annual medicine exports from London

Regions	1700–1704	1720–24	1740–44	1760–64	1770–74
Africa	0	941	381	5,221	7,840
India	1,957	3,895	15,165	160,328	162,624
Europe	2,416	10,473	28,643	26,932	49,787
North America	8,547	16,097	47,413	160,736	319,847
Caribbean	23,743	29,094	101,847	235,072	462,937
Total	36,664	60,500	193,449	588,289	1,003,035

Notes: Averages are in English pounds weight calculated for five-year samples and rounded to the nearest pound.
Source: CUST 3/1–82.

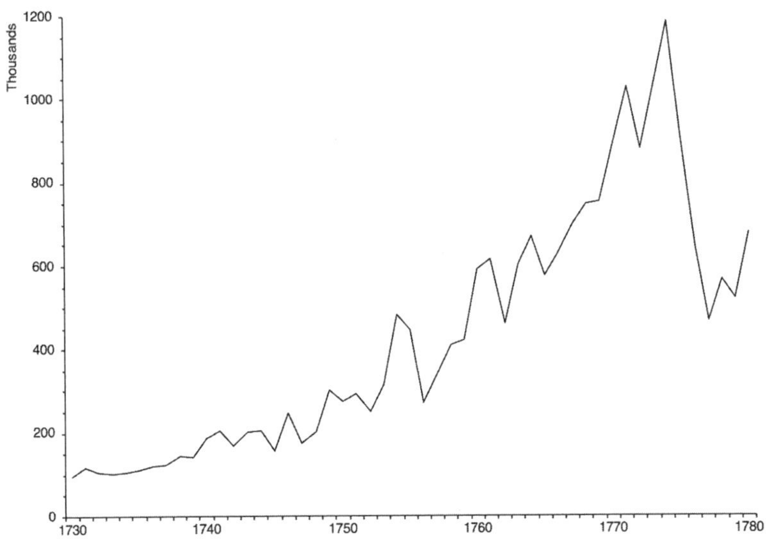

Figure 2.1 Medicine exports from London, 1730–80. Exponential growth began in the 1730s and continued into the 1770s. Exports are calculated by weight (in thousands of pounds). CUST 3/30–80.

export of other manufactures that typically went mainly to Europe. North America's and the Caribbean's demographic growth made those regions desirable markets among exporters of medicines. Medicine exports to some transoceanic locations rose almost 300 percent since the 1720s, while those to Europe stayed comparatively stagnant. Exports experienced even more dynamic gains from the 1740s into the 1760s before peaking in the early 1770s at more than a million pounds per year (see fig. 2.1). The medicine

Table 2.2 Growth rates for medicine exports, 1700–1774

Regions	Growth rate	Total exports
Africa	12.25%	128,677
India	8.85%	3,737,955
Europe	3.81%	1,608,290
North America	5.29%	5,392,277
Caribbean	4.68%	8,672,537
Total	5.17%	19,539,736

Notes: Exports have been organized by region based on locations given in the customs ledgers. Growth rates (k) are calculated by weight (pounds) and rounded to the nearest hundredth of a percent. Total exports are in English pounds.
Source: CUST 3/1–82.

trade still featured enough volatility to make returns unpredictable from year to year, depending on prices or interest rates, for instance. Nevertheless, London's medicine exports grew at a rate of more than 5 percent, surpassing that of total manufactures (at just above 1 percent) exported from London during the same period (see table 2.2).[12] Though British trade in the eighteenth century was less centered on London than it had once been, the city continued to be the hub of the medicine trade, and its commerce grew in absolute terms. The overseas medicine trade reflected, and even outpaced, trends among the other colonial export trades. While textiles accounted for the majority of British exports, medicines comprised an increasing fraction of the total. Contrary to seeing eighteenth-century commerce as somehow fixed, the medicine trade reveals foreign trade patterns still developing.[13] It would be easy to attribute these patterns largely to political economy, but the exponential growth of the medicine trade by midcentury owed to a lot more than taxes.

What contributed, then, to the upwards trajectory of medicine exports beginning in the 1740s? While significant relative increases in exports are apparent from the 1720s to the 1740s, the largest increases on both relative and absolute scales occurred during the 1740s and 1750s. By this time, changes in the medical marketplace and tax policy had reshaped the bulk production of medicines in Britain, while the populations of Britain's Atlantic colonies, military and civilian, free and enslaved, had continued to rise, offering more opportunities for sales. These decades also saw prolonged warfare on numerous fronts. Customs data illustrate that wars could indeed bring windfalls for medicine exporters, yet wartime had an ambivalent effect on the overseas medicine trade. On the one hand, imperial wars destabilized exports and closed foreign markets. They made overseas debt more difficult to collect especially when it was guaranteed by nothing more than personal security after a flood of easy credit in the preceding peacetime. Recouping debts from colonial merchants and planters became even more difficult if metropolitan credit dried up or insurance rates soared, making agricultural produce less valuable. Wars and financial crises could also diminish the confidence necessary for long-distance commerce.[14] On the other hand, imperial wars boosted demand for medical supplies from garrisons and military hospitals as the injuries, morbidity, and mortality of the battlefield took their toll. In these times, British state institutions and trading companies proved to be reliable buyers and providers of credit even when other customers proved less so.

Looking closely at what merchants of medicines, like others involved in transatlantic trade, did to enable their goods to circulate and generate returns illustrates the day-to-day practices beneath the customs statistics and tax policies that facilitated a flourishing medicine trade. Winning a contract to supply the East India Company (EIC) differed from sending a medicine chest to a shop in Choptank, Maryland, or a plantation on Barbados. Some partnerships proved capable of these various tasks, shipping medicines to the Caribbean, India, and mainland North America. The codependence of the medicine trade, armed forces, plantation complex, and EIC will be explored at greater depth in the next chapter. For now, however, considering the strategies Corbyn and Bevan employed to send medicines around the Atlantic world illustrates their approaches to information and debt that helped make exchange possible. The practices of merchants of medicines mattered to the outcomes of long-distance trade and to its consequences. Each instance presented distinct calculations of risk versus reward and required different approaches that when taken together encouraged the distribution of manufactured medicines on a global scale, in the process reshaping contemporary understandings of medicines and the bodies on which they acted.

Managing Information

To facilitate exports, merchants of medicines managed incoming and outgoing streams of information like their peers in other branches of long-distance trade. Transoceanic merchants have been credited with laying much of the Atlantic commercial system's infrastructure over the seventeenth and eighteenth centuries.[15] Keeping a business operative, let alone expanding, across such distances required writing letters and conserving them on a scale unnecessary in personal letter writing. A flood of letters attempted to ensure that items were bought or sold at the desired location, time, or price and that the proceeds arrived as promised. Merchants of medicines exchanged notes about insurance rates, shipping times, tax policy, and local demand. In general, people of certain occupations and social strata were encountering information at a higher volume and frequency during the eighteenth century, which meant new challenges of identifying, categorizing, evaluating, and cataloging relevant details: on prices, currencies, weather, or reputations, for example.[16] Sometimes these streams worked at cross purposes, but other times they enabled a steady export of medicines from London and amounted to a concerted effort to shape expectations of medicines' value. Thomas Corbyn and Joseph Gurney Bevan used form letters, packaging, and secrecy to introduce standardized logic into their foreign correspondence that continued to hinge on interpersonal trust despite some of their informational tactics. The management of information as a business technology in the medicine trade not only helped expand a transatlantic purchasing base, but also raised concerns about the quality of the goods themselves that ultimately informed an epistemic shift in understandings of medicines as interchangeable commodities within the rhythms of long-distance trade.

The scale of merchants of medicines' epistolary networks bears repeating. Merchants, by some counts, maintained more extensive letter-writing networks than any other group in the early modern period. They used letters to cultivate personal and contractual ties when transactions could not occur in person.[17] Merchants of medicines were no different. Their networks spread across the Atlantic world as they developed greater production and distribution capacities. In the first decades of the eighteenth century, Joseph Cruttenden traded with a handful of correspondents in Massachusetts, New York, Antigua, Jamaica, and Barbados from London. Cruttenden had served two apprenticeships before gaining his freedom of the Society of Apothecaries in 1687. Without a laboratory of his own, however, his opportunities for increasing his profit margins through manufacturing remained limited. Instead, as a shareholder in the Laboratory Stock, Cruttenden relied on the apothecaries' laboratory for many of the medicines he exported.[18] He also conducted trade without the aid of kin or religious connections abroad, from which other traders benefited, instead exchanging letters with agents whom he trusted to sell his goods year after year. Despite lacking some of the characteristics often identified in successful long-distance traders, Cruttenden avoided bankruptcy during his long career, though he also never achieved significant wealth.[19] Corbyn, in comparison, constructed a thicker network across the Atlantic a quarter century later during the pivotal

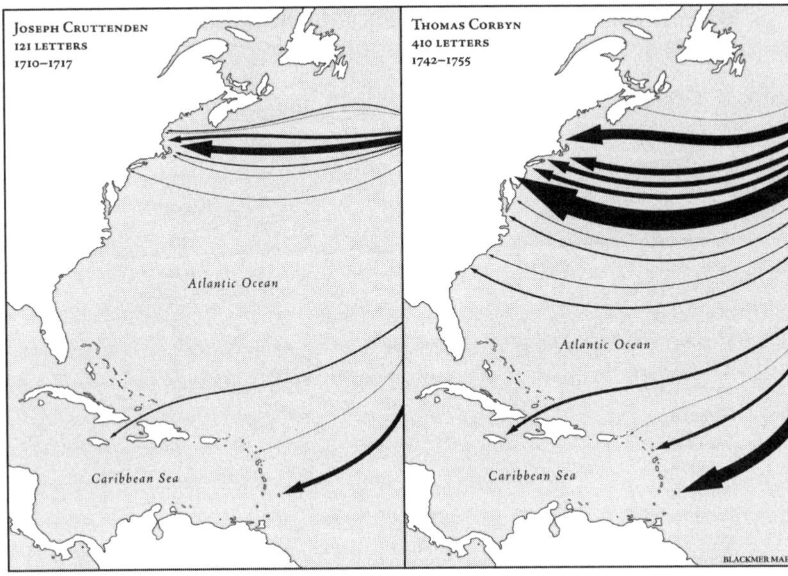

Figure 2.2 Transatlantic correspondence in the medicine trade, 1710–55. The map (*left*) shows Joseph Cruttenden's overseas correspondence. From 1710 to 1717, the apothecary Cruttenden wrote 121 extant letters to contacts across the Atlantic world, including 72 to Massachusetts and 28 to Barbados, for example. The map (*right*), by contrast, shows Thomas Corbyn's overseas correspondence almost three decades later. From 1742 to 1755, Corbyn sent at least 410 letters as part of his long-distance medicine trade, with 58 to Massachusetts and 76 to Barbados, for example. The arrows are weighted by the number of letters to that destination (10 letters is equal to one desktop publishing point). (*Left*) Steele, *Atlantic Merchant-Apothecary*. (*Right*) MS 5442, Corbyn & Co. MSS.

decades for medicine exports of the 1740s and 1750s (see fig. 2.2). With only a decade's experience he sent medicines in bulk to more places than Cruttenden did with more than triple that (see table 2.3). Corbyn's network displayed a few advantages over Cruttenden's that enabled it to reach more people across a broader geography at greater frequency. Corbyn's superior manufacturing capacity and capital stock permitted riskier behavior, namely larger shipments and lengthier terms of credit. He also benefited from co-religionists in ports from New England to Barbados and at English private banks that provided short-term credit. As a result, Corbyn could construct more robust long-distance infrastructure to reach bulk markets around the Atlantic world. Doing so, however, was no easy task.

At such scope and scale, one could no longer expect to meet a contact face to face in the medicine trade. Corbyn, for example, admitted "having no knowledge of any person . . . but by letters" on the island of Tortola when

Table 2.3 Transatlantic correspondence in the medicine trade, 1710–55

Destinations (from north to south)	Cruttenden's letters (1710–17)	Corbyn's letters (1742–55)
Nova Scotia		2
New Hampshire	7	
Massachusetts	72	58
Rhode Island		36
Connecticut		30
New York	3	28
Pennsylvania		116
Maryland		4
Virginia		9
North Carolina		4
South Carolina		10
Jamaica	9	17
Tortola		1
Antigua	2	19
Barbados	28	76
Total	121	410

Notes: Corbyn's letter book also records correspondence with contacts in Ireland and across continental Europe (as well as some without geographical metadata), which is not included in this table. Cruttenden's letters contain a single one to Madras that likewise is not shown here.
Source: Steele, *Atlantic Merchant-Apothecary*; MS 5442, Corbyn & Co. MSS.

he sought to "settle a correspondence [to the island] for sale of Druggs & Compound Medicines" in the 1740s. Routines for encouraging cooperation among strangers thus assumed greater significance in ensuring the distribution of medicines.[20] In the absence of trusted contacts abroad, Corbyn solicited recommendations from those already connected to him by religion, family, or business to find new trading partners.[21] Ties of blood or faith provided some readymade trust. Corbyn relied on Quaker merchants to sell his medicines to local apothecaries and planters, though other times he simply sent letters unannounced to potential customers.[22] Likewise, Joseph Gurney Bevan sought reports about "commercial character" while searching for a contact in Berlin during the 1770s. These background checks were not intended as signs of disrespect but, Bevan explained, "a precaution [as] the distance of place demand[s]."[23] There existed a range of possible customers for exported medicines, each with their own benefits and drawbacks depending on the circumstance. General merchants acting as middlemen, plantation owners and doctors, apothecaries with regional shops, wealthy land speculators, and itinerant rural practitioners all appear in the paper work of the London pharmacies. William Hunter's (1729–77) apothecary shop in Newport, Rhode Island, contained laboratory equipment and the current books for making medicines, including Quincy's *Dispensatory* and Lewis's *New Dispensatory*. It also boasted an impressive range of imported items that mirrored, though not at the same scale, the content of Corbyn's warehouses. Nearly 2,000 miles south in the slave society of Barbados, Joseph Gamble Jackman (d. 1786) kept an apothecary shop in Speightstown furnished with drawers, bottles, and various other equipment. Both of these men, at different ends of the provisioning trade, received medicines imported from London.[24]

Once a contact responded, Corbyn sent additional letters to learn about the local markets and set guidelines for future transactions before exporting any goods. Standardized portions of his letters asked correspondents for "particular" answers about the prices of articles they purchased and from where they received them. Corbyn desired to know which items sold well, which sold poorly, and at what prices. Did medicines usually arrive directly from London or from local middlemen? If the latter, he spied an opportunity to establish a direct trade bypassing those intermediaries. The recipient's name and location, his or her recommender, and a list of items were these letters' only variable parts. The rest came from an easily copied form he kept in his letter book.[25] In this way, Corbyn could save time gathering information from potential customers even when he did not know them very well or they did not have much experience in the medicine trade. His letters simplified his correspondence to extract the types of information he valued. If Corbyn's contacts were to communicate between themselves, they would find that he implemented a more-or-less consistent strategy across his networks. This strategy was not always successful, since many letters did not result in new connections or went unanswered entirely, but it was facilitating more contacts and larger sales by the 1750s.[26]

Medicines' packaging also mattered in a burgeoning trade. Sailing times from England to Boston, the Chesapeake, or Barbados varied from as short as five weeks (for Boston) to nearly ten (for Barbados) depending on the winds and currents.[27] Much could go wrong with a cargo in a damp hold for more than fifty days. Precise packaging helped mitigate the risks of things people could control like leaks and spoilage, if not the privateers and storms they could not. It also added another layer of protection against tampering as medicines passed through many hands in long-distance trade. Seventeenth-century apothecary shops featured a variety of bottles, boxes, pots, and jars, but ceramics remained more cost-effective than glass for storage. By the early eighteenth century, advances in British glassmaking made the exportation of medicines in glassware cheaper and safer. Merchants of

Figure 2.3 Glass medicine bottles for export, ca. 1700s. By the eighteenth century, many medicines were packaged in glass containers of various sizes and shapes conducive to bulk long-distance trade. Glass often proved more durable than ceramic and seemingly offered advantages to discerning quality and selling in smaller quantities for a range of customers. Image courtesy Wellcome Collection.

medicines turned to glass vials for exporting in bulk small doses of manufactured medicines, which provided greater protection to the commodities in transit and enabled simplified sales once they arrived. The inventories of early American pharmacies, for example, show large quantities of British glassware. Corbyn typically used green or white flint glass vials, either long and slim or squat and round, for exportation. The green vials came in a variety of sizes ranging from a half ounce to eight ounces and were cheaper than those made from clear glass. Long, slim green vials have been found in Alexandria, Virginia, for example, imported from Plough Court pharmacy in London. In enabling Corbyn to pack bulk shipments in discrete containers to suit "great & small customers without any trouble to ye Seller," glassware mirrored contemporary solutions to the problem of small change that dogged currency in the early modern period. Medicines packaged in divisible units offered easier resale at a range of scales for a range of customers: in small doses to patients or practitioners and in larger quantities to merchants or planters. Packaging in this manner did for medicines what small denomination coins did for money: it permitted local transactions that occurred below the level of high-value, bulk overseas trade but were, nevertheless, essential to the wider flow of colonial trade.[28]

When purchasing a single dose or many, customers did not need to know much about the products they received because merchants of medicines often sent instructions on how to use them. At Plough Court, Silvanus Bevan insisted that the names of the medicines be inscribed upon every vessel shipped from the pharmacy. Corbyn too provided detailed descriptions of shipments' contents.[29] Careful labeling, bottling, and instructions not only offered a means to elude adulteration, but they also seemingly enabled customers to more simply achieve desired health effects from medicines. Corbyn included instructions in shipments to convey basic details about the proper applications and doses for many popular remedies, theoretically eliminating the immediate need to consult a practitioner or medical text before ingestion. Many people would still have conferred with local healers or family members, in addition to making decisions about imported medicines depending on their own habits, wealth, status, location, or freedom. Supplying instructions, though, enabled Corbyn to establish transatlantic connections with people other than middlemen or medical practitioners—directly with planters in the Carolinas and Virginia, for example—who did not have much experience with medicines. Circulating as they did around the Atlantic world, these instructions endorsed Corbyn's particular products while also extolling the virtues of British medicines more generally.[30] Merchants of medicines thereby spread information about the uses and effects of many common remedies that made those medicines seem more familiar to customers within an intellectual framework that stripped medicinal plants of much of their indigenous context in favor of a European one.[31]

As Corbyn gathered and distributed information to serve his overseas trade, he also limited the availability of certain details about his medicines. Recalling the secrecy surrounding recipes in London's laboratories, information about the quality and composition of medicines remained elusive in the Atlantic world. Corbyn sent two different invoices with exports. One, listing items and quantities but omitting prices, should be shown to potential customers, he insisted. The second, intended for Corbyn's trading partners, included each item's prime cost and an estimate of its local value. Retail shoppers, patients, and other buyers would see only the types of goods and amounts available, whereas Corbyn's trusted vendors would benefit from the "Customary & Market Prices" he supplied.[32] Placeholder or hidden prices helped exporters account for short-term variation in production, distribution, or other incidental costs by leaving prices flexible. This contrasted with other commodities' prices that were usually established through bargaining at the time of sale.[33] Joseph Gurney Bevan, like Corbyn forty years earlier, also refused to share catalogs containing prices. He explained in a letter to Benjamin Rush (1746–1813), the Philadelphian physician and politician, that a list of prices "occasions a disadvantageous comparison against the dealer who chiefly considers quality."[34] The prices of drugs fluctuated too much for a price list to be effective, whereas sharing the prices of his medicines, Bevan cautioned, "might induce a comparison with the prices of other dealers" whereby "mistaken conclusions might be formed."[35] Bevan's desire for secrecy in overseas trade reflected a larger concern about comparisons between goods rather than simply one about prices.

The quickening pace of the transatlantic medicine trade, however, prompted frequent comparisons that conflated markers of value, quality, and efficacy. To Bevan many medicines would look similar on an invoice or in a green glass vial but held distinct values and anticipated effects based on their compositions, which could differ greatly even for nominally the same item. That recipes remained inconsistent and few checks on purity existed offered plentiful chances for mistakes, deliberate cost-saving substitutions, or more nefarious adulterations in the manufacture of medicines, especially given the erosion of regulation in London and the lack of oversight in the colonies.[36] Testing proved inconsistent in the absence of chemical methods, germ theory, or widespread standards. It remained embodied, dependent on the physical senses. Sensory evaluations had always been important to medical practice and assumed greater importance in a globalizing medicine trade. A trader required some understanding of the taste, smell, and texture of medicinal plants to purchase them abroad, for example. Printed texts circulated by the close of the eighteenth century to help with this task.[37] Yet the physiological component to identification also complicated comparisons between manufactured medicines due to their frequent similarity. That comparing remedies remained subjective and embodied made it accessible to many, but also opened the door for confusion. Exporters found ample opportunity to alter the particular composition, and therefore cost, of their products. Corbyn could adjust his own recipes or those from the *London Dispensatory* to manufacture slightly different versions of a medicine, one sometimes "at one third the Cost" of another.[38] Yet the medicines could all appear the same to patients, with some preferring the taste, smell, or appearance of one or the other. Evaluating a remedy's genuineness or quality—that of any commodity, really—required expertise and experience that most consumers would not yet have developed with imported medicines. Nevertheless, the proliferation of printed and manuscript descriptions coupled with increasing availability encouraged people, expert and not, to purchase medicines, though many would fail to tell genuine from adulterated or fresh from spoiled products. Medical practitioners who ordered goods in large quantities like wholesalers, and wholesalers who offered healthcare like medical practitioners further complicated the landscape of medicine distribution.[39] These developments made it nearly impossible to accurately compare medicines in the Atlantic world.

Whereas the turn of the eighteenth century had seen debates about the purity of drugs, merchants of medicines worried more about the adulteration of compound medicines by midcentury. Adulteration not only threatened profits but also a remedy's very efficacy since many medical writers believed that impure remedies, whether adulterated, counterfeit, or rotten, would not produce desired results.[40] It followed that a medicine should work as intended if it were made well, prompting Corbyn to assure a customer that "Thou may depend on Mine to be genuine."[41] Because the concept of *caveat emptor* ("buyer beware") applied in medicine as it did in other fields of contract, English law (civil and criminal) proved ineffectual in combating the circulation of fraudulent or adulterated medicines, particularly following the abrogation of requirements to inspect drugs in 1708 and the weakening of corporate oversight over the medical marketplace around the same time. Some legal protections stemming from common contract law existed for patients in eighteenth-century England, but they generally applied to cases of attendance or advice, rather than sales of medicines.[42] The idea that efficacy depended at least somewhat on proper composition attached moral stakes to financial considerations about the quality of products. A more accessible medicine trade, however, intensified competition over the price and quality of medicines, making it harder to identify bad medicines marketed as genuine, according to participants. Unscrupulous manufacturers—"too few are honest enough to observe [the proper methods of making medicines]," Corbyn lamented—could easily replace a costly ingredient with a cheap substitute and thereby undersell those who used higher quality ingredients.[43] The "great adulterations used by some wholesale Apothecaries & Chymists" were the only reason, Corbyn argued, that they offered lower prices than he did.[44] To make matters more complicated, adulterations could be easily overlooked. Corbyn crowed that he "could make 100 perCt proffit by adulterations" and "def[ied] one Man in 50 to detect [it]," but assured customers that his medicines contained only the purest, proper ingredients.[45] Ingredients represented one of the few material ways to appraise the value of medicines, especially those with similar uses, and distinguish them from spurious ones. As such, the perception of ingredients' quality shaped the values and anticipated efficacies customers projected onto the medicines.

Merchants of medicines' attempts to control information about the medicines they exported, then, stemmed from a concern about commensurability in addition to a desire for profit. On the one hand, they discouraged comparisons between medicines by concealing prices and routinizing correspondence—put another way, they limited accurate description, which seemingly runs contrary to the kinds of object descriptions identified as essential to early modern long-distance trade.[46] Yet, on the other, Corbyn and Bevan realized that with excessive standardization their products would appear illegible, and perhaps even inefficacious, to consumers. They touted the quality of their products without revealing too much proprietary information and promoted the idea that they were indeed like everyone else's, though usually superior. For Corbyn, since his medicines were supposedly of high quality and, by extension, effectiveness, they should be more valuable. Nonetheless, he struggled to realize the prices he sought from American contacts who often found his medicines unduly expensive. In some cases, Corbyn paid for shipping and insurance himself to lower the costs to his contacts and thus the prices they could offer others without hurting the products' reputations.[47] He believed that his medicines would "speak for themselves" with use and justify their prices. How could he "return so many Thousand pounds in England and America with good Repute," Corbyn asked Elijah Collins in 1754, if he were selling bad medicines?[48] Business success now testified to quality, rather than quality dictating sales. Bevan, like Corbyn before him, also turned to quality to deflect accusations of "having extravagantly exceeded another druggist in prices." He reasoned that both personal credit and an item's quality should be considered when determining its value, rather than price comparison alone, which was "not a fair way to estimating the dealings of different persons," in his opinion.[49] Neither efficacy nor value, in other words, could yet be entirely reduced to a single marker despite the abstracting power of long-distance trade.

None of these informational strategies could protect overseas traders from the uncertainty of personal relationships inherent in long-distance commerce. Esther White (1700–77) seemed to Corbyn like the ideal trading partner. She had met him in person, shared connections to local Quaker meetings, and possessed more than a decade's experience in the medicine trade. In 1739, White moved to Wilmington, Delaware, where she began selling medicines on her own account. After meeting Corbyn on a trip to England, White began regular business with him in 1745. She soon sold Corbyn's medicines on commission, remitting payments in bills of exchange and a range of gold and silver coins. In this arrangement, a typical one for many merchants, Corbyn shouldered much of the risk of unpaid debts or les since White did not pay in full for the goods.[50] She benefited from vn price lists (instead of Corbyn's) and the glass vials Corbyn provided ild prosperous retail and wholesale trades, much to the ire of several male vendors in Philadelphia.[51] White had secured Corbyn's trust, which he noted in a letter to Thomas Lightfoot writing that he "[had] a great regard & value for her sincerity & integrity" and in several personal letters to her.[52] At the end of 1750, however, Corbyn's books showed a significant balance overdue him for the medicines, glassware, and clothing sent to White and her husband over the preceding five years.[53] Despite his concerns, Corbyn offered her substantial credit to continue their transatlantic trade for several more years.[54] Returns continued to disappoint, leading Corbyn to suspect White was selling his medicines at wholesale prices instead of at the higher retail prices he desired. Whatever the cause of the account imbalance, Corbyn could not reconcile his perception of White's misconduct with the lofty "Expectation & Dependence [he] had on [her] beyond Common Traders."[55]

Without accurate sales information, Corbyn could not determine whether he accrued any profit from his trade with White. Delays also tied up Corbyn's money, caused goods to spoil, and threatened to undermine his other relationships, provoking lamentations of "as life is uncertain" one must "get old accounts balanced, especially beyond sea."[56] Financial planning, especially in partnerships, required the careful recording of assets and regular profit and loss calculations.[57] As financial literacy became more common on both sides of the Atlantic, inventory taking and account balancing became annual routines for many merchants of medicines.[58] Simply keeping a ledger, however, did not ensure effective or honest recordkeeping, as Corbyn had discovered. From a distance he could not peer into White's books. According to Corbyn's records, Esther and her second husband, John White, owed him almost £500 for medicines, shoes, gloves, and haberdashery sent from 1745 to 1753. Seeking a clearer picture of what goods the Whites still held, Corbyn instructed Robert Foster, a Philadelphia merchant, to examine their ledgers and organize them "as we do in England," rendered in a single value of sterling. Doing so confirmed Corbyn's suspicions of mislaid profits and broken trust. He did not want to lose Esther White as a trading partner, but she would no longer receive the privilege of trading on commission as long as her account did not meet his standards. From Corbyn's perspective, such an arrangement would shift the bulk of the risk to the correspondent and minimize information gaps. He could then offer credit extensions as incentive for careful accounting and punctual communication after she had purchased the goods.[59] Principles of probity, punctuality, and plain dealing—aspects of the Quaker belief structure that

Corbyn espoused—did not necessarily ensure that transactions proceeded as planned.[60] Nor did he always practice those principles with profits at stake. That "[money] affect[ed] him very much," which Corbyn once admitted to White, shaped his approach to transatlantic trade.[61]

It was difficult to know anything with certainty in this world. The medicine trade, like other trades, remained fraught with information asymmetries that prompted abstraction of the goods moving within it. Exports were increasing by midcentury, and medicines appeared in newspapers, inventories, and books across the Atlantic world without differentiating who had made them in a reflection of the commensurability facilitated by long-distance trade. Corbyn's and Bevan's informational tactics made their medicines seem similar to the others for sale around the Atlantic world. Meanwhile, merchants of medicines also advanced the idea that their products could not be compared strictly along economic lines, but instead possessed common salutary effects based on more intangible qualities. Sometimes not knowing, as in Esther White's case, hurt Corbyn's business, but other times it benefited from a lack of information when secret prices or recipes rendered his medicines more commensurable. As medicines circulated the Atlantic world, they contributed to another problem illustrated by Corbyn's concern with Esther White's accounts: debt.

The Problem of Debt

Corbyn's struggles to realize returns from his trade with Esther White illustrate that the timeliness of returns and collection of overdue accounts remained principal concerns of many merchants, especially given the lengthy voyages involved in their businesses. Exporters fretted about the length of time their money stayed overseas and whether they would ever see tangible returns from goods they sent abroad. Overextending credit could threaten merchants of medicines' manufacturing, which required significant liquidity to purchase ingredients. If one's credit dried up because of uncollected debts, the pharmacy's other activities could contract or even grind to a halt. Colonial customers, in turn, worried about the terms of credit they received and how to remit payment. Abundant credit, such as Corbyn extended to his overseas customers, enabled a flurry of transactions but also generated significant debt in an Atlantic world choked with credit and starved for specie. Without a universal medium of exchange, how would these debts be settled across distances? The closest thing to one, silver, remained in short supply; while other options—bills of exchange, local currencies, and colonial produce—all depended on trust to hold a transferrable value. Credit operated in the grey area between a relationship and a contract. Long-standing relational forms of credit coexisted with less personal instruments, including promissory notes, bills, and bonds, which together facilitated exchange between people who did not know each other well. Elite merchant houses typically reduced their risk in these cases by dealing with only a few experienced and trusted contacts.[62] Merchants of medicines, by contrast, diversified their transatlantic trades, with varying results, as medicines provided another possible store of value for transatlantic trade.

Despite encouraging early returns from his overseas trade, the War of Austrian Succession (1740–48) posed several challenges for Corbyn and his partners, even as it offered new opportunities to supply scores of sailors and soldiers. Wartime insurance had risen at least 15 percent on goods shipped to Boston, adding to Corbyn's costs, which had become more difficult to cover as his colonial correspondents carried considerable sums of unpaid debt during these tumultuous years. In October 1746, for instance, returns were "so uncertain" that Corbyn feared he would not see any for more than two years. In hopes of expediting returns, he desired overseas customers to repay debts in specie or merchandise—namely cotton, skins, silver, gold, or indigo—rather than in bills of exchange. If the goods sold well in London, thereby providing him usable credit or currency, Corbyn would not charge commission fees on receiving those goods as long as the senders continued to order medicines from him.[63] There was little he could do to force repayment though, especially when correspondents could not access specie or bills of exchange because of the high exchange rates between local New England currencies and the pound sterling. Corbyn hoped that Parliament's reimbursement of Massachusetts and Rhode Island for their colonists' contributions to the 1745 capture of Louisbourg (the French fortress on present-day Cape Breton Island) would lower exchange rates and thereby facilitate returns.[64] Corbyn also accepted a variety of European currencies, such as Dutch Guilders, from his American customers.[65] Within several years of the war's conclusion in 1748, most of Corbyn's correspondents in New England had regained their good standing with him and "pay well," he noted, though he still hesitated to extend plentiful credit overseas for several more years.[66]

When trade slowed due to periodic credit or cash shortages, Corbyn repurposed his transatlantic infrastructure to invest in other colonial commodities. The war years offer several examples. In 1746, he proposed a voyage to purchase sugars in Antigua and wrote to the merchant Israel Pemberton Jr. (1715–79) in Philadelphia about purchasing a ship together with John Hunt, the tobacco merchant in London, to expedite remittances from

North America. Later the same year, Corbyn sought partners to purchase a different ship in Boston to carry Caribbean tar and oil to London where he would serve as commission agent for the goods to be "turned into money," he hoped.[67] Diversification, as opposed to specialization, could offer a path to success for British merchants, though it remained risky and presented its own pitfalls. That some merchants of medicines could now engage in such behavior following those in the plantation or slave trades reflects their integration into Atlantic commercial systems by midcentury.[68]

An effort to corner the market in guaiacum is perhaps the best example of Corbyn's attempts to invest in colonial produce. Guaiacum, commonly known as lignum-vitae, refers to a genus of tall, flowering plants native to tropical and subtropical regions of the Americas, particularly in the Antilles. The wood is hard and heavy, a dark greenish-brown, that emits a pleasant odor when burned. "In Venereal Cases it [was] said to do wonders," so much so that medical writers referred to it as the "*Holy-wood*." The tree's resin or gum was believed to be a specific for gonorrhea, as well as salutary for gout and other joint ailments.[69] The price of guaiacum gum rose as it gained popularity in Europe and as the plant's natural habitat faced destruction from plantation agriculture.[70] For more than a decade guaiacum had been "plenty & cheap" though recently supplies had begun to dwindle, Corbyn observed in 1753, as the trees were cut down to make room for coffee plants on Jamaican estates. Coffee had been grown in Jamaica since at least 1728, developing into a significant export crop by the end of the century. The ongoing deforestation and soil depletion made guaiacum so "scarce & dear" it could only be purchased in London with cash at prices more than a third higher than in recent years.[71] To Corbyn it also presented an investment opportunity, reflecting his understanding of botanical commodities as objects of speculation as well as raw materials for medicine manufacturing.[72] In June, Corbyn wrote to Edward Penington (1726–96), a Philadelphia merchant in the plantation trade, with a proposal to buy as much guaiacum as they could find at below-market rates. Corbyn had heard that plentiful trees still grew on the island of Saint Martin and desired Penington "to find a proper person going there" to purchase high-quality gum at lower prices than could be had at Jamaica. Corbyn also relayed precise instructions for packing the gum to avoid adulteration or spoilage in transit.[73]

Corbyn did not intend to use all the guaiacum for making medicines. To turn a profit on the gum required purchasing it cheaply before others with similar intentions drove up prices in the Caribbean and then selling the gum in London while demand for it remained high in the medical marketplace. Not only was this a time-sensitive scheme, it also depended on information asymmetry at a larger scale than Corbyn's typical overseas trade. He asked Penington "to keep [the gum] private & if possible out of your Bills of Entry, lest it should happen to be known here before thou canst ship it," a level of secrecy that would have amounted to smuggling.[74] Before Penington could act, however, the price of guaiacum fell by nearly 50 percent in Atlantic markets and continued to decline through the end of 1753 and into the next year. What Corbyn had predicted as a long-term rise in guaiacum prices turned out to be a short-term fluctuation. Though Penington never made the voyage to Saint Martin, Corbyn, anticipating prolonged high prices, had already purchased wholesale quantities of guaiacum in London and Philadelphia.[75] Needless to say, this did not turn out well for Corbyn who was left with guaiacum he had bought at the height of a bubble and now faced selling for a loss. Corbyn's adventure in commodity arbitrage embodied the daily efforts of exporters to realize returns from medicines they had sent overseas on credit.

Despite such a miscalculation, Corbyn's business continued to grow. In general, the amount of credit in the transatlantic medicine trade increased throughout the 1740s and 1750s. These decades saw Corbyn offer more credit, and on lengthier terms, to both expand his networks and accommodate overdue wartime debts. In the early eighteenth century, colonial legislatures had passed bills to shelter debtors from the claims of creditors. In response, the Colonial Debts Act of 1732 eased colonial debt litigation and made the personal property of debtors in the American colonies open to seizure and sale for the settling of debts. These laws gave creditors the ability to confiscate land, in addition to chattel, livestock, and slaves, to recover costs, effectively transforming land and people into substitutes for money. Human collateral had already been used to both secure debts and raise cash and credit in some colonies since the seventeenth century. Planter's bond became a more effective, and trustworthy, debt instrument, which increased the prevalence of bonds and mortgages in long-distance transactions. By the end of the 1740s, then, British traders could be more confident of receiving the proceeds of sales made on credit.[76] This newfound confidence coupled with rising production capacity to support bulk distribution based on more abundant credit.

Merchants of medicines remained uncertain, however, of when and how much they would receive in return for their goods given the variability of the long-distance medicine trade. Charges on trade, such as high commission rates, duties, and transportation costs, and a reduction in the legal maximum interest rate in 1752 also limited the size of returns.[77] Interest on outstanding debts alleviated some of the risk, but remained inexact and difficult to enforce. Any interest collected rarely rewarded long delays during which time exporters' money and goods sat beyond their reach. The promise of more payment later was often no substitute for less money sooner. When Caribbean correspondents carried substantial debt, for instance, Corbyn refused to send additional medicines until they supplied at least some bills or specie.[78] During a long term of credit, paper money could depreciate; customers could commit fraud, theft, or fail to pay interest; and medicines could rot in a damp storeroom. Medicines and money languishing abroad did little for Corbyn besides spoil and lose value, which made expediting sales and returns all the more imperative. He experimented with extending only six months' credit to some of his correspondents, usually medical practitioners, in an effort to settle debts more quickly. To Caribbean merchants and planters, Corbyn offered twelve months when he could "make an advantage of the ready Money" they possessed and sometimes even proposed eighteen months' credit in hopes of more lucrative future returns from wealthier customers.[79]

Accepting a range of remittances proved an opportunity to more easily move value and take advantage of price differentials without resorting to undue speculation. A bill of exchange, for instance, could be drawn against a credit balance with a metropolitan merchant house or against the anticipated value of colonial commodities, though the bill's value remained inaccessible until it was accepted in London. Merchants of medicines benefited from connections to private banks that could discount bills—purchase a bill earlier than its due date for less than its face value, that is, at a discount—on a scale that individuals typically could not. This manner of discounting expanded liquidity in a number of trades and also financed much of the slave trade during the eighteenth century.[80] Sugar was a popular option for returns in the medicine trade. It could usually see a profit in London and was already well known to merchants of medicines as a drug they used in their medicinal preparations.[81] Commodities, in contrast to bills, effectively invested a trader's returns while overseas. Corbyn applied proceeds from the sale of Caribbean produce to correspondents' accounts to satisfy preexisting debts and acted as a commission agent by offering lower fees if a consignor took medicines instead of bills or specie when goods sold.[82] Medicines thereby flowed back across the Atlantic from the sale of sugar and other planation crops in London. Accepting commodities, though, introduced its own risks because their value ultimately depended on a series of transactions. Corbyn typically received commodities only from long-standing customers who settled accounts punctually. To obtain sterling or credit in London, Corbyn would have to receive, store, and sell the goods, and then credit the customer's account with the proceeds, subtracting commission, freight, and customs. Or, Corbyn would become the owner of the goods at a colonial port and assume the entire risk of the return voyage before selling them on his own account in London. In 1749, Corbyn estimated that even though the Antiguan sugars he received from Robert James sold well, after paying all the charges he lost 15 percent more than if James had simply sent bills.[83]

In many ways, exporters remained at the mercy of what colonial agents could or would send. Thomas Richardson (1680–1761), a general merchant in Newport, Rhode Island, received medicines and clothing from Corbyn. After moving to Newport in 1712, Richardson shifted from selling British imports to the West Indian trade, sending Rhode Island goods south in exchange for sugar and other agricultural products. Nevertheless, he sold Corbyn's medicines locally, perhaps shipping them to Caribbean plantations as well. Ann (Newberry) Richardson, Thomas's wife, also specialized in importing medicines.[84] The medicines typically sold at nearly tenfold markups when they did not lay "heavy on hand," meaning when they sold quickly, according to Richardson.[85] The longer he waited to remit the proceeds, and thereby held Corbyn's money, the more Corbyn demanded bills or silver to compensate for lost interest.[86] If Richardson sent bills, Corbyn allowed returns at below current exchange rates, a politeness he did not extend to every customer.[87] Some merchants of medicines hesitated to charge interest on accounts in arrears for fear that doing so would discourage the debtor from settling the account. Instead, they marked up prices more than 50 percent from prime cost to account for that risk, but, even then, some complained that they made less than 10 percent profit on long-distance trade.[88] Having lost confidence in receiving bills or specie from Richardson, Corbyn requested New England lumber in March 1751 and authorized him to send beaver skins if the prices seemed more advantageous. Though beavers had become scarce in New England by this time, their furs for more than a century had been a common way for Native traders in the region to acquire European goods and, in effect, served as a medium of exchange. Furs posed a particular challenge because they required long chests "with plenty of Tobacco dust between each" to avoid damage from folding or worms during voyages. Corbyn did not entirely trust Richardson to select well-cured specimens that would fetch good prices either.[89] Sometimes he dismissed the colonial products Richardson offered to send, such as soap, whalebone, and lumber, whereas at other times he requested specific things, such as deer skins, bees wax, or cotton, believing they would "yield a handsome proffett" in London.[90] Even more profitable could be a slaving voyage.

Richardson's home, eighteenth-century Newport, was the center of the New England slave trade. Over a thousand slave trading voyages, more than half of those launched from North America, departed Rhode Island before the abolition of the transatlantic trade in 1807. Sugar and molasses produced by enslaved laborers in the Caribbean arrived in Rhode Island for distillation into rum, which was exported to West Africa and exchanged for captives sold in the Caribbean to propagate this cycle of exploitation. In 1764, Newport alone boasted twenty-two distilleries.[91] With its bustling maritime economy, the city supported sizeable local and regional medicine trades of which William Hunter, a physician and apothecary, availed himself. Residents relied on medicine imports from London, though a lack of regulation also meant that many could buy, sell, or make remedies locally.[92] Hunter's early life before appearing on American shores in 1752 at the age of twenty-two or twenty-three remains shrouded in mystery. He was said to have fought among the Jacobites at the Battle of Culloden in 1746 before studying medicine first at Edinburgh under Alexander Monro (1697–1767) and then at Leiden. After settling in Newport, Hunter outfitted the slave ships that lay at anchor in the harbor and the men who financed them with medicines he received from London and made in a small laboratory.[93] He forged relationships with prominent merchants, politicians, and slave traders during his career, including Aaron Lopez and the Brown family, and married the daughter of Godfrey Malbone, one of the most successful merchants and land speculators in New England at the time.[94] Hunter's ties to the transatlantic slave and medicine trades permeated his daily life. Besides the contents of his shop, his estate included mahogany furniture and Mark and Quarts, his slaves.[95]

The slave trade offered a way to pay for imported medicines from a place where the local currency lacked fungibility. In a letter to a Glaswegian merchant from 1762, Hunter warned that any money left in Rhode Island would lose value; he thus advised the merchant "the sooner you get your money out of this country the better." Value stored abroad in local currency created anxiety for British traders because those monies often depreciated compared to the pound sterling, while the interest intended to mitigate this risk could be collected only infrequently. The small colony of Rhode Island offers a case in point. A surfeit of paper money issued in the first half of the eighteenth century triggered severe inflation over the decade preceding Hunter's letter. To make matters worse, when Rhode Island did not follow Massachusetts in returning to a silver standard in 1750, its bills of credit could no longer circulate in its neighbor to the north, and its paper currency continued to lose value relative to the price of silver until 1763. Based on these circumstances, a "guinea voyage . . . is still the only true way of making remittances" from Newport, Hunter concluded. The merchant would be better served purchasing rum with local money to exchange for captives along the African coast who could then be sold in the Caribbean for sugar, which would finally be sold in London to satisfy the original debt and hopefully turn an additional profit. Hunter claimed to have used this method to remit sums to London in exchange for his medical supplies.[96] A ship captain familiar with the slave trade would have been simple for Hunter to find since slavers had sailed for years from the port financed by many of Newport's leading families. Each ship would have required medicinal supplies provided by local vendors like Hunter. Though slaving voyages did not always return a profit on the books, the sale of slaves gave New Englanders much-needed specie and bills of exchange (in other words, liquidity) to send as remittances to British creditors.[97] Hunter's opinion of the slave trade was hardly a minority one among merchants of medicines. A number of partnerships arose at the intersection of the medicine and plantation provisioning trades.[98]

Trading a decade later, Joseph Gurney Bevan typically declined taking goods on commission or as remittances. He demurred from serving as commission agent for commodities he did not fully understand, preferring "to stick to [his] own business of a Chemist & Druggist." He sold only colonial goods, including Barbadian sugar, as direct repayment for medicines, as opposed to Corbyn's strategy in the 1740s and 1750s of accepting a range of consignments.[99] Bevan generally turned away from speculative trading practices as the medicine manufacturing industry matured in London. As this happened, merchants of medicines could give better terms of credit to their overseas customers because they received longer credit from their suppliers. They also supplemented their capital stocks by borrowing on bond or negotiating bills for short-term liquidity. Such credit activity among traders has been recognized as creating an infrastructure for the industrialization and export growth that followed in the nineteenth century.[100] Even then, medicine exporters found it difficult to recoup stubborn debts, including from kin and co-religionists.[101] This was an age-old problem, but the distances and quantities involved made attempts to resolve bad debts more onerous and time consuming as the century wore on.[102]

Financial developments in the late seventeenth and early eighteenth centuries had added significant tools to the system of public credit, but neither remade exchange nor obviated its social embeddedness.[103] Because promises usually could not alone guarantee repayments, exporters also turned to bonds, mortgages, and other forms of legal recourse when dealing with delinquents. The first line of arbitration came via the epistolary networks that established the trust necessary for transaction to happen. When his written exhortations failed to induce timely payment, Thomas Corbyn turned to the transatlantic Quaker community before withholding goods or taking legal action. In 1748, John Easton, a Quaker merchant in Newport (and Elijah Collins's nephew), owed Corbyn the proceeds from a sale of medicines in 1742 that he had used for his own ends instead of remitting. The money had depreciated over the six years between the sale and Corbyn's action again Easton, while the 5 percent interest Corbyn charged did not balance the value lost compared to if Easton had paid promptly. Easton not only was late in his payments, but he had also, in effect, stolen Corbyn's interest. Corbyn desired that the Newport Monthly Meeting force Easton to remit the original sales value of more than £550 without adding any interest. While less than what Easton fully owed, this sum would have been better than nothing at all and may still have allowed for some profit. Corbyn rallied Friends and "Considerable Merchants" to his cause by evoking the Quaker ideal of "strict Honour" in trade.[104] Yet, in 1754, more than ten years after the sale, Easton had not sent anything despite numerous promises to pay. The Meeting proved unable to settle the affair, so Corbyn instructed Thomas Richardson to proceed with a sale of ten acres of Easton's estate to satisfy the debt.[105]

Arbitration from Quaker institutions seldom compelled individuals to pay in a timely manner, Corbyn discovered. Informal efforts to deal with insolvency in the transatlantic Quaker community proved largely ineffective.[106] Legal means did not always occasion a prompt or desirable outcome either. Corbyn could terminate trade with those he suspected of spending his money after debts had come due, which he noted was too "common a practise for Merchants in Jamaica," but doing so could still fail to produce a return and risked the end of a lucrative connection to the plantation complex.[107] Only after exploring several other informal options did Corbyn settle accounts by litigation. Legal action, especially at a distance, could take years to resolve, which strained a trader's cash flow in the meantime. Even with a local planter acting as his attorney, Corbyn waited more than eighteen months before he saw returns from some medicine shipments to Barbados in the 1750s. Facing such delays without guarantee of a favorable settlement depleted Corbyn's coffers and required borrowing in London to cover the interest on his own loans.[108] In another instance, Corbyn authorized William Logan (1717–76) to serve as his "true and lawfull Attorney" in the province of Pennsylvania, granting him power to "Sue for and by all lawful ways and means Recover" sums due Corbyn with "nothing in Law or Equity excepted or reserved," including arresting and imprisoning debtors.[109] As a result of his diverse approach to debt, or perhaps simply due to luck, Corbyn managed to avoid bankruptcy during his career in transatlantic trade.

Yet for others around Corbyn, "mistakes & unnecessarily entering into this Foreign Trade," he recounted, led to financial ruin.[110] John Hunt, years after Corbyn had hoped to purchase a ship with him, fell on hard times at the conclusion of the Seven Years' War when another merchant, John Pleasants, absconded to Virginia with a large sum of interest due him. Though Hunt had been a prosperous tobacco merchant, Corbyn now feared he would have to "Mortgage or Sell some of [his] Lands" to balance his accounts. Hunt gave up his foreign trade but, even then, lacked the liquidity to answer his creditors as he waited in vain for the return of almost £4,000 from Pleasants. Quaker ties did not compel Pleasants to pay, and four years later, in 1768, the affair remained unsettled. Meanwhile, Hunt's debts had grown. Mistakes in long-distance trade also threatened his reputation within the British Quaker and mercantile communities. Hunt's creditors insisted on his staying in London until he could find enough money to fulfill his obligations. If Hunt attempted to go to Virginia to deal with Pleasants himself, Corbyn wrote, his creditors would sue him for insolvency and "take the Affairs out of his hands, which will Cause the greatest reproach that has happened in Our Time." Corbyn understood that his success in overseas trade had been fortuitous as he reflected on Hunt's case and others like it in his correspondence with other merchants.[111]

A rhetorical emphasis on plain dealing did not necessarily generate more timely repayments for Joseph Gurney Bevan either. In the 1770s and 1780s, he increasingly turned to mortgages to recoup debts from correspondents connected to the plantation complex. Mortgages were by then well known to Caribbean planters who relied on them for long-term credit to purchase slaves and equipment, clear land, or endure disease, crop failure, warfare, and storms. Often, mortgages were provided as the result of mounting debts or, other times, as surety for a loan unrelated to one's crumbling finances. When a debt came due, a planter would enter into a bond for that amount with a merchant to receive more time to pay it off. If the bond remained unfulfilled after the allotted time, the merchant or his agent could attain a legal judgment against the planter. To delay action on the judgment, the planter could give the merchant a mortgage on some portion of his property. Two acts of Parliament on colonial mortgages passed in the early 1770s illustrated the contemporary hunger for mortgage money to develop new agricultural land in the West Indies. Mortgages became more common in the medicine trade as they did in other transatlantic trades.[112]

When mortgage progressed to foreclosure, merchants of medicines came to own property in the West Indies that posed fresh challenges of long-distance estate management in hopes of seeing returns in London and brought the medicine trade even closer to the plantation complex. In foreclosure, the final step to this debt process, the creditor assumed ownership of the mortgaged property, such as a sugar plantation. Sometimes, the new owner would have to take out another mortgage on the property to gain some liquidity after the opportunity costs of the capital tied up in the legal proceedings for such long and varying amounts of time.[113] In 1782, for example, Bevan received George Pugh's estate at Greenwich Hill, near the town of Montego Bay, Jamaica, (estimated at £1,100) through a mortgage to discharge Pugh's extensive debts. This arrangement enabled Pugh, a medical practitioner, to procure goods from Bevan for a store he intended to open nearby in Savanna la Mar. Rather than keep the property, Bevan instructed his representative to facilitate a sale but insisted he take no bonds in payment. After so many years of waiting for returns, Bevan desired immediate money or produce, noting that if it "should sell for less as in moderate certainty is better than more in expectation."[114] In Jamaica, metropolitan creditors faced a variety of legal obstructions and charges, as well as a planter government that enabled debtors to pay off debts with debased currency or overvalued island produce as legal tender in some instances.[115] To avoid some of these obstacles, Bevan engaged the mercantile house of Birkbeck, Blakes & Company to handle his future trade with Pugh: corresponding, shipping items on commission, and receiving returns.[116] The entrance of sugar factors into the medicine trade points to a major reason many merchants of medicines countenanced the risks of transatlantic trade. The incorporation of medicines into plantation health regimes offered significant markets of potential patients to be supplied from overseas.

Corbyn and Bevan tried personal appeals, diversification, and legal action to mitigate the effects of debt in the eighteenth century. Despite a diversified approach, tallies on the credit side of Bevan's ledger mounted by the close of the century. This was not necessarily a sign of success if that abstracted value never materialized in London.[117] Corbyn's and Bevan's experiences suggest that religious and family ties alone could no longer sustain bulk medicine exports, though neither could bills, bonds, and mortgages. The extension of easy credit introduced risk that would, in time, limit the expansion of medicine exports as more debts went unpaid following a series of credit crises and conflicts in the coming decades. For the time being, the high volume of trade this credit enabled returned profits in a variety of forms to merchants of medicines. Though their long-distance trading strategies enabled merchants of medicines to access markets of bound labor across the British empire, neither Corbyn, Bevan, nor anyone else possessed a monopoly on the transatlantic distribution of medicines.

British medicines comingled with myriad commodities as they became commonplace in the Atlantic world. Merchants of medicines acted like other transatlantic merchants exchanging a variety of commodities by necessity in the absence of a universal medium of exchange. This diversification contributed to transforming a manufacturer of medicines into a merchant of medicines. Like many others, they sought to trade with distant strangers who could connect them to lucrative markets of settlers, soldiers, and slaves. Their trade came to depend on markets of unfree migrants who often comprised the principal labor forces in the British Atlantic colonies.

Conclusion

Despite their best efforts to establish routines and manage relationships, Thomas Corbyn and Joseph Gurney Bevan often faced long delays and complicated decisions. From their letters, succeeding in the long-distance medicine trade seemed like a daunting task. Yet medicine exports and the assets of merchants of medicines rose significantly during this period, suggesting that at least some of their strategies generated returns. When faced with wartime risk and credit shortages, or simply the quotidian challenges of overseas trade, merchants of medicines devised ways to remedy the ills of distance. The decades of the 1740s and 1750s witnessed a sharp uptick in medicine exports from London due to a changing medical marketplace, colonial wars, new populations of patients, and the strategies employed by exporters. Both Corbyn and Bevan employed information asymmetries, accepted a range of remittances, speculated, and acquired mortgages often with success, though also with notable failures. Due to such activities, manufactured medicines flowed alongside other commodities to port cities, plantations, military camps, and rural trading posts. The business practices underlying these exports remained fluid throughout the century, but with large enough scale, commercial flexibility, favorable power dynamics, and luck, one could turn a profit as Corbyn, Bevan, and others did. Credit, in all its forms, proved essential to the distribution of healthcare.

The process of long-distance exchange began to alter understandings of the medicines that traversed the Atlantic world. Faced with the task of offering goods of nominally superior value in a system that prevented accurate comparisons, merchants of medicines established a sense of equivalence by obscuring price information and evoking the intangible qualities of their products. Accepting a variety of colonial produce as payment for debts also enabled medicine exporters to introduce their goods into the inventories of merchants and planters across the Atlantic colonies. Doing so brought them closer to the slave trade and plantation complex. Soon, medicines circulated as stores of value beside—and were exchanged for—sugar, furs, tar, and people. Even though concerns about efficacy lingered, in many cases they did not impede exports since manufactured medicines increasingly went to groups of unfree migrants with little choice in the medicines they were offered or forced to ingest. For treating such patients, convenience, interchangeability, and utility soon mattered more than quality. Selling ever-larger quantities of medicines to plantations, trading companies, or the armed forces required an expectation of efficacy that transcended the limitations of geographical or bodily variation that, for centuries, had determined the application of remedies.

THREE

The Possibility of Unfree Markets

British military fiscalism encouraged a bulk medicine trade to the edges of empire. Broadly speaking, empire enabled the collection of raw materials from and distribution of manufactures to colonial settlements where slaves, servants, soldiers, sailors, and trading company employees confronted British medicines, often irrelevant of consent. While unfree migrants can be understood as patients in the sense that they managed their health, interacted with practitioners, and were expected by planters, officials, or slave traders to ingest remedies provided for them, the benign language of healthcare quickly runs into the violence of slavery and other forms of unfreedom. Recognizing this tension, the distinctive experiences of enslavement, impressment, or servitude can illustrate the patterns of unfree migrants' encounters with healthcare that formed the foundation of the long-distance medicine trade. Markets for medicines did not simply arise from a profusion of goods or new production techniques in the eighteenth century but depended on institutions that organized labor at scales requiring simple, portable, and widely applicable remedies for tropical ailments. As medicines found use alongside and instead of other body-altering treatments, they shaped underlying assumptions about bodies and efficacy across the British empire: sometimes in contradictory ways.

By some accounts, the eighteenth century differed little in terms of European medical thinking from earlier periods, but it saw the spread of an ontological view of disease that supposed a more interchangeable body for the purposes of treatment. Illness had been understood to derive from a person's constitution, which meant care could vary, for instance, with geography or temperament, variables that would need to be put aside to explain using a medicine manufactured in London for a fever in Bridgetown and one in Madras.[1] The medical needs of overseas empire prompted the adoption of an ontological approach that had been visible in one form or another for centuries. Theory and practice converged in the "specific," a medicine that could have the same effect on anybody in any location suffering the same affliction and, therefore, offered the prospect of simple, widespread treatment. For example, two people beset by ague (a malarial fever) could both expect a similar treatment of Peruvian bark, even if they differed in sex, age, or skin and dwelt in different places. Accepting that medicines could act in such a way, however, was not so much a deliberate refutation of the physiological methods of individualized healthcare than a reaction to the failure of those methods to provide what the Admiralty, East India Company, and planters sought: portable medicines, produced in bulk, and disconnected from particular places or vernacular traditions to solve the omnipresent challenge of manpower across an empire rife with sickness and death.

The use of exported medicines in greater quantities during the eighteenth century reflected that manpower had become an economic concern, which transformed healthcare, by extension, into both an economic and a political-economic problem. Military officials, Company directors, or plantation owners could hardly ignore the health of those people they depended on for agriculture, commerce, and warfare. Capital from these pursuits underwrote the distribution of British medicines to colonial locations where they were imposed on diverse landscapes of healing. In hospitals, plantations, barracks, and ships, remedies classified as specifics found wider use than contemporary medical texts recommended, as cures for everything from a tumor to a toothache.[2] The turn toward widely applicable medicines appears less intellectually rigid considering patterns of colonial practice but no less crucial for implying that different bodies would react similarly to medicines.[3] Keeping bound laborers alive in the colonies seemed to require an ontological approach that helped popularize a particular kind of medicine but also medicines more generally. Merchants of medicines soon supplied military camps in New England, sugar estates in the West Indies, and Company settlements in South Asia where, in the face of overwhelming mortality, whether the products were actually effective was almost beside the point.

The medicine trade, then, owed much of its expansion to the exploitation of men, women, and children across the globe, especially in the plantation complex. Everyone born under the institution of slavery lived through a medical experience. The enslaved experienced few distinctions among curative, intimate, or violent forms of touch and little privacy or consent in medical matters, though such matters also offered opportunities to contest the power exercised over their bodies.[4] While medicines could represent hope for convalescence, they also carried reminders of trauma as another way to inscribe the violence of slavery onto some and not others in the Atlantic world.[5] For many, symptoms would ebb and pains shift though likely never fully dissipate, but the memory of the objectification medicines represented remained long after the bottles were emptied. As practitioners and planters embraced simple treatments for tropical ailments, they opened the possibility to view all persons, free and unfree, across the British empire as alike in the face of disease, a possibility soon foreclosed by hardening categories of bodily difference. What persisted instead were markets for British medicines. By the close of the eighteenth century, colonial power had come to depend on a bulk medicine trade even as the very systems extending British military fiscalism threatened to destabilize one aspect of the racial thought underpinning those systems.

For Every Disease, a Specific Treatment

The expansion of colonial trade and warfare prompted new considerations of what was desired from medicines and, by extension, what they could do. The 1690s offer a paradigmatic moment when such considerations began to shift toward a more capacious application of medicines. During the Nine Years' War (1688–97), English soldiers and sailors had faced a health crisis. An abundance of sick men overwhelmed the army's medical service in Ireland, while disease disrupted successive expeditions to the Caribbean. Naval forces had to abandon an attack on the French island of Martinique in 1693 after fever decimated sailors across the fleet despite medicines furnished by the College of Physicians. Soon after, in 1695, several ships undermanned due to disease sank on their return voyages from another expedition to the region. The Lords of the Admiralty then began to reevaluate the Royal Navy's medical supply to avoid repeating the recent misfortunes. They sought more economical and reliable remedies. The college's medicines, which followed humoral and place-based logics, did not fulfill the Admiralty's desire to more efficiently treat the diseases sailors suffered in a range of locations. Physicians prescribed complicated individualized treatments according to patients' constitutions and circumstances, whereas officials wanted simpler remedies, in other words, ones that would act like specifics. When the college refused to accommodate this request, it lost a lucrative contract with the Admiralty. Other merchants of medicines, including the London Society of Apothecaries, jumped at the opportunity to provide the kinds of medicines needed by institutions that oversaw large groups of people. Nevertheless,

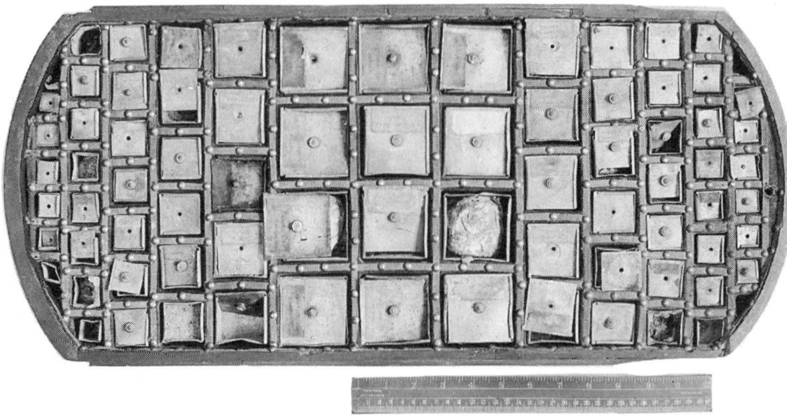

Figure 3.1 Naval medicine chest. An early modern medicine chest demonstrating the range of drugs aboard some ships before the shift toward bulk, chemical medicines in the eighteenth century. Image courtesy Wellcome Collection.

competition for contracts and high mortality would continue to characterize military medicine. Shortages of medical supplies among troops stationed abroad compelled further research into specifics with a profusion of publications on the topic emerging after 1740. Medical news traveled widely and shaped public opinion about the government. Observers around the Atlantic read about the casualties suffered by British troops at Havana in 1762 when more men died in two months of peace than during the years of combat across North America that preceded it, for example. The weight of therapeutic choices combined with other material pressures to shift expectations about medicines more generally.[6]

Physicians, along with many others at the time, could hardly comprehend treating alike the scores of military personnel and slaves who populated the eighteenth-century empire. As the Admiralty's decision to purchase medicines it hoped would act like specifics suggests, the requirements of managing groups of people motivated new expectations for manufactured medicines. The forced and voluntary migrations of the early modern world confronted multitudes with unfamiliar disease environments and brutal labor regimes. These dislocations necessitated adjustments to preexisting therapeutic frameworks. For example, Britons disembarking in settlements across the Indian and Atlantic Oceans considered whether their expectations about European medicines would apply to their new surroundings. For others, the challenge of adapting healing practices would be more difficult, though far from impossible. The concept of the specific that came into fashion in European intellectual circles in the seventeenth century helped assuage this worry for some as it, at least theoretically, assured that a collection of medicines could be used to treat the sick, whether an enslaved laborer on an Antiguan sugar estate, a soldier at Madras Hospital, or a sailor ashore in Boston. The preference for medicines that could act similarly across distances also reflected pragmatic developments particular to the British long eighteenth century: an influx of chemical remedies and new botanical drugs, the grouping of the sick in institutions, surgeons' and apothecaries' rise to prominence, and the circulation of popular medicine texts, in addition to the aforementioned voluntary and forced migrations.[7] These trends found expression in imperial institutions to propel conceptions of diseases and the bodies they affected.

During the late seventeenth and early eighteenth centuries, the classical understanding that internal imbalance caused illness—what has been termed the physiological view—ran up against the opinion that diseases had essential qualities—the ontological view. While ostensibly opposed, these two views coexisted in medical manuals and people's decision making

Figure 3.2 Eighteenth-century depictions of disease. The prolific satirist Thomas Rowlandson (1756–1827) depicts ague and fever as creatures menacing a patient, illustrating public recognition of an ontological view of disease. *Ague & Fever*, colored etching by Thomas Rowlandson after J. Dunthorne, 1788. Image courtesy Wellcome Collection.

about their bodies for centuries. Some ailments, such as plague or venereal disease, had long been seen in an ontological way, which had roots in folk practices emphasizing a take-this-for-that approach. Religious doctrine had also at times presented diseases as ontological entities aside from classical medical physiology.[8] A generally ontological view now presented advantages for imperial healthcare. No longer was disease simply an interior matter; it was becoming an entity unto itself that threatened the body from without. The body gradually began to lose its uniqueness and become interchangeable for treatment purposes since all bodies were subject to a similar group of ailments in an ontological frame. Put another way, if a disease no longer owed to the particular conditions of someone's constitution, it follows that groups of bodies afflicted with similar symptoms could be treated by similar remedies. Such logic supported the use of medicines in far-flung locations among diverse peoples, even if increasingly seen through emergent racial categorizations. Surgeons in the South Sea Company turned to the same medicines for European sailors and African captives during long slave trading voyages, especially in the face of deadly disease outbreaks. In an effort to prevent mortality from a "bloody flux [that] raged among the Negroes," the *St. Michael's* surgeon "gave ym ye Same Medicines" he gave the crew while they lay at anchor off Madagascar in 1726. Despite the medicines, several people died before reaching Buenos Aires where the captives would be sold.[9] Despite its usefulness to healthcare practices being implemented across imperial institutions in the early eighteenth century, an ontological view did not completely replace other ways of thinking about disease and the body that attributed illness to inner flows, unhealthy environments, or celestial events, for example.[10] Yet the allure of medicines that could treat diseases in anyone, anywhere, supported this conception of disease, which reciprocally provided theoretical backing for treatments that were less personalized and thus more easily manufactured and distributed in bulk.

Thinking of diseases in this way would have seemed familiar to the broad swath of Britain's population steeped in the country's rich folk and self care traditions or those who could not afford to pay physicians for individualized advice. Most people at home or abroad rarely experienced the kinds of advice provided by a proper physician, instead turning to care provided by surgeons, apothecaries, or family members. Apothecaries and surgeons, in particular, provided the majority of formal healthcare (outside of domestic healing), while similarly pragmatic and empirical practitioners predominated overseas or in the military as well.[11] By the eighteenth century, more people could also hope to purchase remedies in addition to growing, foraging, or making them themselves.[12] Advertisements, inventory lists, self-care texts, and plantation manuals directed potential customers' attention toward medicines generally conforming to an ontological view, while herbals and gardening manuals spread information about the medicinal properties of plants. In laboratories, workshops, and kitchens across Britain, chemical remedies and foreign drugs entered people's daily health routines. By the late eighteenth century, medical drug imports in England were fifty times greater than they had been only two centuries earlier. Broadening markets for remedies manufactured in England from a range of imported ingredients decoupled medicines from their places of origin and cultural associations, replacing them with new expectations of efficacy and application. Manufactured medicines' portability, in particular, offered a way to avoid complications posed by the idea that remedies naturally arose in the same location as the diseases they treated.[13] That these changes occurred during a moment of quickening consumerism expedited medicines' proliferation now as necessaries instead of luxuries, and sparked the global search for other specifics. Keeping medicines at home was nothing new, but purchased medicines circulated more commonly and in greater quantities as people across the socioeconomic spectrum turned to commercial remedies to relieve common afflictions, rather than by more personalized diets or regimen alone.[14]

Both physiological and ontological approaches to health coexisted across the breadth of European medical practice, but the ontological view's convenience encouraged its wider adoption in the early eighteenth century. Its proponents espoused their views in a variety of printed texts from European medical institutions. Thomas Sydenham (1624–89) declared that diseases could be classified by their symptoms just as plants could be grouped by their attributes. Investigators, such as Sydenham or Robert Boyle, felt that showing the existence of specifics would prove that material stuff (atoms or corpuscles) animated diseases as well as treatments. Herman Boerhaave (1668–1738), a chemist, botanist, and physician who taught at the university of Leiden, likewise recognized that diseases could be distinguished one from another, and soon classifying diseases—nosology—became the subject of works including Linnaeus's *Genera morborum* (1763) and François Boissier de Sauvages de Lacroix's *Nosologia methodica* (1763). Many of the founding members of the medical school at Edinburgh attended Boerhaave's lectures on chemistry.[15] By the mid-eighteenth century, the kinds of allegory used to describe disease and the body in the Renaissance had been replaced by more material terms that followed mechanical laws.[16] Drugs such as opium and Peruvian bark became well known as specifics that were incorporated into a range of medicines produced for export. Specifics also challenged long-standing assumptions of humoral medicine because they typically were not evacuant medicines (purgative, emetic, or diuretic).[17]

News of specifics traveled farther afield via the medical texts colonial practitioners received or the instructions included in medicine chests. John Denison Hartshorn, an apprentice surgeon, physician, and apothecary in Boston, recorded the books his master Silvester Gardiner ordered from London in March 1755. They included works by Boerhaave as well as six copies of Robert James's *Pharmacopoeia Universalis* that discussed various guaiacum products as specifics for gonorrhea and a "Tincture against the Phthisic" as a specific for "hectic Fevers." While Boerhaave acknowledged some medicines as specific to a particular ailment or organ, he also scorned others portrayed "as an infallible Specific" or that "promise[d] a specific Cure."[18] The term *specific,* in other words, remained quite unspecific. The distribution of European pharmacopoeias, nevertheless, enabled colonial practitioners to prepare British medicinal items and helped foster overseas demand for them. An abridged version of Nicholas Culpepper's *Pharmacopoeia Londinensis* printed in Boston in 1708 and followed by a full edition in 1720 was the first medical book printed in English America. Texts like Culpepper's often took a more pragmatic and ontological approach that addressed concerns relating to distance or scale in healthcare.[19] Before long, British medical texts appeared in the Caribbean and Asia addressing the medicinal needs of long-distance commerce, including James Grainger's *An Essay on the More Common West-India Diseases* (1764) and John Clark's *Observations on the Diseases in Long Voyages to Hot Countries* (1773).[20]

Their functional promise encouraged writers to proclaim certain plants and preparations as specifics for a range of ailments, disconnecting the term from its strict theoretical precepts. According to William Salmon, balsam de chili was "one of the greatest Specificks in the Cure of the Palsie, Scurvy and Gout . . . transcending all other Medicaments." Later in the same 1698 text, he celebrated a spirit of sal ammoniac as "a Specifick" for urine stoppages. Salt of tartar and rose water were "indeed a Specifick against the Dropsy," he reported, while balsam of antimony could be "given as a specifick against the Dropsy, gout, Kings-Evil, and obstruction of the Courses in Women." Salmon's frequent invocation of the term *specific* to underscore the efficacy, and by extension commercial potential, of certain remedies reflects a practical, even opportunistic, understanding of what constituted specifics and what they could do.[21] Some practitioners marketed medicines of their own devising as specifics for "curing all sorts of Agues" or tried to depict seemingly anything as a specific for one ailment or another.[22] These kinds of phrases reappear in many sources. Specifics were popular in contemporary medical texts and with English consumers, though not everyone accepted their legitimacy right away.[23] As new pharmacological trials identified additional applications for specifics, the idea of a remedy's specificity gradually transformed into one of more general usage that influenced the ways these medicines could be deployed outside the European medical marketplace.[24] Since there existed no definitive set of specifics, in print or otherwise, many common medicines could be interpreted as having extensive efficacy under an ontological view of disease, informed as it was by medical attention to large groups within the context of overseas empire. A broad conception of specifics also proved expeditious to healthcare in the armed forces.

Specifics offered an opportunity to quell pressing concerns about manpower. Rather than by its administrative capabilities alone, the British state's credibility during the eighteenth century's imperial wars hung on its ability to provide what appeared to be adequate medical care to its armed forces. Disease came to symbolize a failure of leadership and threat of disorder, making expenditures on medicine all the more necessary as illustrated by the certificates issued to sailors noting the nature of their ailment and subsequent treatment.[25] The commercial and political rivalries between European states that motivated military conflicts across the globe in the seventeenth and eighteenth centuries drew large numbers of men to Asia, North America, and the Caribbean. Though Whig administrations avoided major European military engagement for twenty-six years after 1713, the specter of war propelled the distribution of medicines in times of conflict and of peace. Britain's ability to wage war across four continents depended on the welfare of its armed forces as part of the buildup of central power, particularly as the contest for overseas empire (in the form of territory and trade) with France gained momentum in the 1740s. Soldiers and sailors typically lived in dirty, overcrowded quarters exposed to the weather, and often with rotten rations. Under such circumstances, disease accounted for a higher portion of military mortality than injuries sustained in battle and carried public stigma absent from other military deaths. This realization prompted officials to emphasize healthcare in military planning even as costs soared, national debts mounted, and European economies sagged under the burden of colonial warfare. Average annual expenditure on the armed forces increased from £5.46 million annually during the Nine Years' War to an estimated average of £18.04 million per year during the Seven Years' War nearly half a century later. The number of military personnel had also risen during this time to nearly 170,000, representing another opportunity for merchants of medicines to sell their products in bulk to populations regularly needing care.[26]

As the colonies became theaters of war, they provided new groups of patients requiring versatile medicines at scale. During the Seven Years' War (1756–63), British regulars sailed overseas in record numbers: more than 20,000 to North America, 7,500 to the Caribbean, nearly 4,000 to India, along with around 80,000 on ships. Medicines produced in Britain soon followed, arriving up and down the Atlantic coast, from Halifax in the north to Savannah in the south.[27] Between 1688 and 1815, the Royal Navy mobilized roughly 500,000 men, volunteer and impressed, in Britain's wars against France and Spain. Impressed sailors comprised anywhere from one- to two-thirds of this total, putting them (a very distant) second to enslaved Africans among forced labor groups in the eighteenth-century British empire. In contrast to slaves, whose legal bondage was hereditary and permanent, impressed sailors could go free when wars ended and received wages for their work. Overcrowded, poorly ventilated, and vermin-infested men-of-war exacerbated mortality rates that neared 50 percent from disease alone for sailors on Caribbean naval expeditions. Such staggering losses from tropical diseases exacerbated the navy's considerable manpower issues. Efforts to improve sailors' health became a major part of plans to maintain the British navy's competitive advantage.[28] Elsewhere, practitioners in Boston prepared medicine chests of drugs, compound medicines, and utensils for an expedition to Niagara in 1755. They also treated sick and wounded seamen who returned to harbor and sent "Surgeons Necessary's" to navy vessels anchored in Nova Scotia.[29] An extensive business existed for new or refitted chests in a variety of contents, sizes, and styles for the "East & West Indies, The Army, Navy and Shipping."[30] Medicine exports quickly followed once British forces captured a territory from a European rival, such as in the cases of Martinique and Guadeloupe. After the invasion of Martinique in 1762, 4,284 pounds of medicines reached the island before it was given back to France after treaty negotiations in 1763. Similarly, Guadeloupe received an average of almost 13,000 pounds of medicines yearly during the British occupation of the French sugar island from 1759 until 1763. Official ledgers read as chronicles of British territorial gains and losses based on which colonies did or did not receive medicines in a given year.[31]

If one was fortunate enough to get one, a contract to supply an institution such as the navy could guarantee income for merchants of medicines as the trade grew more competitive by the mid-eighteenth century. For example, the Society of Apothecaries saw substantial profit from its sales to the Royal Navy, its navy stock returning larger dividends to shareholders as the century wore on.[32] The desirability of this business spurred rivalries among those who hoped to secure a contract with the navy. In 1742, Silvanus and

Timothy Bevan of Plough Court pharmacy proposed to supply the Sick and Hurt Board's Caribbean hospitals with medicines from their inventory in Jamaica, offering a price 35 percent less than what the contracted apothecaries charged or 20 percent less than what their own laboratory in London charged.[33] The Bevans' attempt to wrest some naval supply from the apothecaries proved unsuccessful, though not long afterwards, in 1755, William Cookworthy received a contract to serve the hospital ship *Rupert* docked at Plymouth in Devon. The master and wardens of the Society of Apothecaries tried to block Cookworthy from doing so, but he supplied the navy on small contracts throughout the Seven Years' War. Despite such competition, the apothecaries continued to sell medicines to the armed forces well into the nineteenth century.[34]

Endemic disease made healthcare a daily concern among British troops. Fevers, particularly malaria and yellow fever, already had a long history of inflicting severe losses on military expeditions and settlement schemes in the Americas.[35] Specifics, in the form of various tinctures, teas, and pills, offered the possibility of treating some of these diseases, but in practice they often failed to match the lofty expectations set for them in print. One day in January 1756, John Denison Hartshorn made his usual rounds to check on the troops who had just returned to Boston tired, wounded, and sick after the capture of Fort Beauséjour on the Isthmus of Chignecto in Nova Scotia. The volunteers complained of a litany of ailments, including difficulty breathing, aches, chills, deafness, diarrhea, cough, weakness, and swollen tongues. To those diagnosed with fever, Hartshorn administered several medicines, including a snakeroot tea believed to act as a specific for certain fevers. While some of his patients survived for more than a week, others, such as twenty-five-year-old Ebenezer Styles, succumbed quickly.[36] Hartshorn noticed that even as some of the soldiers improved, they passed on a fever to the nurses, often their principal caregivers. In a matter of days nineteen were sick. Though Hartshorn did not have time to "write the cases of each," he noted giving the women "the same treatment as the soldiers" for "the contagion." Soon, it claimed him as well. On May 18, 1756, Hartshorn recorded in his diary feeling "prodigious tired" after visiting patients all day. He never wrote in the diary again, dying shortly thereafter, presumably of the fever, at the age of twenty.[37]

Chances at recovery seemed to depend less on the kinds of individualized evidence that had once influenced prognosis than on the theoretical potential of specifics to transcend distances and human variation. Customers often had to look to Europe to fulfill their desires for familiar medicines because Britain's Atlantic colonies were not self-sufficient, especially in terms of health-related items. In some cases, drugs popular in Europe were native to the Americas—guaiacum, sarsaparilla, and sassafras for the treatment of venereal disease, for example—but production of most compound medicines continued to occur in European centers of overseas trade.[38] Since no large-scale medicine manufacturing yet existed in North America, colonial vendors had to rely on transatlantic trade to maintain their inventories. Apothecaries, druggists, and physicians in Atlantic port cities communicated with exporters in London to acquire medicines that they then put to a variety of uses.[39] Wealthy colonists also ordered medicines or purchased them locally from intermediaries. William Logan, the Quaker merchant in Philadelphia, typically received "Apothecary Ware" from Plough Court for his family's "pretty large" health requirements, prompted by his opinion that medicines were "often wanting" outside the city.[40] With the quickening pace of exports, British medicines used in the manner of specifics with fewer place-based requirements for efficacy could be bought at prices and in quantities that made them convenient for faraway customers with a range of needs. They could be prescribed by colonial practitioners regardless of where the patient had fallen ill or lay suffering. For example, Peter Renaudet, a medical apprentice in New York City, used a stomachic tincture described in the *Pharmacopoeia Edinburgensis* to treat the swollen legs of a gentleman from Philadelphia who had become sick in Jamaica and then sought treatment in New York in 1742 or 1743.[41]

The promotion of an ontological view of disease in the eighteenth century made it easier to envision soldiers, sailors, servants, and slaves across the British empire as consumers of medicines. Specifics theoretically offered a solution to the pressing challenge of managing groups of people in disparate geographies with extreme mortality rates. Traders, officials, and plantation owners sought quick, portable, and efficacious medicines stripped of preexisting cultural associations, which both reflected and reinforced an acceptance of and, indeed, enthusiasm for specifics. Remedies expected to behave in such a way had for a long time been an instrument of the poor and rural in Europe, but an ontological approach spread from the vernacular to the elite and imperial because it offered a convenient way to think about colonial healthcare. The wider application of specifics, often broadly defined and manufactured in London, also proved mutually beneficial to agents of empire and merchants of medicines. Plantation doctors and East India Company surgeons relied on them to treat enslaved laborers and soldiers on almost opposite ends of the globe. Medicine exports from London grew exponentially after 1700, driven largely by remedies produced for markets of unfree migrants in the armed forces, plantation complex, and trading companies where those products served as medicines of desperation. They could offer a sense of hope or an opportunity to advance political-economic agendas, but those who turned to them often had few alternatives whether that was in the slave society of Jamaica or company-state of Madras.

British Medicines in the Company-State

The English East India Company (EIC), first incorporated in 1600, offered another intersection of health and empire. Similar to the armed forces, it moved people across the globe who required streamlined therapeutic options and thereby prompted familiar questions of maintaining a workforce in foreign environments: questions yet again answered with British medicines. That treating a sick or wounded person in the Company's service remained cheaper than training a replacement institutionalized a reliance on imported medicines and led to experimentation with various treatments in the eighteenth century.[42] Officials responded to disease pressures and labor requirements by implementing a centralized system for distributing healthcare to the variety of patients in Company settlements across South Asia.[43] Merchants of medicines competed to supply those settlements with medicines each trading season at significant expense to the EIC. To make decisions about from whom to purchase medicines, Company officials evaluated the efficacy of medicines in India and tested new treatments in hospitals across the region. Company surgeons could exert some measure of influence on medicine orders and thus promoted a preference for British medicines that would shape standards of care and infrastructures for distributing them across the company-state.

Expansion of the EIC's territorial sovereignty in India during the eighteenth century set the stage for increasing demand for medicines from the troops and employees, both European and Indian, dwelling in its settlements. The alliance of capital and the state that developed during the 1690s enabled the EIC's evolution from a maritime trading company into a sovereign company-state over the ensuing decades. By 1700, it had become a colonial proprietor with a network of outposts, factories, and plantations across Asia and the South Atlantic. Conflict with the successor states of the Mughal empire after 1707 coupled with the extension of European wars to India drove officials to send more troops to Company outposts in the 1740s and 1750s. Victory at the Battle of Plassey in 1757 during the Seven Years' War and acquisition of the Mughal office of *diwan* (revenue collector and administrator) in eastern India in 1765 pushed the commercial institution further toward governing its South Asian territory as a sovereign state.[44] As they gained more territorial responsibilities, Company officials faced the challenge of keeping healthy the scores of functionaries, laborers, sailors, servants, and soldiers required to maintain their rule.[45] This undertaking quickly became an expensive one as adoption of a more expansionist policy led to the first regular forces stationed in Bengal and Madras Presidencies by midcentury.[46] The EIC's army had swelled from 500 soldiers in 1750 to almost 3,000 in the 1760s. Royal troops also supported Company forces on the ground. Frequent shortages of European troops led officers to recruit Indians into sepoy battalions, who numbered more than 25,000 in Bengal alone by the late 1760s, as well as to renew efforts to forestall sickness among Europeans.[47]

The relative availability of medicines prompted concern in Company settlements and on voyages to reach them. Crowded quarters and lengthy passages resulted in high mortality aboard Company ships. When the average voyage from England to India lasted almost four months, diseases had ample time to proliferate among the passengers, though "a sufficient stock of medicines" was believed to reduce mortality.[48] In terms of treatment, Company surgeons relied less on bloodletting by midcentury than mercury-based specifics to treat fevers and fluxes among the growing number of patients under their care.[49] That many of these medicines arrived from Europe stoked anxieties about a lack of medical self-sufficiency within the Company but did not quell ordering from London. The surgeons "required so much beyond what we have hitherto been furnished with" noted officials at Fort William (Bengal Presidency), they feared "the supplies [of medicines] hitherto sent [were] by no means equal to the demand" in November 1766. During the Company's campaigns of territorial expansion, military officials, such as Robert Clive (1725–74), and surgeons alike worried about delays in the delivery of medicines to troops in the field.[50] Settlements faced acute shortages when shipments were damaged or destroyed during the passage from Europe, such as in April 1766 when medicines were "badly needed" at Madras Hospital in Fort St. George (southwest of Fort William in Madras Presidency) after the anticipated supplies were lost at sea.[51] Such shortages were often described in life-or-death terms because the settlements featured high morbidity and mortality rates, which officials hoped could also be mitigated through proper hygiene, discipline, and treatment.[52] The Seven Years' War, in particular, taught administrators that disease would determine the EIC's success in India, sparking decades of dynamic growth in medicine exports to the region.[53]

To keep employees laboring in the EIC's service, Company ships carried medicines from England, around the Cape of Good Hope, and across the Indian Ocean. The largest shipments generally occurred during times of war when manpower concerns could more easily justify higher expenses to the Court of Directors.[54] India experienced rapid growth in medicine imports during the eighteenth century according to official records, despite the considerable distances and risks involved (see table 2.2). Though they received medicines from England for decades, the principal Company settlements—Fort St. George, Bengal (Fort William), and Bombay—saw major increases in shipments of manufactured medicines after the start of the Seven Years' War in Asia. Smaller Company outposts in the region, including Fort Marlborough on the island of Sumatra and Fort St. David along India's Coromandel Coast, also received medicines from London. Not only did the size of shipments increase, but medicines began to comprise a larger portion of merchandise transported aboard Company ships. Another significant spike in medicine exports occurred during the EIC's conflict with the Kingdom of Mysore and its French allies known as the Second Anglo-Mysore War (1780–84).[55] During peacetime, by contrast, Company officials sought to limit medicine exports because they were "of considerable expence" among the supplies sent to the settlements. The Royal Navy's Sick and Hurt Board also complained about the "inconvenience" of caring for its troops in India. Efforts to impose "the strictest economy" when administering medicines did little to prevent employees from seeking imported medicines for their health, nor India from becoming one of the fastest-growing destinations for British medicines, trailing only the Caribbean in amount of medicines received from London in the second half of the eighteenth century (see fig. 3.3).[56]

Healthcare ostensibly followed a formal routine in Company settlements. Surgeons provided most of the medical services at sea, in port, and in the field, while also engaging in private trade and moneylending to generate supplementary income. Employees could receive advice or medicine without fee as part of their service, though they had to pay for treatment of venereal diseases outside Company auspices. In March 1766, for instance, hospital expenses at Fort William were set at a rate of eighteen *sonaut* rupees a man per month, or about eighteen pence per day, which was six pence less than allotted at His Majesty's hospitals elsewhere in India. Each trading season, Company surgeons compiled lists of the items they desired from England for these duties. Medicines for ordinary employee care arrived via the EIC's infrastructure, but other medicines for private trade arrived separately. Examples of orders from the Mocha factory show the breadth of drugs, compound medicines (pills, syrups, tinctures, spirits, unguents, electuaries, elixirs, oils), and chemicals sent via such channels. Often, however,

Figure 3.3 Medicine exports from London to regional destinations, 1730–80. Medicine exports expanded during the eighteenth century with the most dramatic gains visible to the Caribbean and India amid this general pattern. Exports are calculated by weight (in thousands of pounds). CUST 3/30–80. For context on compiling this data, see McCusker, "Current Value," 613, 615, 618; McCusker, "Weights and Measures," 604, 612.

these routines remained more notional than realistic due to shortages, rule breaking, and other exigencies. Company surgeons turned to local vendors for medicines deemed "essential to the proper treatment of the Sick and wounded," while they also reportedly used products meant for the Company service in their own businesses. The frequency, and necessity, of sourcing country medicines, as many seemingly local articles were called, reflected the challenge of providing care at a distance to the increasing numbers of employees and troops administered by the company-state.[57]

Facing rising costs and inadequate supplies in the 1760s, the directors sought to reform the EIC's long-distance medicine delivery. Their approach was as much about infrastructure as it was about health. They feared that without oversight surgeons redirected Company medicines into their private trades in a way that contributed to the soaring medicine expenses. To curb such perceived abuses and, by extension, lower costs, the directors monitored medicine use in their settlements and more carefully evaluated future orders.[58] These practices mirrored other attempts by officials in the armed forces to collect information on the condition of troops in different theaters of the Seven Years' War. Such data, such as rates of sickness, influenced British strategy and medical theories on surviving in foreign climates.[59] The strategy seemed to work in the short term since it also created a contest among merchants of medicines, who by this time had more manufacturing capacity, for the contracts to supply the Company settlements. Each year saw new attempts to lobby the Court of Directors for a share of the season's medicines. In 1760, Silvanus and Timothy Bevan provided 60 percent, Alexander Johnston and Company 30 percent, and the London Society of Apothecaries 10 percent of the annual supply.[60] Yet, in 1766 the apothecaries received a nominal monopoly for the trading season based on the perceived dependability of their medicines in faraway places, recalling the similar shift in the Royal Navy more than half a century earlier. The directors admitted that the apothecaries' prices "exceed those of other persons," but there was a "certainty of being supplied by them with the best sorts" that motivated their decision—a sentiment shared by the surgeons and officials abroad who requested medicines from the apothecaries' laboratory. The Society of Apothecaries saw significant returns from this trade, which continued into the next century.[61]

The promise of portable, effective medicines aroused hopes for improved healthcare in India, as it did in the Caribbean, but such hopes also depended on the EIC's recently installed medical surveillance. The directors demanded more precise recordkeeping of medicines, drugs, utensils, instruments, and hospital patients in the settlements.[62] To better manage orders and limit supplies from outside Company auspices, they instructed employees to "report to us from time to time the quality of those [medicines] sent to your Presidency." If any were found "not [to] answer our expectations," the directors solicited observations about the particular defects so they could determine "whom to blame in case of failure in Quantity or Quality."[63] That surgeons in India could test medicines sent by suppliers before the directors chose whom to contract for the next season incited competition among merchants of medicines. It also institutionalized a preference for British medicines.[64] Officials in London marked outgoing medicines according to who had made them so that Company surgeons could record observations on the items' "goodness and quality" once they reached their destinations. Suppliers emblazoned their names on the outside of packages and on each bottle, pot, or parcel to identify their products among the others.[65] Company surgeons did as they were instructed, often grousing about the medicines they had been sent in a given trading season, but nevertheless seeking more from London.[66] Observations could be gathered through this infrastructure because of the expectation that British medicines would act in predictable ways on Company employees, who had limited options for tending their health.

Perhaps as a result of these observations, the directors returned to purchasing medicines from multiple manufacturers only a few years after granting a monopoly to the apothecaries. Other druggists and chemists, including Samuel Hannay, James Bogle French, and Timothy Bevan, provided medicines into the 1770s, although surgeons in India often preferred those sent from the Society of Apothecaries that still provided medicines for the Royal Navy each year.[67] Several merchants of medicines lobbied the directors about the "exorbitant" prices and inferior quality of other people's products in an effort to recover their share of the Company medicine trade. In 1773, for example, Philip Hurlock urged them "to reconsider your late resolution," pleading that he had "really aimed to give all the satisfaction in [his] power" when he had previously provided medicines.[68] Timothy Bevan also wrote to the directors in 1773 expressing his concern at losing a share of the trade. He noted Plough Court's more than thirty years of sales to the EIC and reminded the directors that the Committee of Shipping had commended him in 1758 for "'serv[ing] the Company several years with great Reputation.'" In all his years of service, Bevan argued, "no cause of complaint [had ever] been brought against [the] quality" of Plough Court's medicines. To support his claims, Bevan included documents testifying to the "goodness of our Medicines" from merchants who exported them across the Atlantic.[69] Nevertheless, such evidence did not persuade the directors to abandon their business with the Society of Apothecaries. Company settlements continued to receive much of their imported medicines from the apothecaries' laboratory even in years when several manufacturers filled the orders.[70]

The settlements proved an opportune setting to assess the efficacy of treatments developed for places with similar concerns about labor and health. On hearing reports of a "Nostrum for the Cure of the Dysentery & all inward Bleedings" in the West Indies, the directors hurried to "[give] this Medicine a Tryal, in hopes of its preventing the Loss of many Persons in India where such Numbers are carried off by that dreadful Disorder." They sent bottles of the medicine to Company surgeons in Madras and Bombay with instructions to use the medicine and report the results thereof.[71] Surgeons at Fort St. George gave Ruspine's Styptic Solution "a fair trial" in 1786 to help the directors "judge of the beneficial Consequences to be expected from the uses of this Styptic; and how far it may be proper to give suitable encouragement to the Inventor by sending annually such Supplies of it to our different Settlements."[72] They also tested local cures for tropical diseases and disorders. The directors made clear that surgeons should serve as conduits of medical information "for the public good" from India to London where official judgment on any findings would be rendered.[73] Sailors, slaves, soldiers, prisoners, and the poor usually had little choice about participating in these tests or agency in their treatment at the hands of Company surgeons, plantation doctors, or other medical practitioners. Though the objectives of such experimentation varied depending on the location or subject, countless unfree migrants endured it with varying degrees of coercion and severity as the search for new cures continued across the empire.[74]

Vulnerable bodies in Company settlements, as they would in different circumstances across the Atlantic world, offered opportunities to sell medicines and validate their usefulness to the imperial project. The significance of India as a market for exported medicines continued to grow, a trend acknowledged by officials at Madras who hoped to "increase the demand for European Medicines to an object of commercial importance" by establishing a dispensary there later in the century.[75] The territorial and commercial presence of imperial institutions, such as the East India Company, led more Britons overseas to places where unfamiliar diseases and battlefield injuries threatened their health. It also meant larger numbers of Africans and South Asians exposed to medical systems designed to arrogate their bodies into European frameworks. Some medicinal needs could be fulfilled by local plants, while many seemed to require British medicines. The company-state's capacity to transmit medicines and medical information extended its authority over the bodies of those in its barracks, hospitals, and ships. Merchants of medicines supplying the Company settlements also supplied plantations across the Caribbean where a distinctive set of conditions aligned medicines and empire but also exposed paradoxes inherent in both.

Plantation Medicine

The transatlantic slave trade that forcibly relocated millions of African captives to unfamiliar environments and brutal labor regimes presented stark questions of power and productivity on Caribbean plantations. Nearly constant labor shortages, whether due to sickness and injury, the delayed arrival of slave ships, or a dearth of capital, incentivized slaveowners to provide a measure of medical care to the enslaved. Efforts to maintain health occurred already during the slave trade era, sparking contests over treatments, of which enslaved women often bore the brunt.[76] The Atlantic plantation complex became a lucrative market for the British medicines increasingly designed for conditions of desperation and large numbers. This extension of European corporate medicine did not dull the early modern Caribbean's multifaceted healing landscape where imported medicines contended with other treatments based on what appeared effective or least bothersome. Unlike other healthways, British medicines were backed by systems of slavery and capitalism.[77] They appeared in plantation hospitals, counting houses, and doctors' bags. Most colonial settlements, but those of the Caribbean especially, relied on imported goods, including tools, construction materials, apparel, foodstuffs, and medicines. The extent of monoculture offered little hope of self-sufficiency in the West Indies.[78] The sugar islands soon became the principal destination for British medicine exports. An environment of systemic overwork, sickness, and violence necessitated frequent replenishment of medicines to treat the free and unfree, Afro-Caribbean and European alike. The incorporation of these medicines into plantation health routines simultaneously presented a radical challenge to the racial hierarchies shaping the region and underpinning an empire.

Sickness and dying assumed terrible new forms as the plantation complex transformed the greater Caribbean's demography and ecology. More than three-quarters of transatlantic migrants between 1700 and 1760 crossed the ocean as captives, and many others crossed carrying indentures or labor debts.[79] The installation of plantations for large-scale sugar cultivation across the West Indies spread disease, particularly yellow fever which became endemic to the region. Marshes formed after the removal of forests for canefields on Barbados, for example, provided excellent habitat for mosquitoes, a common vector, if not understood as such at the time. The arrival of monkeys to the Antilles on slaving voyages from West Africa between 1640 and 1690 gave yellow fever another vector: one with no natural predators on the islands and more concentrated prey.[80] The average number of slaves on a sugar estate nearly doubled from the 1740s to the 1780s by some measures. As settlements in tropical areas supported larger populations, observers' concerns about health mounted. The high mortality rates seen on the sugar plantation raised the question of whether to maintain an enslaved workforce through healthcare and reproduction or through purchasing more slaves.[81] Acclimating to the climate and disease environment of the islands typically claimed 15 to 20 percent of recently arrived captives. Squalid living conditions, poorly constructed dwellings, and brutal treatment of the sick exacerbated mortality for the enslaved, whose life expectancy in the West Indies remained less than forty years, with many not surviving two decades.[82] The most profitable crops were the deadliest: sugar above all. Male and female field workers were recorded as sick or disabled for much of their working lives. On Golden Grove estate in Jamaica, the percentage of enslaved people who were sick never fell below 10 percent at a given moment in the 1760s and 1770s, though the figure was usually larger. Other Jamaican estates featured even higher incidence of sickness.[83] Because of high child mortality and low fertility, among other factors, a steady flow of new slaves susceptible to disease and requiring medicines was necessary for planters to maintain enslaved populations in many cases.[84] Death proved generative for commerce as it did for culture and politics under these circumstances. It weighed on people's minds and influenced their economic decisions when it came to health.

With such pressures, a diverse marketplace developed for healthcare across the early modern Caribbean. Captives and free creoles brought African and Amerindian therapeutic practices to the West Indies, which ran up against the medicine of Europeans working on plantations. This varied, but not quite syncretic, healing culture relied on a range of practical experience in contrast to the scholastic models of Galen and Hippocrates that predominated in Europe. Healers of African origin worked alongside Europeans, though few glimpses remain of the hybrid application of medicines. Some ritual practitioners did use manufactured medicines purchased from British, Dutch, or French sources and learned from botanical expeditions. Planters and slave traders boasted inventories of locally made and imported medicines that in practice ignored the strict therapeutic boundaries still in place across much of Europe. An iterative approach pushed the sick and suffering to consult a range of practitioners, enslaved and free. Far from ideological, their choices depended on appearances of efficacy that afforded practitioners and patients some flexibility in a marketplace subject to the hierarchies that ordered Caribbean slave societies.[85] Effective cures could come from a variety of places: imported or local; African, Amerindian, or European. For example, the planter James Knight urged European physicians to study the "many Secrets in the Art of Physick, [that] may be obtained from the Negro Doctors . . . as it may be of great Service to themselves, as well as mankind."[86] Black healing practices could coexist with European ones as different models for explaining disease and the body's workings circulated in the Atlantic world.[87] Patients, in other words, needed not adhere to a single therapeutic system among those they encountered.

While largely unregulated for some, the medical marketplace in the British Atlantic colonies remained circumscribed for others. Unfree black practitioners often treated colonists, who variously admired or denounced (sometimes practically in the same breath) their work. Across the Caribbean, African and Afro-Creole healers outnumbered European practitioners and continued to do so until the end of slavery.[88] Enslaved practitioners provided daily care on plantations despite a range of restrictions, as refracted through planters' letters and diaries. William Hillary, a physician working in Barbados during the 1740s and 1750s, noted the effectiveness of what he interpreted as African treatments for several common ailments, such as using "caustic Juices of certain escharotic Plants" for yaws, a skin infection. These "notable Cures," he lamented, the enslaved "keep as a Secret from the white People, but preserve among themselves by Tradition." Herbal remedies attributed to Africans were also often recommended for venereal disease.[89] So prevalent and ingrained in plantation life were black healers that several colonies passed laws to prevent them from plying their trade outside certain parameters. Planters feared poisoning under the guise of medicine stemming from an overarching paranoia about conspiracy and insurrection. A 1748 Virginia law prescribed the death penalty for any enslaved person who "shall prepare, exhibit, or administer any medicine whatsoever." South Carolina passed a law several years later prohibiting anyone who made medicines from employing free persons of color or slaves in such work. Exemptions could be granted with an owner's consent, revealing the tension between restricting the practice of Afro-Creole healers and depending on it. Several enslaved men and women suffered mutilations as a result of the medicine laws between 1754 and 1771.[90]

European practitioners filled a variety of roles on Caribbean estates by the 1720s, though their numbers remained small overall. They typically provided medicines and attendance on retainer to planters for their families and enslaved workforces. Adam Anderson, a physician and surgeon in Jamaica, noted "'he had the physical care of near 4000 slaves,'" encompassing services often provided separately in Europe by physicians, surgeons, apothecaries, and midwives.[91] These practitioners also served the plantation's dominant power structure by supervising pregnancy and delivery or by recording alleged instances of induced abortions, for example. Surgeons were also said to have amputated the legs of slaves who attempted escape. Enslaved women rarely sought the services of white doctors in an effort to guard what little privacy they had.[92] Similar patterns, but on a smaller scale, were also seen among the mainland slave societies to the north. Imported medicines appear on a 1767 invoice from James Carter to the College of William & Mary for the treatment of the college's slaves. He provided medicines, extracted teeth, and applied plasters to what could have been whipping injuries.[93] Alexander Johnston (1739–87) annually received five shillings per head for treating 345 slaves on two estates of Philip Pinnock's, one of the richest men in Jamaica. Around five shillings seems to have been a typical amount for this work, though Johnston charged planters considerable sums to treat their slaves outside such a piecework arrangement. In 1770, Johnston provided bulk care to around 1,428 enslaved people, in what he called an insurance policy, on nine plantations. That number rose to 1,959 people on eighteen plantations in St. Ann's parish by 1774. The care of so many amounted to the delivery of medicines and little direct attention. Johnston preferred to attend white patients who paid more for preserving their own health, but plantation visits generated substantial returns and facilitated the sale of large quantities of medicines for which he was well compensated. By the end of his life, Johnston owned 3,000 acres in St. Ann's with 80 slaves and 329 animals thanks to his medical practice.[94]

British medicines counted among an eighteenth-century sugar estate's annual expenses. Common ailments included fevers and fluxes, as well as ulcerated limbs sustained as a result of deficient diet, lack of suitable footwear, labor-related incidents, and the routine violence of the plantation. Many proved resistant to available medical interventions.[95] Planters nevertheless stockpiled imported medicines and were expected to administer remedies as part of managing an estate, even if they had no formal medical training. Often, they tasked skilled Afro-Caribbean attendants with much of the daily care. In other cases, overseers and bookkeepers dispensed medicines in a practitioner's absence having read about them in a manual or similar text.[96] The efficacy and portability purported of specifics made applying them a relatively simple task in theory, though forcible means could be used to ensure medicines were ingested. For the enslaved persons subjected to medicines in this manner, the experience was anything but simple.[97] Shipments from Plough Court to Jamaican estates contained medicines framed as specifics, such as preparations of ipecac, opium, cinchona, jalap, rhubarb, and sarsaparilla, to assuage many tropical ailments.[98] Instructions included with the shipments described how items could be applied regardless of a patient's particular circumstance.[99] In such a framework, plantation medical guides recommended importing a narrower range of medicines.[100] John Taylor of Running Gut estate in Jamaica received medicines from the merchants Higgin & Crawford and purchased others on the island, which may have been sourced locally or come via transatlantic channels.[101] Over the years, Joseph Gurney Bevan sent balsams, chemicals, tinctures, and preparations of cinchona and opium (the two most popular specifics) to Jamaica, Barbados, and Antigua. Thomas Corbyn too distributed a variety of these items throughout the region.[102]

Plantation produce underwrote the importation of medicines. The London-based sugar factors and commission agents Long, Drake & Long carried orders regularly from Plough Court to the Swamps and Golden Grove estates in

Jamaica, both of which had considerable enslaved populations by the nineteenth century. These shipments traveled along typical British sugar routes and could be quite sizeable, weighing upwards of 200 pounds in many cases, and included newer chemical remedies (antimonials, mercurials, and nitric acid varieties) as well as botanical preparations. These medicines traveled alongside the litany of provisions imported for a sugar estate and its enslaved workforce: hoes, chains, hammers, shovels, hats, flannel, paint, and countless other goods.[103] In 1765, close to 400 enslaved individuals (valued at £11,852) lived at Golden Grove, which featured cane fields, sugar mills, and animal pens like other large estates. It also had a "Hott House," or quarantine hospital, where several enslaved men and women worked under the oversight of a practitioner named Hayward. Despite reports that the enslaved appeared healthy, mortality continued apace at Golden Grove into the 1770s due to age, injury, and sickness, spurring frequent purchases at the Kingston slave market.[104] Mortality also prompted more medicine orders, for which planters could pay by sending sugar and cotton to London. Such was the case for John Pickering (1708–68), a ship owner, plantation owner, and the most active member of Tortola's Quaker community, who became the island's foremost contact for the transatlantic medicine trade. Pickering was said to have owned about 500 slaves and, at his death in 1768, was considered one of the wealthiest planters in the West Indies. He helped Corbyn expand his trade on the islands by finding "Doctors & considerable Planters" who would purchase Corbyn's medicines while also buying medicines and making a tidy commission himself.[105]

That categories of doctor and planter were often not mutually exclusive encouraged British medicines to find use on plantations. Take, for example, Joseph Gamble (1678–1756), a Quaker physician, who participated in transatlantic literary circles, practiced medicine, and lived at the Fontabelle estate about a mile outside Bridgetown.[106] Through Gamble, Corbyn connected to the wealthy Jackman family on Barbados. Joseph Jackman (d. 1776), Gamble's nephew and also a medical practitioner, served as Corbyn's primary agent on the island during the 1750s.[107] The combination of family wealth and a thriving plantation medical practice enabled Gamble and Jackman to purchase imported medicines in bulk based on what they knew of contemporary medical theory and local experience.[108] All of Corbyn's customers on Barbados were tied to the plantation economy by owning slaves or practicing medicine, often by both. They included both leading families, such as the Skeetes, and smaller landowners. An illustrative example surfaces from Corbyn's business with Jonathan Worrell (baptized 1734), third generation on the island, who amassed considerable wealth from a

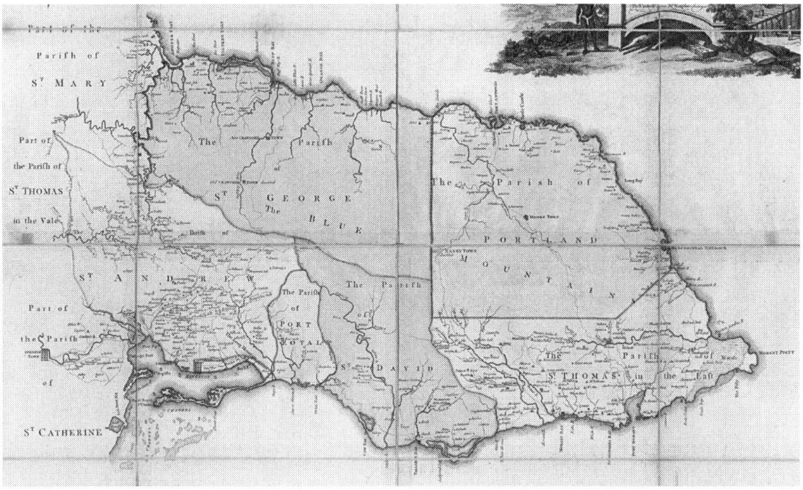

Figure 3.4 The eastern end of Jamaica, ca. 1760s. Estates dotted the landscape of eighteenth-century Jamaica. Note their density especially in the southeastern portion of the island. Each estate would have required medical services, such as the Golden Grove plantation in St. Thomas Parish (lower right) that received supplies from Plough Court in London. Craskell, *This map of the County of Surry in the Island of Jamaica*. . . . Courtesy of the John Carter Brown Library at Brown University.

number of estates, including Sturges, Neils, the Spring, and the Hill. Worrell had briefly studied medicine at Glasgow in 1752 but inherited his father's estate shortly thereafter and returned to the island, so he never worked as a medical practitioner. Nonetheless, he would have learned about specifics and known to administer them.[109] Planters also kept apothecary's scales, weights, and other instruments for preparing medicines from local and imported ingredients. The Newton plantation, for example, received powdered bark to presumably prepare specific remedies for the rampant fevers. The medicine inventory of a sugar estate, then, would have contained a mix of remedies made on site, imported ingredients, and premade medicines.[110]

Colonies with enslaved labor regimes became the principal destinations for London's medicine exports. Antigua, Barbados, Jamaica, Virginia, Maryland, and Carolina, in particular, developed robust markets for British medicines. Burgeoning demand for manufactured wares during the eighteenth century arose partly from rapid population growth coupled with rising living standards and changing tastes among the free, white population in North America.[111] Yet medicine consumption largely falls outside this narrative. Given its maturing economy, the Caribbean emerged as a pillar of the long-distance medicine trade that touched all corners of the Atlantic

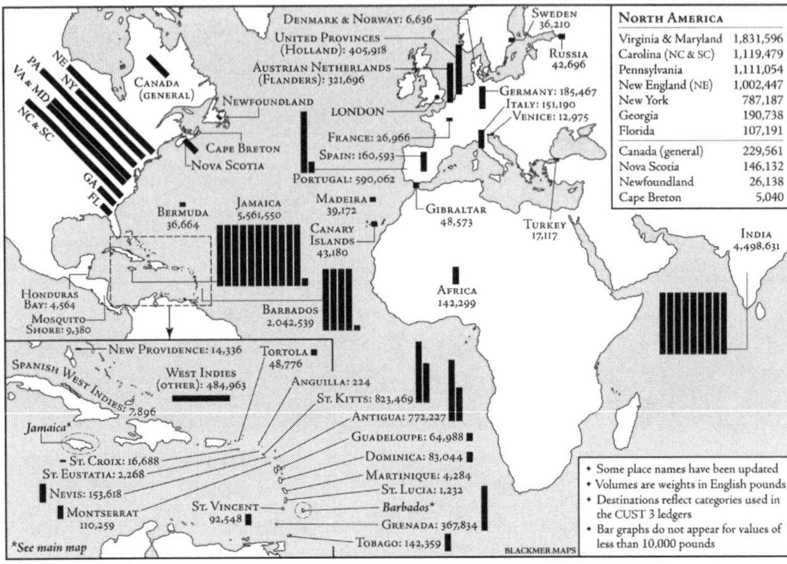

Figure 3.5 Destinations of medicine exports from London, 1697–1780. British medicines flowed across the Americas, Europe, and South Asia with particular emphasis on areas associated with plantation agriculture, the slave trade, and the East India Company. Jamaica, Barbados, and the EIC's settlements received the largest quantities of medicines, though exports large and small generally followed the British armed forces or plantation complex to points throughout the Atlantic world. CUST 3/1–82.

world (see fig. 3.5). The region generally received the largest quantities of medicines and experienced steady growth in imports for much of the century, driven largely by Jamaica. The island accommodated Britain's richest and most prominent imperial subjects, its mightiest naval squadron, and its largest enslaved workforce. During an astonishing period of economic growth from 1740 until 1776, the number of plantations increased by 45 percent and the aggregate value of Jamaica's economy increased fivefold, supported by an enslaved population that almost doubled from about 100,000 to 197,000 people and accounted for nearly 90 percent of the total population by 1788.[112] In 1774, for example, Jamaica received more than seven times the weight of medicines received by Barbados and five times that received by Virginia and Maryland. Medicine imports exceeded in value and weight those of many other provisions to the island, trends that held for much of the eighteenth century.[113] As the century wore on, the Carolinas, Virginia, and Maryland began to catch up to the West Indies in terms of annual medicine imports and surpassed them in the growth

Table 3.1 Medicine exports to Atlantic slave societies, 1700–1774

Destinations	Growth rate	Total exports (in thousands of pounds)
Barbados	3.18%	1,743
Jamaica	5.29%	4,275
Leeward Islands	5.32%	1,460
Carolina	7.97%	1,091
Virginia and Maryland	5.22%	1,818

Notes: Growth rates (k) are calculated by weight (pounds) and rounded to the nearest hundredth of a percent. Total exports are in thousands of English pounds, rounded to the nearest thousand.
Source: CUST 3/1–82.

rate of those imports (see table 3.1). While the connections necessary for bulk export of medicines to the greater Caribbean had been established decades earlier, those in the American South were still developing, reflective of the quickening pace of plantation agriculture in the region.[114] The rate at which medicine exports increased across the North Atlantic exceeded the growth rate of colonial populations, free and unfree, in those locations.[115] Higher per capita medicine consumption reflected a core conundrum of the eighteenth-century Atlantic world: how to maintain health in spite of daily violence and tropical diseases amid a system generating more and more wealth. Strategies included coercion, confinement, hygienic measures, diets, and consumption of more medicines.

In addition to along transatlantic ones, medicines moved via intra-Caribbean flows that could cross nominal imperial boundaries. British medicines exported to other European states, such as Portugal and France, may have been reexported to Brazil or Martinique, for instance, where they also would have been turned on the enslaved. Jamaica served as a major entrepôt for illicit trade, suggesting medicines were likely smuggled onward along routes that do not appear in the official records. In one notable case of interimperial smuggling, South Sea Company ships carried British medicines to Spanish ports where sailors intended to sell them under the guise of the slave trade to circumvent mercantilist trade restrictions. In 1722, Spanish authorities caught Company sailors onboard the *navio de permiso* (annual ship) *Royal George* trying to carry medicines into Cartagena alongside the other merchandise. The medicines may have been on the *Royal George* for its entire voyage or been transferred from a Company slave ship ("paquetboat") that, under the terms of Britain's slave trading agreement with Spain (the *asiento*), could carry medicines in unlimited quantities. The sailors argued that the nine boxes of medicines were for treating African captives in plantations and factories, and thus should not be taxed alongside the general trade goods. They tried to draw a distinction based on the increasingly common practice of allocating British medicines for enslaved laborers. Spanish officials identified the medicines as "European" ones that could fetch "well-known profits" on account of their appeal to colonists; they therefore concluded that the medicines were trade commodities that should be taxed. Examples such as this one hint at the larger interimperial commerce in British medicines furthered by, but not limited to, the slave trade.[116]

Expanding populations across the Caribbean spurred demand for European practitioners who, in turn, employed imported medicines in their businesses. Advertisements in eighteenth-century London newspapers sought "young men bred to physick and surgery" to serve as physicians and surgeons abroad.[117] The West Indies offered a chance for economic gain and perhaps even prosperity to young men otherwise faced with a competitive medical marketplace. Demand for medical practitioners fluctuated in the sugar islands where it remained a challenge to attract qualified ones. Charles Leslie noted in 1740 that Jamaica was "quite crowded with raw unexperienced youths" seeking medical work. Newcomers also found it difficult to gain acceptance among the inhabitants.[118] Nonetheless, the allure of fortune and dearth of prospects at home enticed young men across the Atlantic in search of a career within the plantation complex. Many relied on relationships with merchants of medicines for supplies and medical knowledge. Thomas Downes had struggled to earn a living in London, so he left for Antigua in 1756 with only his surgical instruments. He was licensed to practice medicine and surgery in St. John's parish and soon was working on the estate of Colonel King, where he claimed to have built a successful practice within a year of his arrival.[119] Antigua, though lacking the soil quality of Barbados, attracted English colonists in the seventeenth century who grew cotton, sugarcane, and tobacco.[120] The scale of bound labor used to harvest these crops necessitated a medical program on the island similar to those on Jamaica or Barbados. By early 1758, Downes visited multiple plantations and treated more than 100 enslaved persons, such that he could boast that "I am quite content & consequently richer than Thousands."[121] His prosperity would be brief. Downes typically received medicines from Thomas Corbyn in exchange for sugar. His remittances, however, remained infrequent, and unfulfilled promises to pay became a common theme of their relationship.[122] Downes continued to send Caribbean products, mainly rum and sugar, to London until his abrupt death later in 1758 presumably from disease.[123]

Orders, invoices, and ledgers can also belie the complexity of medicine use in the Caribbean. Merchants of medicines provided treatments theoretically convenient and useful for disparate populations of willing or unwilling patients, but actual application varied according to the hierarchies of race and gender that structured eighteenth-century slave societies and often remain concealed in the business archive. Such records rarely capture the daily intricacies of healthcare in the plantations, households, and urban areas of the West Indies. Merchants, apothecaries, and practitioners sold local medicines and imported ones prepared in Britain, often in large quantities, for a variety of purposes.[124] For example, Speightstown, Barbados, supported a significant free community of color and several apothecary shops. The apothecary Joseph Gamble Jackman, Joseph Jackman's son, owned several slaves who may have worked at his shop in town selling medicines received from Corbyn.[125] Details about what Jackman's apothecary shop represented in Speightstown or who comprised its clientele, for example, remain elusive, but other glimpses of contests over healthcare and its meanings emerge from contemporary plantation records and manuals. Many of the tracts that touted medicines showcased the paternalistic benevolence of West Indian planters, so if accounts mostly endure through the voices of slaveowners, other medical knowledges and practices remain more difficult to uncover. Manuals or account books miss when planters selectively ignored certain advice or when practitioners bent European learning and local experience to the ecological and economic realities of the plantation. Healing practices were caught between disparate interests: to bury supposedly superstitious or folk practices with ones validated by European scientific institutions.[126]

During the eighteenth century, practitioners such as Downes or Johnston struggled to fit British medicines into the health cultures they encountered in the Caribbean. They initially dismissed many Afro-Caribbean treatments but soon saw some ailments as requiring them.[127] The planter Edward Long found several of the "chief medicaments among the Negroes," particularly "about thirty different herbs," to be "wonderfully powerful" in cases "which have foiled the art of European surgeons at the factories," though unsurprisingly he acknowledged little creativeness, expertise, or intentionality among enslaved healers. Others generally agreed with these opinions.[128] Therapeutic practices of folk healers and nurses most likely surpassed other forms of plantation healthcare in effectiveness, though this did not prevent practitioners, planters, manufacturers, and authors from recommending British medicines in large quantities as part of the plantation health regime. If a planter was willing to pay for imported medicines, little incentive existed for practitioners to curtail their ordering, even if enslaved people refused them.

Practitioners encouraged overordering and overdosing since they could charge more for the sale of imported medicines than other services or goods they typically provided. Practitioners and planters also justified overdosing by suggesting that people of African descent possessed a higher tolerance than people of European descent for the strong, heroic medicines that induced bleeding, vomiting, and purging. The medicines a slave received relative to a colonist could vary in dosage and severity despite similar afflictions and prescriptions.[129] The practicality of stocking medicines applicable to free and unfree persons alike encouraged conversations about specifics between merchants of medicines and their overseas customers, which gave theoretical heft to such an approach. Joseph Gurney Bevan discussed the promise of cinchona (Jesuit's or Peruvian bark) with doctors John Moodie and Alexander McIntyre, who operated in partnership as medicine importers in Kingston, Jamaica. Bevan also described various specifics for syphilis in correspondence with Jonas Langford Blizzard, a physician who attended several plantations on Antigua.[130] By contrast, John Quier, a practitioner in Jamaica from 1767 into the 1800s, dismissed harsh medicines as simply easing the minds of those who could pay for them.[131] Yet in many cases there were few alternatives to these medicines of desperation. Though planters drew on Afro-Caribbean healing practices, these approaches were also disregarded for cultural reasons that are covered elsewhere as British medical ideas were imposed on Afro-Caribbean bodies.[132]

The emerging genre of plantation manuals endorsed British medicines for enslaved adults and children of both sexes, as planters and practitioners examined the conditions that afflicted the enslaved. Over time, such texts eroded the universal idea of specifics even as they promoted imported medicines for common plantation ailments. From a manual, planters, their employees, and practitioners could read that British medicines would indeed be effective in the Caribbean and for non-Europeans. They provided detailed instructions on which medicines to order and when to administer them.[133] There is a long history of Britons living in slave societies and then writing guidebooks on management or treatises on disease. James Grainger's *An Essay on the More Common West-India Diseases*, published in 1764, was the first of its kind to offer advice to West Indian planters specifically for managing the health of slaves, and many more followed.[134] Grainger (ca. 1721–66), a plantation owner and physician on the island of St. Kitts, studied at Edinburgh and served as an army surgeon during the Seven Years' War when he would have learned the necessity of bulk, portable medicines for groups of soldiers and encountered a variety of them in the medicine chests supplied to the armed forces. He then joined the Royal College of

Physicians in London where he practiced medicine and met a number of absentee plantation owners. Like Downes, Grainger found it difficult to earn a living in London, so in 1759 agreed to go to St. Kitts to ply his craft and tutor the young heir of a sugar estate on the island. His career flourished in the West Indies, and soon Grainger owned a plantation and advised physicians and slaveowners on treating enslaved laborers.[135] Arising from this experience, his text recommended a variety of remedies for fevers and fluxes, but in the case of diarrhea, for example, it favored local plants and foods to ease the bowels. For yaws, Grainger suggested European medicines, such as mercurial pills from the *London Dispensatory* or green balsam from the *Edinburgh Dispensatory*, alongside several Afro-Caribbean ones using the leaves of the castor bush (*Ricinus communis*) for more severe cases.[136] Works published a few decades later promoted imported European medicines for households of the American South lacking access to a medical practitioner. In these circumstances, remedies needed to stay fresh for long spans and have straightforward applications; thus manufactured medicines couched in the language of specifics offered a convenient solution to the limitations of the kinds of care available in rural areas with unpredictable incidence of disease.[137]

Medicines appeared so convenient partly because they did not interfere with the demands of plantation agriculture while replacing complicated body-altering treatments with simpler ones among people portrayed as interchangeable. With routines of confinement, punishment, and surveillance, medicines comprised a method of healthcare designed to enforce discipline and productivity.[138] Grainger's work shows the influence of the classical canon, but he also touted medicines directed to ailments of the plantation. For fluxes among enslaved laborers he suggested vomits and purges (hallmarks of the Galenic tradition), as well as laudanum (a form of opium), a specific. Some British medicines, such as the "mercurial pill of the Edinburgh Dispensatory" he praised in cases of the liver and spleen ailments that were common among people, both enslaved and free, in the West Indies. Preparations of opium and mercury were easy to transport and administer to groups, as surgeons in the armed forces and EIC had discovered.[139] The climate made some ailments more common in the Caribbean than in Europe, according to Grainger, rendering some long-standing approaches, namely bleeding and purging, ineffective in curing them, but it did not alter the condition of the body so much as to preclude employing well-known specifics and forms of regimen.[140] While practitioners turned to European medicines for chronic diseases, "accute diseases are often so sudden and violent in these climates," wrote the planter Samuel Martin in 1750, "that all remedies come too late."[141] Practitioners employed a variety of measures, including opiates, bleeding, purging, shaved heads, blankets soaked in seawater, and antimony in the face of fever.[142] The opinion that acute diseases could be prevented only with a combination of diet, regimen, hygiene, and medicines gained popularity among observers, while European medicines continued to be celebrated for pleurisies, inflammatory fevers, and chronic diseases as "wonderful cures"—a phrase of Martin's reused by other authors in subsequent texts.[143] Natural histories confirmed a preference for medicines that more directly targeted a particular ailment as opposed to Galenic medicines that targeted an individual's constitution. Some types of bitter pills and clysters from Britain, for example, emerged as popular alternatives for treating dysentery and other distempers, though they burned the skin and irritated the throat.[144] The pain or side effects of various remedies mattered less than mitigating the ailment and returning to work. In times of sickness or for minor offences, enslaved individuals could be confined to the hospital, instead of the plantation's dungeon, as "a place of security, where they suffer a privation of amusements, and are forthcoming to their labour," according to one planter and practitioner. Plantation hospitals contained restraining devices to limit mobility and thereby hasten recovery or deliver abuse. Windows were usually fortified to prevent escape. In these settings, planters, overseers, or practitioners forced medicines, such as calomel (mercurous chloride), antimonial powders, or other purgatives, on enslaved men and women for purposes of control and convalescence.[145]

By the end of the eighteenth century, British medicines had become commonplace enough in the Caribbean for authors to question whether specifics indeed acted on black bodies as they were thought to on white bodies. An inherently different body, useful for supporting white supremacy and sugar production, would eventually supplant conceptions of a universal body during the next century. This outcome, however, was hardly foreordained in the late eighteenth century when planters and slaves alike received medicines expected to act as specifics despite emergent recognition of fixed bodily differences. A universal, or at least practically interchangeable, body proved necessary to the emergence of a global trade in British medicines tailored to institutionalized healthcare across the colonies. The self-styled "Professional Planter," Dr. Collins, a physician and planter on the island of St. Vincent, declared that purchasing manufactured medicines for slaves was well worth the expense to avoid sickness and increase productivity. Collins instructed other planters to maintain stores of foreign drugs and European medicines, and referred them to merchants of medicines in London for those supplies. To that end, he also provided a list of medicines to order annually in bulk and directions for how most effectively to administer them on an estate of up to 300 enslaved persons. Local botanical drugs could be substituted for imported ones—and often were—but only "at an expense of time and labour," he warned readers of his *Practical Rules for the Management and Medical Treatment of Negro Slaves in the Sugar Colonies* (1803). Convenience remained paramount.[146] Grainger, nearly forty years earlier, had also recommended that "every owner of an estate ought to have" medicines and various ingredients "sent him annually from England."[147] Now, amid rising calls for abolition, Collins again emphasized medicines as one facet of how healthcare could sustain a workforce and planter profits, though he provided a poor guide to those experiencing the consequences of that debate every day.[148]

On publication, Collins's became the most comprehensive and widespread manual for the medical treatment of slaves in the West Indies. Like others, it acknowledged the necessity of medical care for the enslaved while reinforcing their value as chattel through such care. Beyond improving the management of sugar plantations, Collins had another goal with his text: to draw distinctions between individuals based on the ways they responded to environment, labor, disease, and medicine in the Americas. Recalling earlier tracts, he differentiated between the severity of ailments suffered by those of Afro-Caribbean and European descent. Dysentery, according to Collins, affected black people more severely than white people "so that the two varieties of men seem to pass out of life by two different outlets; the one by fluxes, and the other by fevers." Such a statement amounted to an assertion of physiological difference marked by skin color in terms of disease susceptibility and treatment effectiveness even if Collins lamented that "the knife of the anatomist, however, has never been able to detect [different internal organization]."[149] Theories of constitutional race could pose a dilemma for the long-distance utility of specifics; but, as in previous centuries, the effectiveness such medicines offered persisted in certain contexts because they helped maintain subjugated and healthy workforces across an expanding empire. Comparing Europeans and Africans in such a way nevertheless contributed to the ongoing process of establishing biological characteristics as a basis of human difference, even if many in the field of European medicine still adhered to a general concept of human uniformity at the end of the eighteenth century. Race gradually emerged as an essential medical category through efforts to compose a coherent tropical medicine.[150] For his part, Collins still recommended European medicines for enslaved populations, meaning merchants of medicines could continue business as usual supplying the plantation complex. Whether planters accepted at face value every word of assertions like Collins's, such statements lent medical authority to expectations that diasporic African bodies differed innately from white bodies in the context of American slavery.[151] They also normalized a logic of overdosing and forced medication, practices benefiting the long-distance trade in British medicines that persisted after ameliorative legislation in 1787 and the legal closure of the transatlantic slave trade in 1807.

As a result of their experiences, writers in the West Indies increasingly framed their prescriptions as outside of, but no less efficacious than, the tenets of institutional European medicine. Grainger had advocated incorporating British medicines into plantation health regimes in the 1760s, and four decades later Collins encouraged planters to import medicines while cautioning that treating slaves required a departure from "the rules laid down by European practitioners," including from "the refinement practised with respect to white patients."[152] The nature of this departure can be interpreted several ways. It reflected the adoption of portable medicines suited for large populations and exported in bulk to the greater Caribbean. Across the region, enslaved men, women, and children comprised a mass market of patients, though one barred from many typical consumer behaviors. This market nevertheless returned significant profits to Europeans involved in the medicine trade. The opinions of Grainger and others reflected the convenience of British medicines intended for groups of unfree migrants even as they were distinguished from others who fell ill, suffered, and died in tropical locations. Collins's reference also reflected the preclusion of some neo-Hippocratic theories of human variation and the universal body by more fixed categories of bodily difference.[153] The plantation offered openings for new tenets of medicine to diverge from European precedents in a region of significant wealth, mortality, and hierarchy. Under such conditions, medicines could embody opposing things: pain and palliation, punishment and resistance, sameness and difference. American slavery provided markets for the medicine trade that catalyzed the shift from a universal to a racialized body as part of the longer process of institutionalizing tropical medicine across the European empires. The relationship between disease and difference during the long eighteenth century culminated in a connection between physical variation and racialized medical characteristics that took shape only once bulk medicines had become fixtures of plantation health regimes.

Conclusion

In the long eighteenth century, the distances separating where medicines were made and where their intended patients lived prompted several questions: would remedies spoil before reaching Bridgetown or Madras, and would they work on non-European bodies? Medical infrastructures in the Atlantic and Indian Oceans propagated an imperial approach to healthcare that affirmatively answered the first question. Over time, unfree migrants comprised markets for the bulk export of British medicines that provided answers to the second question. Faced with maintaining the health of soldiers, sailors, and slaves in far-flung places, surgeons, planters, and officials sought simple, portable medicines that treated tropical diseases in a less individualized way. Though borne of desperation, reliance on exported medicines also derived from an ontological view of disease that emerged from European intellectual circles and vernacular practices. Specifics seemed to provide a solution to the chronic labor shortages that gained renewed urgency with the demographic, economic, and territorial changes of the eighteenth century. Losses of manpower in the colonies could hurt Britain's prosperity, which led commentators to attach economic significance to medicines in discussions of political economy.[154] Medicine exports from London expanded dramatically, especially to the armed forces, East India Company, and plantation complex. These exports consisted of specifics as well as other medicines made possible and popular by the idea of specifics and the infrastructure of empire. Many of London's largest manufacturers relied on overseas trade, which prevented others from accessing the profits of institutional contracts.[155] Medicines shipped to the plantation complex, in particular, tied successive generations of merchants of medicines to the West Indies. Many healers—domestic, folk, or otherwise—could not reap the material wealth generated by this association between medicines and global commerce. The infrastructures devised by bureaucrats, Company officials, merchants of medicines, and planters to connect supply to demand transferred wealth to some and forced medicines onto others.

An imperial medicine trade also transformed conceptions of human difference. Pragmatic health needs in the colonies coupled with the ontological framework of specifics presented an opportunity for conceptualizing a universal body that responded similarly to different diseases and treatments regardless of origin, location, or physiology. Unfree bodies provided markets for this kind of simple, portable, and broadly applicable approach. A desire to control those bodies eroded the radical potential of specifics even as they became widespread. Instead of a remedy that treated a certain disease regardless of the patient's condition, as Sydenham had written, a specific became defined more generally as "such medicines as are more infallible than any other in any particular disorder" in the *Edinburgh Medical and Physical Dictionary* published in 1807. This quotation also appeared in a number of other publications around the same time, suggesting just how pervasive but also how empty the notion had become at the start of the nineteenth century.[156] In other words, sameness had served a purpose to boost medicine exports, but there was also a use for difference in an empire. Medicines had been such a successful vector for the former that they became a vector for the latter. Responding to the broader pressures facing the armed forces, company-state, and plantation complex, European observers began to see medicines affecting diasporic Africans and South Asians differently than themselves, gradually undermining the ontological basis of specifics.[157] Manufactured medicines nevertheless remained a standard throughout imperial institutions given their established convenience as markets, exports, and profits continued to grow. That infrastructures had developed to accommodate such medicines and rely on them rendered a once essential part of their efficacy, the universal body, no longer necessary for the expectations medicines carried moving forward. Medicines continued to infiltrate local economies around the Atlantic world and provoke debates about the cost and effectiveness of imported remedies and the labor they supported. They also provided opportunities for importers to diversify their businesses, as was the case in the New England woods.

FOUR

Pine Trees and Profits

While medicines shaped healthcare regimes in slave societies and trading companies, they offered a slightly different set of options across New England, a colonial space that illustrates another facet of a globalizing medicine trade. Thousands of miles from the plantations and factories of the West and East Indies, Silvester Gardiner (1707–86) watched as his sloop the fifty-ton *Kennebeck* sailed out of Boston harbor for the first time on a "Clear & Pleasant" day in May 1753. Soon, the forty-five-ton *Industry* joined the *Kennebeck* in his service, and together the two vessels plied the Atlantic coastal waters carrying food, medicine, potash, and lumber between Boston, Newfoundland, Maine, and Virginia.[1] These voyages knit together Gardiner's various trades and lined his pockets. When the Continental Army confiscated what they could of Gardiner's estate in 1776, they found more than £2,000 of medicines, an impressive stock given a single medicine chest sold for around £20.[2] Gardiner represents what a successful Anglo-American distributor of medicines could accomplish not necessarily in healing but in wealth accumulation as a result of the medicine trade. While medicine offered a good start to that end, Gardiner diversified into the typical colonial investments: land speculation, provisioning, and shipping. Gardiner had begun his career in the 1730s as a formally trained surgeon before entering the medicine trade less than a decade later. Soon, he was speculating in land and distributing colonial goods in his own ships. How Gardiner did so is as much a story of good fortune as it is of shrewd tactics and privilege, and a common one in which colonial merchants amassed significant wealth at the expense of those who rented the land. For Gardiner, receiving British medicines in the North American colonies appeared less a calling than a way to gain wealth and status, which offers a few clues to the extractive processes and motivations behind an expanding medicine trade.

The maturing commercial infrastructure unlocked opportunities for Anglo-American traders to tie the medicine trade to other major regional trades of the era, namely in provisions and timber, thereby diversifying their approach to transatlantic commerce. The Atlantic economy by the mid-eighteenth century had grown more coherent, connected, and complex, though it remained unpredictable.[3] Men and women distributed medicines they received from abroad or prepared themselves along pathways set by land speculators, settlers, and others motivated by the promise of colonial natural resources. Their actions represented another way to disseminate manufactured medicines and the profits therefrom. In Gardiner's case, his trade benefited from the imagined and real bounty of the Kennebec River Valley in Maine and Boston's position at the nexus of military, agricultural, and shipping interests. Medicines became one among many commodities supplied to northern settlers, slave traders, military expeditions, or rural doctors from Boston with several economic, environmental, and political consequences. Returns from these enterprises underwrote an emergent American medicine industry, while the infrastructure supporting them concentrated profits from northern New England among absentee landowners in Boston and eventually provoked backlash during the imperial crisis.

The New England Medical Landscape

Like similar gains elsewhere across the British empire, the New England medicine trade was growing in the eighteenth century. Men and women with transatlantic connections and some business acumen could assume prominent roles in the colonial medical marketplace, helped by its relative lack of formal credentialing, an insufficient supply of doctors, and the shortage of large-scale medicine manufacturing. By midcentury, traders in port cities imported British medicines and made some of their own to sell in regional and continental markets. Already fourteen apothecaries sold medicines in Boston during the 1720s, a number that rose over the coming decades as members of the medical community developed capacity to import and distribute goods.[4] Distinct from other spaces of the medicine trade, New England presented several conditions conducive to commercial diversification that contributed to an expanding transatlantic medicine trade. A noteworthy surgical career advantaged Silvester Gardiner in the colonial medicine trade, where success depended on skill, who and what one knew, and more than a little good luck and status.[5] Surgery positioned him well to conduct trade in New England, though a surgeon in London would not have experienced a similar prospect. Gardiner received formal education in surgery and learned enough pharmacy to later instruct his apprentices in preparing medicines. When John Denison Hartshorn signed apprenticeship papers with Gardiner in 1752, he promised five years of service in exchange for the opportunity to learn "Physick, Surgery, and the business of an apothecary." Gardiner's other apprentice at the time, William Jepson, kept Gardiner's "Rules to be remembered in Trade" in his diary.[6] The networks colonial practitioners built represented another interface between medicines and patients as well as one between producers and consumers separated by long distances.

A practical surgeon-based healthcare prospered in New England as it did in other colonial spaces. Gardiner's training enabled him to bring an ontological approach to the medicine trade, but his was also pragmatic in its concern with expertise and profit. As a child, he investigated the plants around his family's estate in South Kingston, Rhode Island. This early interest, as well as his patrimony, would take him first to Boston and then to Europe in pursuit of a medical education. Gardiner's surgical career began in earnest at fourteen when he began an apprenticeship to John Gibbins (d. 1760), an English physician in Boston. After turning twenty in 1727, Gardiner journeyed to Europe to continue his training. He spent the next eight years (1727–34) in the surgical halls of Paris and London, notably studying under William Cheselden (1688–1752), a specialist in removing gallstones and bladder stones. From Cheselden, Gardiner learned a new lateral method of cutting for bladder stones, a procedure known as lithotomy, which would later bring him success as a practitioner.[7] Gardiner married the daughter of his former master, Ann Gibbins, in King's Chapel, Boston (where he would later become a vestryman) during a visit home in 1732 and opened his own practice focusing on obstetrics and surgery on his permanent return to Boston in 1734. He quickly joined the ranks of New England's medical elite since few European-educated medical practitioners operated in the North American colonies in the eighteenth century, and fewer still attempted lithotomies: by some estimates, only one out of nine practitioners had had any formal training at the time.[8] Gardiner's training thus distinguished him from other surgeons in New England and facilitated his connections across the Atlantic.

Lithotomy, the cutting out of bladder stones, remained a rarely performed procedure due to its risks and expense for the patient, but it occupied a prominent place in the minds of the medical public as one of only a handful of major surgical procedures performed by practitioners in Britain, Europe, and the American colonies before the nineteenth century. Its risks were high, with failure often meaning death for the patient and a damaged reputation for the practitioner, though the promise of renown and economic reward awaited the successful surgeon.[9] Given his alleged quickness with the lateral method—he could perform a lithotomy in under a minute, an astonishing feat at the time—and success rates, Gardiner's surgeries often drew a crowd. In October 1741, for example, the *Boston Weekly News-Letter* and *Boston Evening-Post* featured accounts of Gardiner's successful extraction of a stone from Joseph Baker, age six, "without Distraction or Dilaceration of the Parts, which too frequently kill the Patient."[10] An audience attended the Baker surgery, and seven practitioners traveled to observe Gardiner operate on Samuel Brown on October 15, 1754. By some estimates, at least fourteen men, women, and children underwent apparently successful lithotomies in the American colonies before 1773. Gardiner is known to have performed five of these surgeries.[11] Mortality rates for lithotomies in Europe ranged from 8 percent to as high as 45 percent and averaged above 30 percent. Although it is impossible to define mortality rates for Gardiner's practice, the extant records suggest that none of his lithotomy patients died immediately following surgery, though not all of his surgeries were entirely successful. Such was the case for Colonel Amos Turner of Scituate, Massachusetts, who endured the first lithotomy in North America in 1738. Though the initial surgery was successful, Turner died the following year. An autopsy revealed additional bladder stones that Gardiner had seemingly missed. Gardiner lost a suit against the Turner estate and did not receive any pay for the lithotomy, though his expertise emerged unscathed.[12] Gardiner also treated funguses, let blood, amputated limbs, and set broken bones for patients. Though he was recognized as the best-known surgical lithotomist in New England, given the risks and expense of major operations and competition for services, surgery alone could not support Gardiner's aspirations financially. By 1744, he had begun selling a variety of drugs and medicines to his patients.[13]

Gardiner combined previously specialized roles in medicine: the surgeon, the physician, the apothecary, the shopkeeper, and the wholesaler. While such breadth was not uncommon in the colonies, the scale at which Gardiner did so was. He prepared his own medicinal recipes and kept abreast of the latest practices from London while seeing patients and performing surgeries.[14] *Doctor* remained a term loosely applied in New England and often self-proclaimed, as there existed little professional organization, certification, or regulation before the late eighteenth century, which afforded many the opportunity to try their hand at medical practice. Practitioners could often fulfill a variety of services because occupational boundaries remained fluid and care was pragmatic and iterative. Practitioners mixed

A Medical *Society in* Boston New-England, *with no quackish View, as is the manner of some; but for the Comfort and Benefit of the unhappy and miserable Sufferers by the excruciating Pain, occasioned by a Stone in the Bladder, do Publish the following Case.*

IN *Boston, Octo.* 8 1741, *Joseph Baker* Æt. 6, was cut for the Stone in the Bladder, according to Mr. *Chefelden*'s late Improvement of the *lateral* Way, by Mr. *Gardner* of *Boston* (who had some Part of his Education in the Hospitals and Infirmaries of *London* and *Paris*) in Presence of the *Medical Society* and others without reserve.

This Boy by his Complaints seems to have had the Rudiments of this Stone from his Birth, and by degrees the Paroxyms of Pain became so violent, that Death or the Operation were unavoidable. This Stone was extracted in the *lateral* Method without Distraction or Dilaceration of the Parts, which too frequently kill the Patient in a few Days or Weeks; or if the Patient escapes with his Life, the Urine continues to issue involuntarily by the Wound, during an uncomfortable Life. Several such miserable Instances we have had in this Province for want of Skill and Discretion in the Operator.

This Stone when extracted, was of a lenticular Form, somewhat elongated, imitating the converging Part of the Bladder, its surface *instar Lapidis arenosi*, but more hard and compact, the Circumference 4.75 Inches; its centeral Thickness 8.5 Inches; the longest Diameter 1.5 Inch; the shortest Diameter 1.25 Inch.

The Dressings were soft easy and simple, with a Milk Diet to mitigate the scalding of the Water in the Wound: The 4th Day the Urine began to trickle the natural Way; the 11th Day it passed the natural Way in a full strong Stream; 15th Day no more Urine by the Wound; and makes Water only 3 or 4 times in 24 Hours; after the 23d Day, no more Dressing, the Wound being well cicatrized and the Boy at Play about the House.

Thus in about 3 Weeks a compleat and easy Cure for the Stone in the Bladder, was affected by the *lateral* Way.

Figure 4.1 Notice of a successful lithotomy, 1741. Descriptions of surgeries he performed circulated in Boston newspapers during Silvester Gardiner's career, often with details emphasizing his surgical expertise. *Boston Weekly News-Letter*, 13 Nov. 1741, [2]. From NewsBank, Inc., and the American Antiquarian Society.

various techniques, remedies, and theories in what they did based on a range of informal and formal training. Surgeons sold medicines, physicians handled patients, and apothecaries made house calls, for example. Domestic household medicine also persisted in eighteenth-century New England despite broader trends of commercialization.[15] Fewer opportunities existed to make medicines in bulk, so those entering the medicine trade in New England had often worked as practitioners, whereas those in London had apprenticed as apothecaries or merchants. An apprentice to Gardiner was responsible for many tasks. Hartshorn collected debts, drew up accounts, copied letters, took notes, rolled pills, filled bottles, packed chests, ran errands, collected news, and treated poorer patients including some sailors, servants, and free people of color.[16]

By contemporary accounts, Gardiner was the leading apothecary in colonial New England, selling "all Sorts of Drugs and Medicines, both Chymical and Galenical" to a variety of customers across the region, including members of Boston's mercantile, political, and religious elite. He prepared remedies for Governor William Shirley's daughters and the merchant Thomas Paine, for example. Gardiner displayed patent medicines, compound medicines, and botanical drugs received from abroad alongside his own medicinal preparations that incorporated a range of specifics at his shop under "the Sign of the *Unicorn* and *Mortar*" on Boston's Marlborough Street.[17] Such wares enticed potential customers and visually assured them of Gardiner's large, well-prepared stock of the "freshest and best of all Kinds of Drugs and Medicines," recalling the organization of Thomas Corbyn's shop in London. This presentation depended not only on the perceived quality of what Gardiner and his apprentices prepared in Boston, but also on that of medicines and drugs he received from London.[18] Gardiner gained the confidence of his customers and fellow practitioners to provide high-quality, reliable products due in large part to his surgical work. Medical practitioners shared information on a variety of topics, including treatments, news, and anatomy. Other surgeons sometimes sought Gardiner for his expertise, which encouraged subsequent medicine sales.[19] Gardiner also sold materials in bulk to rural practitioners, in some cases so they could set up apothecary shops of their own and presumably continue to purchase medicines from him.[20] John Wheeler of Rehoboth, Massachusetts, for example, purchased medicines from Gardiner for at least sixteen consecutive years; some of these were recipes tailored to individual needs, and others were premade, imported medicines.[21] Gardiner's most prosperous clients remained, however, Boston's commercial and political grandees who underwrote the importation of medicines as merchants and planters did elsewhere around the

Atlantic world. Distributors, such as Gardiner, benefited from the military and slave trade economies that shaped northern colonial ports in both subtle and unsubtle ways during the eighteenth century.

British medicines appeared on the shelves of specialty retailers as well as in some of mid-eighteenth-century Boston's most prominent shops. John Loring promised "a Compleat Assortment of Medicines, both Chymical and Galenical" at his shop fronting Cornhill Street in Boston and offered to supply "Country Practitioners and others . . . on the most easy Terms."[22] Across the river in Charlestown, Isaac Foster Jr. advertised a similar selection of British medicines "cheap for Cash or short Credit" at his shop with the "Sign of the *Pestle* and *Mortar*."[23] Peter Roberts, a self-described surgeon and apothecary, imported drugs and medicines from London and prepared "Doctor's Boxes for Ships & private Families" including "proper Directions" for using the items.[24] Beginning in the 1760s, Jane Eustis advertised several patent medicines and tinctures alongside fabrics and garments at her shop opposite the Boston Town House. Eustis described herself as a milliner, selling a variety of imported lace, necklaces, hoops, stays, gloves, mittens, thread, silks, and clothing beginning in 1755.[25] Women advertised medicines in print throughout the colonies.[26]

Since no large-scale drug manufacturing yet existed in North America, colonial practitioners and traders relied on transatlantic trade to maintain their stocks. From 1720 to 1774, the value (and weight) of recorded annual medicine exports to New England increased more than eightfold. The region surpassed others, with the exception of some areas of the Caribbean, in importing medicines, which were often resold via coastal trade networks.[27] Gardiner received orders from London at least twice yearly throughout the 1750s, with five arriving in 1757 alone.[28] When ships met misfortune during the Atlantic passage, Gardiner faced shortages of drugs and chemicals. He had "not an ounce of Spirit Nitri" remaining in Boston and lost 800 boxes of Lockyer's Pills after the *Elizabeth*, captained by John Bradford, was blown off course on route from London to Boston in December 1759 and then taken by French privateers near Antigua several months later. Though far from the only British ship to be lost or captured in these war years, news of Captain Bradford and his crew appeared in print up and down the coast in the spring of 1760.[29] Gardiner's shipments of medicines and the materials required to make them usually came from four London wholesalers: Kilby, Barnard & Parker; Harrison & Ansley; Hughes & Whitlock; and Cluer Dicey & Company.[30] While much of Gardiner's drug stocks came from Europe, he also purchased items such as castor from Native traders.[31] Others in New England had imported medicines before Gardiner began to do so, but he and his contemporaries traded at a larger scale than their regional predecessors: across the Atlantic, along the coast, and into the countryside in times of peace and of war.[32]

By 1760, Gardiner had become one of Boston's highest-volume retailers thanks to local settlement patterns, military contracts, and connections to long-distance trade. He kept a large stock on hand, and his income in these years exceeded that of an ordinary practitioner by perhaps sevenfold.[33] Gardiner's apprentices kept busy preparing orders for rural medical practitioners, military regiments, merchant seamen, and settlers across Connecticut, Maine, Massachusetts, New Hampshire, New York, and Nova Scotia. One 1769 sale to Robert Southgate of Hardwick, Massachusetts, totaled just over £23, nearly twice what most residents earned in a year.[34] Since medicines often went unnoticed by customs officials or unrecorded on shipping lists, much of the transatlantic and coastal commerce remains unseen in the archive.[35] That Gardiner's sales could be quite large, however, underscores the variety and quantity of items he supplied to his customers. He also treated sick and wounded seamen and assembled medicine chests for navy and merchant vessels. An extensive business existed for assembling new chests and refitting old ones for the "East & West Indies, The Army, Navy and Shipping." Such "cags and chests of medicines," filled with an assortment of remedies, had become a common feature of British voyages, settlements, hospitals, and military maneuvers across the globe. Gardiner supplied medical stores for both Massachusetts expeditions to Niagara and Crown Point (Fort St. Frédéric on Lake Champlain) organized by Governor William Shirley in 1755. Of the hurried preparations, Hartshorn wrote, "it was Sunday [June 15, 1755] and all of us staid at home all day to stow the medicines for the Governour's regiment," though neither expedition accomplished much compared to the high hopes for them (and in contrast to the Nova Scotia expedition of 1755). "Trooping and training in town," in Hartshorn's words, brought opportunities to sell medicines, as did the arrival of men-of-war in Boston harbor in the spring of 1756. Gardiner also sent medicines to the navy hospital at Annapolis Royal in Nova Scotia.[36]

New England supported numerous participants in the medicine trade by midcentury. Since the 1740s, Thomas Corbyn had supplied a handful of contacts across the region, including John Greenleaf in Boston and Jabez Bowen in Providence, Rhode Island, who sometimes competed with Gardiner.[37] Greenleaf's business grew in the 1750s, supported by medicines from Corbyn & Company and Hugh Hughes (later of Hughes & Whitlock who supplied many in the area) and rising demand from Massachusetts's sizeable military and mercantile communities. Greenleaf received smaller shipments from London than Gardiner, but nonetheless carried on a transatlantic trade in drugs, chemicals, compound medicines, medical texts, glassware, and surgical instruments that expanded during the 1760s. He paid for these goods with a combination of bills drawn in London and specie, relying on connections to Boston merchants, including John Hancock, and to the Society for the Propagation of the Gospel in Foreign Parts to do so.[38] Greenleaf then supplied medical practitioners across the colonies of Connecticut, Massachusetts (including Maine), and New York, who would have used the products in their local practices. His clientele in the Boston area, including Joseph Warren, Samuel Tufts, James Otis, and Hezekiah Chaffee, often displayed Whig leanings, in contrast to many of Gardiner's who numbered among Boston's Anglican, loyalist, and mercantile elite. Greenleaf also provided medicines for Joseph Warren's almshouse, which generated significant yearly sales for him from 1766 to 1772.[39] Medical commerce retained strong personal connections that shaped its contours, though in the coming decades it would threaten to split along political lines.

Like Gardiner and Greenleaf, Martin Brimmer (1742–1804) imported medicinal items from London in the 1760s and 1770s. His account books show the ease with which he moved between Massachusetts money and pounds sterling to conduct this business with Harrison & Ansley and Hughes & Whitlock. Brimmer sold medicines to local practitioners of various political leanings and socioeconomic backgrounds, including Gardiner. Brimmer and Gardiner, however, typically did not exchange money, only certain medicines, since they both had their own transatlantic suppliers. Brimmer conducted a brisk business with practitioners inland who depended on distributors in port cities for British medicines and paid with a combination of cash, notes, and bonds.[40] To reach these customers, Brimmer advertised extensively in newspapers. He informed the *Boston Evening-Post's* readers that he desired mail-order business to sell a "general Assortment of Drugs and Medicines," announcing that "Country Practitioners and others will be as well supplied by Letter as if present themselves." In his advertisements, Brimmer promised to deliver "a fresh Supply [of medicines] every Spring and Fall" and to accept short credit in addition to cash at his shop near the top of Seven Star Lane in Boston. By 1768, Brimmer not only sold drugs, compound medicines, and patent medicines, but also assembled medicine chests for the ships leaving the harbor in a nearly endless stream.[41]

Together, traders like Gardiner, Eustis, Greenleaf, and Brimmer extended credit, established connections, and set expectations that expanded the infrastructure for importing and distributing medicines in New England. While similarly connected to transoceanic flows and concerns, this infrastructure at first appears less tethered to the institutions driving healthcare provision in the Caribbean or South Asia. Gardiner did not have an equivalent in Kingston or Madras, for example, due to the particularities of the medicine trade in those places, though some commercial partnerships in the West Indies approached his level of diversification. New England featured fewer bound laborers and military personnel than other locations that received British medicines in bulk, though indentured and enslaved workers were certainly not absent from eighteenth-century New England.[42] Several common demographic and intellectual trends encouraged local conditions conducive for markets for medicines to develop across the empire, namely increasing populations (especially of unfree migrants), new forms of mercantile organization, regular warfare, and a pragmatic approach to treatment. Boston merchants tapped into these trends and profited from them through shipping. From a regional perspective, the New England medicine trade depended in other perhaps less direct, but no less important, ways on plantation agriculture, military fiscalism, and settlement projects that, in turn, benefited merchants and medical practitioners.[43]

This was a regional and a continental development, rather than simply a Boston one. To the south, a robust medical community developed in Newport, Rhode Island, that similarly depended on ties to the military and slave trade. In the eighteenth century, the town supported several apothecaries and other practitioners who served local patients and supplied the ships anchored in the harbor.[44] Born in Scotland into a clerical family, John Halliburton (d. 1807) set up his medical practice in Newport around 1750 after serving as a maritime surgeon on a British frigate.[45] William Hunter, who also had military experience with medicines, arrived around the same time and began seeing patients in Newport and Providence, though neither man imported medicines to the same extent nor covered as wide an area as Gardiner. Hunter treated local tradesmen and grandees, including Aaron Lopez, members of the Brown and Malbone families, and the politician Jabez Bowen Jr. By the time of his death in 1777, Hunter had amassed a substantial stock of drugs, chemicals, and medicines, in addition to equipment for preparing them (scales, weights, mortars). His shop contained bottles of different shapes and sizes, midwifery instruments, and some "elaboratory" equipment, while his house featured fashionable decoration.[46] Such objects indicate a thriving medical practice and medicine trade, as do his accounts. Hunter's account with the merchant and slave trader Aaron Lopez reveals the range of surgical activities, visitations, and medicines he provided for Lopez, his family, and his slaves over more than a decade from 1763 to 1776. In January 1769, for example, Hunter gave Cudjo, an enslaved man in

Lopez's house, "a phiol of Pectoral Mixture" and, several days later, "a box of Anodyn Pills," while Cudjo's brother received "a large plaister for his stomach." In another instance from November 1772, Hunter dressed the sore hand of a patient he identified as "yr [Lopez's] Negro Man Fortune." Sometimes Hunter charged Lopez for visitations and in-person care, while other times he sold Lopez medicines in bulk, such as a box of pills or several bottles of mercurial liniment. He also supplied medicine chests for Lopez's slaving voyages. A single "Medicine Chest for the Coast Guinea" cost more than months of other goods and services.[47] Gardiner too had connections to Newport. He sent medicines to John Tweedy (d. 1787), an apothecary whose wife, Freelove-Sophia Tweedy (1714–68), helped tend their shop.[48]

Familiar concerns about quality arose in this burgeoning regional trade. Fears of counterfeiting, whether of paper money or medicines, crept across New England, sometimes embodied by the same person. John Hallowell, a self-proclaimed "practitioner of physick," had turned to medical practice in the late 1720s after being caught, though not convicted of, printing paper money. Two decades later in 1742, he figured prominently in the death of Sarah Grosvenor, a resident of Pomfret, Connecticut, who to end a pregnancy had ingested a powdered abortifacient he provided. The substance given Sarah may have come from John Tweedy who had already been accused of selling adulterated medicines on multiple occasions. Practitioners alleged his medicines were of poor quality or diluted such as to have little or even harmful effect, especially if the connection to the Pomfret case is to be believed. Following Sarah's death, Hallowell sued Tweedy for selling him bad medicines. Though Tweedy denied the charges, several witnesses, including the elder Jabez Bowen, himself an active importer of medicines, examined Tweedy's medicines and confirmed their spuriousness. After various countersuits, the Court of Equity sided with Hallowell that Tweedy had indeed sold counterfeit medicines.[49] The affair, however, did not prevent Tweedy from continuing, and even expanding, his apothecary business over the next decade. During the 1750s, he received medicines from Gardiner and may also have used "Chymical & Galenical Preparations" made by Corbyn to whom he was recommended by Thomas Richardson. In a letter to Tweedy, who was not a Quaker, Corbyn acknowledged the opportunity for deception "to the poor patients damage" through adulteration, but quickly reminded Tweedy that "our Maxim in Trade hath been to deal upon Honour."[50] Corbyn's words underscore the blurry line between adulteration and manufacturing where a substitution of ingredients or modification of their quantities could turn palliative to poison and debt into credit.

Meanwhile, Gardiner began prioritizing trade over medical practice, but he had not entirely given up surgery. In May 1758, for example, he wrote to his now former apprentice, William Jepson, that he had successfully "extirpated a very large Tumour" weighing more than 6 pounds "from a Boys back, hips & thigh."[51] Gardiner abandoned performing risky surgeries, such as lithotomy, only a few years later. He professed no fondness for the operation after nearly twenty years undertaking it, describing cutting for the stone as "the most difficult operation in Surgery."[52] Gardiner declined to perform a lithotomy in the spring of 1764, citing his "great Hurry of business" and his advancing age.[53] Jepson performed several lithotomies in Connecticut during the 1760s using Gardiner's training in the lateral method. He removed a "very rough Stone weighing 2 Ounces 7 three Quarters" from Mr. Bradley of Tolland in April 1767 and another of about an ounce from a ten-year-old boy in New Haven later that month. Gardiner counseled Jepson on the procedures but refused to assist him since he had begun "declining (and have been some time) all kinds of practice."[54] He thought medical practice would "interfere very much with . . . other business" and devoted more attention to making and selling medicines.[55]

Though Jepson started off as Gardiner's apprentice in surgery, he too saw the profits to be taken from the medicine trade. Gardiner had trained his apprentices in commercial as well as surgical practices, which was unusual for the time.[56] After his apprenticeship concluded in 1757, Jepson entered into partnership with Gardiner to sell medicines from a shop in Hartford, Connecticut. Gardiner furnished the stock and Jepson transacted all local business per the terms of their agreement. Gardiner supplied drugs at the prices they cost him to import and medicines he compounded at the prime cost of their ingredients.[57] Jepson, despite his training, remained rather deficient at preparing medicines himself. Gardiner urged him to hire laborers and rely on his "negro boy . . . to make most of your Plaster and unguents Tinctrs and Elixirs," which would "Save me a great deal of trouble."[58] Gardiner offered to purchase another enslaved person for Jepson in the summer of 1759. Not long afterwards, he sent Jepson "a good natured well behaved Irish man Named Timothy Malony" bound for two years to help with the business.[59] Bound labor was not uncommon in the New England medicine trade. Daniel Lathrop and his younger brother Joshua relied on enslaved laborers, wage earners, and apprentices to maintain their apothecary shop, gardens, greenhouses, stills, and warehouses in Norwich, Connecticut. Enslaved people often did similar work to their free counterparts across New England, including preparing medicines, such as in Jepson's and likely Hunter's shops, and may also have done so in Caribbean apothecary shops.[60] Gardiner generally left Jepson to conduct business without much oversight as long as he provided annual accounts, and for a few years the partnership generated a profit. Yet sending payments to London for their medicines remained a challenge that pushed Gardiner and Jepson to find other means of moving value across the Atlantic.[61]

The potash trade offered the partners a solution to their problem. Potash, a potassium salt derived from plant ashes, was the eighteenth century's foremost industrial chemical, used in processes ranging from flint glass manufacture and textile dyeing to something more familiar to Gardiner, medicine production. Abundant forests and Parliament's exemption of American potash from import duties in 1751 positioned New England as a significant source of the salt. Though prohibited from private trade in the partnership, Jepson established a potash works at Middletown, Connecticut, before the agreement concluded in 1765. These were hopeful years for the American potash trade. From Massachusetts to Virginia works sprung up to supply Britain's thriving cloth industry. Even when potash failed to return much of a profit to investors, it proved a useful commodity that usually could be sent directly to Britain without incurring a loss, which attracted the attention of merchants like John Hancock who were always looking for ways of moving value without losses. Exports rose dramatically after 1763 such that by 1775 Britain received almost two-thirds of its imported potash from North America. Gardiner initially hesitated to involve the partnership's capital and his own proceeds in the venture, but after judging that a sample of the salt "look'd good" he agreed to participate, wishing Jepson "Joy at your Success in the Potash works." Gardiner would then receive potash from Jepson's works or other regional producers and send it to London where it was sold to settle the partnership's overseas accounts.[62]

In 1767, Gardiner counseled Jepson to "keep closely to [his] own business for [he] will find more Profit in that [than he] will by changing or running into others," meaning one must specialize and be patient to maximize returns.[63] In practice, however, Gardiner rarely followed his own advice. He had turned from surgery to selling medicines and assumed risk to take advantage of changing patterns in colonial warfare, backcountry settlement, and plantation provisioning. The regional medical community had also become more crowded and competitive. This risk soon paid off as profits from the medicine trade enabled Gardiner to join the ranks of Boston's mercantile elite and invest in northern land. Emergent commercial infrastructure not only provided a conduit for British medicines to reach men and women across New England but also transformed the region's environment and system of land tenure.

Investment Downeast

Following the success of his surgical and medicine pursuits during the 1730s and 1740s, Gardiner joined a group of speculators hoping to revive an old claim to northern lands. The object of their fantasies, the Kennebec Valley, had lured investors, schemers, and settlers to the Maine woods for a century. Maine offered seemingly vast tracts of land to exploit and remained a part of Massachusetts until 1820. The Kennebec River travels more than 150 miles from Moosehead Lake to the Atlantic coast over waterfalls and through dense hardwood forests. The river offered those who have peopled its banks fish, furs, and timber, but also hardships. In the winter it could be cold and desolate, and during flood times, wild and overpowering.[64] For Gardiner it represented a better chance of acquiring real assets (land and hard money) compared to selling medicines, in light of the risks inherent in long-distance trade and the limitations imposed by trade laws. As a proprietor and then the Kennebeck Company's moderator, he planned settlements, distributed provisions, managed transportation, and dispersed credit for the company's and his own benefit. Provisioning and land speculation presented lucrative opportunities for mixed economic activity and remained the primary options for large-scale capital investment in New England. Potash or medicines, by contrast, could generate only so much revenue without the corresponding expansion of colonial resources and markets for them. Investing in land gave Gardiner access to provisioning and lumber trades where he learned the often tortuous routes to receiving returns from long-distance trade, as many others around the Atlantic world already knew. Such diversification nevertheless knit regional markets for a variety of goods, with medicines reshaping New England economic life in the eighteenth century.

Land speculation already had a long and complicated history in northern New England by the middle decades of the eighteenth century. Countless groups had attempted to buy land to exploit its resources and collect rent from settlers based on claims of varying legitimacy over the years. While the previous century saw much armed conflict in the region, the Maine frontier experienced a period of relative peace between Wabanaki leaders and colonial elites after Dummer's Treaty formally ended Father Rale's War (also known as Dummer's War and a host of other names) in 1727.[65] Large landowners and speculators—the Great Proprietors—then turned their attention to the farmers and settlers who inhabited the land they sought and often opposed their claims. The proprietors buttressed their claims to contested titles by recognizing Native rights to unsold tracts when it suited their aims, in contrast to and often against the expansionist policies of the ordinary colonists—a temporary form of equilibrium, perhaps. Yet, with the arrival in 1741 of a new Massachusetts governor, William Shirley, for whose daughters Gardiner would later prepare medicines, the prevailing balance of frontier relations faltered. Shirley supported claims by the Muscongus and Kennebeck Companies that did not rest on accepted Native land sales but on old letters patent from European governments, reversing decades of policy. The Anglo-Wabanaki alliance of Dummer's Treaty soon collapsed, and a period of renewed violence followed in Maine, exacerbated by the new land claims and intrusions on Wabanaki sovereignty.[66]

The Kennebeck Company further upset the fragile equilibrium in 1749 after Samuel Goodwin, a Charlestown housewright, found the missing Plymouth patent. The patent, a 1629 royal grant to William Bradford and the other members of the Plymouth Company (known also as the Virginia Company of Plymouth), was one of four seventeenth-century royal patents authorizing vague claims to mid-Maine lands. The document had been lost decades earlier and followed a convoluted and likely apocryphal journey before ending up in Goodwin's hands. The rediscovery of the original enabled an association of Boston merchants to claim approximately 3,000 square miles around the Kennebec River at a time when the geopolitical situation seemed favorable for development of that area. The largest shareholders among this group organized themselves as the "Proprietors of the Kennebeck Purchase from the late Colony of New Plymouth" to remove any confusion with the former Colony of New Plymouth following a Massachusetts law in 1752 permitting speculative land companies to incorporate. The "Plymouth Company," "Kennebeck Purchase Company," or simply "Kennebeck Company," as it was typically called, continued to develop its grant until 1775. Company shares quickly gained value and desirability as development occurred across the claim after the 1748 treaty of Aix-la-Chapelle eased violence in the region. The proprietors operated on a larger scale than most land speculators in northern New England. In addition to building two forts (Frankfort and Fort Western), establishing a dozen towns, clearing much acreage, and settling hundreds of families, they acted as a self-interested force in politics, intimidated opponents, and clashed with agents of the Crown over forest rights. Local opposition to the Kennebeck Company from towns, such as Brunswick, Wiscasset, Woolwich, and Georgetown, and growing resentment from settlers prompted the proprietors to adopt a litigious and deceitful policy of defending their claim.[67]

Gardiner entered this realm of money, land, and politics when he attended his first company meeting as a proprietor on December 6, 1751.[68] The medicine trade had furnished him with business acumen, familiarity

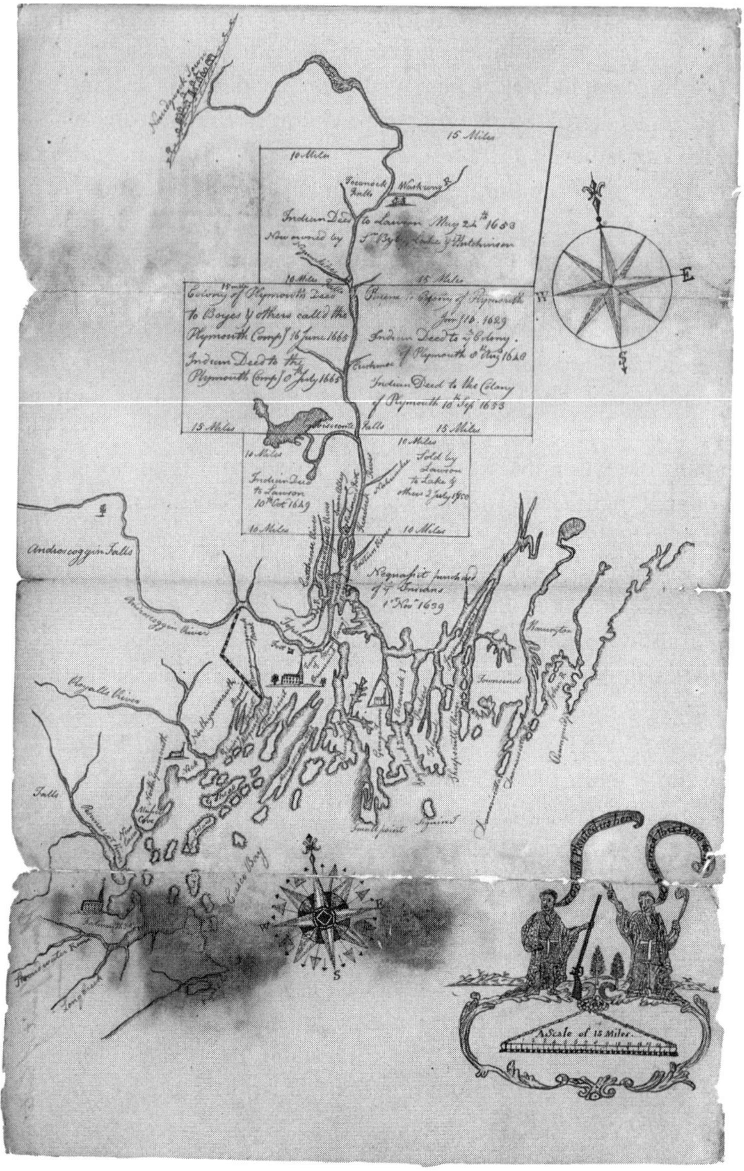
15 Miles
10 Miles
Indian Deed to Lawson May 22th 1653
Colony of Plymouth's Deed to Boyes & others call'd the Plymouth Compy 16 June 1665
Indian Deed to the Plymouth Compy 8th July 1665
Indian Deed to ye Colony of Plymouth 8th Aug 1648
Indian Deed to the Colony of Plymouth 10th Sep 1653
Falls
Indian Deed to Lawson 10th Oct 1649
Sold by Lawson to Lake & others 3 July 1650
Negnasit purchased of ye Indians 1st Nov 1639
Androscoggin Falls
Androscoggin River
Royalls River
North Yarmouth
Townsend
Smallpoint
Sequine
Casco Bay
A Scale of 15 Miles
E
W
S

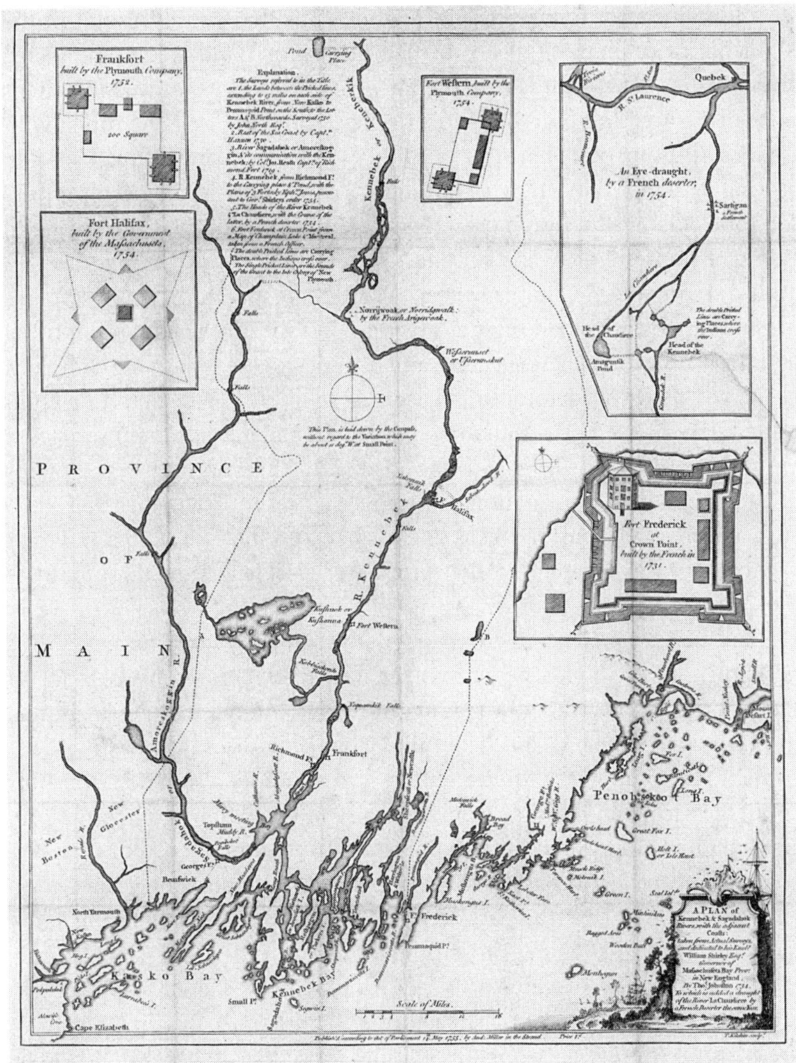

Figure 4.2 Kennebeck Company claims, ca. 1750. The manuscript map (*left*), created by Thomas Johnston in 1753, shows the lands granted to the Kennebeck proprietors on both sides of the Kennebec River. The printed map (*right*), also created by Johnston around 1755, was used in litigation between the Kennebeck proprietors and the proprietors of the Township of Brunswick. In contrast to the manuscript map, it depicts a more expansive view of the Kennebeck proprietors' claims as demarcated by the dotted lines reaching to the shore. (*Left*) Plan of part of the Eastern Shore, 1753, map 56-2-1, Kennebec Papers. Image (#11977) courtesy Collections of Maine Historical Society. (*Right*) Kennebec and Sagadahok rivers plan, 1755, map 56-2-2, Kennebec Papers. Image (#37703) courtesy Collections of Maine Historical Society.

with long-distance transport, and relationships that proved useful to the Kennebeck Company; it also provided him capital and credit. He hosted lavish parties, played cards with members of the Quincy family, and had his portrait painted by John Singleton Copley. Some evenings he dined with Charles Apthorp, one of Boston's wealthiest merchants and a churchwarden at King's Chapel, as well as a slave trader and fellow Kennebeck proprietor. John Adams, the future president, described Gardiner as possessing "a thin Grashopper Voice, and an affected Squeak; a meager Visage, and an awkward, unnatural Complaisance."[69] Gardiner's speech may not have impressed Adams, who had built a career on oratory, but he cut an imposing figure as Copley's likeness suggests, and he quickly rose within the company. He served as its moderator and a leading member of the standing committee responsible for formulating investment and development strategies. From these posts, Gardiner corresponded with surveyors and ship captains to gather as much information as he could about the Kennebec Valley. To such a degree did Gardiner act as the link between the proprietors and their agents that John Temple, another proprietor, declared him "the mouth of the Company."[70]

Gardiner assumed greater oversight of the company's trading activities as he learned about the Kennebec over the next several years. Unlike many of the proprietors, he visited the region during the summer to perform annual inspections of his land and determine what provisions, such as pig iron, corn, or rye, to purchase for the settlements. Gardiner also made decisions about the local lumber trade from his house in Pownalborough.[71] By 1760, Gardiner had established the Gardinerston Plantation at the Kennebec River's head of navigation and confluence with Cobbosseecontee Stream on land granted him over the past decade. These grants, and any profits to be had from them, depended on settling families on the land. The last mile of the Cobbossee featured a 130-foot drop, which he used to power two sawmills, a felting mill, a potash factory, and a gristmill. By some accounts, the gristmill was the only one for 50 miles for settlers to grind their corn.[72] Because of their relative isolation, settlers in the Kennebec represented markets for goods and services sold by the proprietors in addition to sources of rents. Gardiner supplied food, building materials, dry goods, medicines, and arms to settlers; mobilized their labor to build churches, dams, and mills and clear farms; and coordinated communication and shipping. This development provided the proprietors a source of rents and produce, and it created a market for manufactured goods.

Gardiner became one of the Kennebeck Company's largest sources of capital for the region's development. The company already owed Gardiner

Figure 4.3 Portrait of Silvester Gardiner, ca. 1772.
John Singleton Copley (American, 1738–1815), *Dr. Silvester Gardiner*, ca. 1772, oil on canvas, 50 × 40 in. Gift in honor of the seventy-fifth anniversary of the Seattle Art Museum, 2006. Catalog no. 125. Photo: Susan Cole.

more than £1,300 in 1754 and continued to be indebted to him into the 1770s. After another proprietor failed to build a sawmill and gristmill, Gardiner purchased the rights to the project, had the mills built, and received 1,000 acres of land, in addition to the profits from the mills, as compensation.[73] Issuing loans and investing in local infrastructure proved lucrative for Gardiner. At first, the Kennebec Valley lacked the infrastructure necessary for the kinds of regional trade and long-term settlement the proprietors envisioned.[74] In 1752, Gardiner volunteered to build a sloop, the *Kennebeck*, at his own expense to establish regular transport between Boston and the Kennebec for the company. The proprietors had hired a ship to carry families to the region in 1751, but this approach proved too costly to continue. Gardiner already organized the transport of people and goods, so the other proprietors accepted his proposal to also manage the shipping. In return, Gardiner received 400 acres and access to the sloop for his own business alongside the company's.[75] The *Kennebeck* sailed between Boston and the Kennebec at least once a month; it voyaged as far north as Newfoundland, though not always without incident.[76]

Regardless of whether Gardiner received them through his medicine trade, partnership with William Jepson, or company business, provisions offered him an effective, if not circuitous, route to tangible returns in an otherwise uncertain colonial trading scene. From 1752 to 1757, Gardiner sent supplies, usually grain, meat, arms, building materials, and medicines, worth more than £1,000 to the Kennebec for which the company compensated him with money or land.[77] Gardiner used many of the goods he received as returns from his medicine trade in New York and Connecticut to provision the Kennebec settlements, while some he kept for medicine manufacturing. Hog's lard, for example, went toward making unguents and plasters; pork went "to supply [his] eastern Settlers."[78] Connecticut's laws pertaining to bills of credit made it complicated for Gardiner to receive proceeds from his medicine trades there directly, so he relied on remittances in "the Produce of that Colony" via Jepson. Beginning in April 1751, bills from Connecticut, New Hampshire, and Rhode Island could no longer circulate in Massachusetts, which made regional trade more difficult given the lack of specie in New England. Without the equivalent of a single regional currency, Gardiner preferred Jepson to exchange Connecticut bills for provisions to avoid holding depreciating paper currency. Only later in the 1760s could Connecticut bills again circulate in Massachusetts at par with the local currency.[79] As a result, Gardiner's returns often sat with Jepson for five or six months as he purchased and shipped foodstuffs from southern New England. Such delays hampered Gardiner's ability to make timely payments to his suppliers in London. And when returns did come, they often brought other logistical difficulties. To avoid paying high duties, Gardiner asked Jepson to omit his rum from a bill of lading—in other words, to smuggle it to Boston.[80]

Gardiner's provisioning trade soon stretched beyond New England. When the Kennebec River froze in the winter, Gardiner sent his ships south to Carolina and Virginia where his agents purchased crops for the upcoming year. The ships left Boston laden with molasses and sugar, and they returned carrying tobacco, grain, lard, and pork. Gardiner held his understanding of continental markets in high esteem and believed that he could secure goods at lower prices than could his fellow proprietors, writing that they had "acted like fools" for buying pork at too high a price without first recognizing the "great quantity of Pork in Carolinas and Virginia" in 1762.[81] These activities required more than just acumen though. Gardiner wrote about needing ready money to an extent during the 1760s that he had not known over the previous ten years.[82] Tensions flared among the proprietors on the topic of provisions that had been scarce in the Kennebec by the end of the Seven Years' War. Farmers across New England and the southern colonies subsequently planted more so-called cargo crops to meet this demand. Consequently, Gardiner predicted a plentiful harvest and falling prices in the summer of 1763. He hoped to purchase extra corn at low prices for the Kennebec where, he noted, the inhabitants "cant yet from the newness of their Farms raise their own corn."[83]

On the other end of this provisioning trade, certain New England trees and their byproducts contributed to the maintenance of British military and economic power, such as through shipbuilding, dyeing, and medicine making. Donald Cummings, a physician who owned portions of a mill, provided Gardiner with an estimated 40,000 boards each season in exchange for loans and medicines.[84] Cummings, in turn, depended on Gardiner for a medical advantage in Biddeford—a burgeoning mill town that in 1766 was "Crowded with Doctors," he reported—and without Gardiner's supply, Cummings lamented that he would be "Quite out of Medicine."[85] Cummings often apologized for his unpaid debts and implored Gardiner to send him more medicines because, he noted, he "Injoys Practice anough Provided I can keep in yr favour."[86] Gardiner sold lumber from Maine to customers across Massachusetts who, in many cases, already partook of his medical care. Samuel Watts purchased cordwood, shingles, and "merchtble boards" in addition to "medicins & attendance" from Gardiner, accruing quite a large debt from 1758 to 1772.[87] Gardiner's agents determined "exactly what Sort of Lumber" customers wanted and had trees cut to specific dimensions in advance of shipment.[88] After the conclusion of the Seven Years' War, Gardiner's ships transported white pine boards, with one sloop able to carry up to 15,000 of them, from the Kennebec to the Connecticut River Valley where they reportedly sold well. Proceeds of these lumber sales, like those of medicines, would then be sent to Boston, usually as pork, wheat, corn, or rye for provisions.[89]

With their large diameters and heights, white pines in particular held enormous economic and military value across the Anglo-American world.

Europe faced extensive deforestation in the early modern period, so the old growth forests of North America and especially those of New England offered an alternative source of masts for the Royal Navy. The best masts for high tonnage ships came from a single straight tree. White pines were bigger and lighter, though less resinous and flexible, than the Baltic firs that had been the primary mast trees in England, encouraging the navy to adopt New England white pine as its new standard. A host of legislation, collectively known as the Broad Arrow policy after the mark emblazoned on the trunks, protected the precious large trees for the king and thereby sought to ensure the navy's advantage over its continental rivals. At first, the policy reserved trees over a diameter of 24 inches, but the threshold was dropped to 12 inches in 1772 to include more trees. Compensation for mast trees remained contentious in the northern woods since pines suitable for masts numbered only a few per thousand and government bounties barely covered the cost of cutting and hauling them. Crown agents, squatters, and absentee landowners fought over rights to the trees given the reward they could bring. The Broad Arrow policy proved almost impossible to enforce due to the size of the area covered and the lack of Crown agents available for the task.[90] The 1760s, however, brought renewed determination to try as settlers and proprietors across Maine increased their lumbering operations in violation of the laws. A 1762 fire destroyed many of the region's finest pine tracts after sweeping eastward from New Hampshire for nearly 50 miles before burning itself out on the shores of Casco Bay. People fleeing the conflagration moved into eastern Maine and set up new lumber camps from Falmouth to Machias. Lumbering settlements soon spread up the Kennebec and Penobscot Rivers. Thomas Scammell, a surveyor for the Crown, saw rampant violation of the pine laws when he visited the region in the early 1770s to record the output of the new mills. Inland, he found more trespassers, but also stands of old growth pine trees, one of which had a circumference of more than 17 feet, he claimed.[91]

That the timber trade was experiencing a moment of flux proved opportune to the Kennebeck proprietors. In 1766, their lumbering operations provoked the ire of John Wentworth, surveyor-general of the King's Woods and royal governor of New Hampshire. For the next seven years, Wentworth sparred with the Kennebeck Company for control of the lucrative traffic in white pine from the area. This was not the first conflict between a Crown representative and the proprietors over mast trees. The Kennebeck proprietors had clashed with Wentworth's uncle, Benning Wentworth, the previous surveyor-general, in the 1750s. For years they petitioned to reform the Broad Arrow administration and sued to disrupt the enforcement of the policy, which often had little support in colonial courts.[92] Proprietors also evaded enforcement. The implementation of the Stamp Act in 1765 made it more difficult for Gardiner to clear boards in Boston on account of the naval office there. He worried how the officers would implement the act, so he decided to avoid it altogether by shipping lumber directly between the Kennebec and New London where there were not yet naval offices.[93] For years the proprietors had also flouted the Broad Arrow policy with the wide floorboards visible throughout their Pownalborough Court House, built in 1761 along the Kennebec River. The three-story building features 45-foot supporting beams, indicative of the timber wealth of the area and the skill with which it was built, and a courtroom 45 feet long with large windows on the second story above space used as a tavern on the first. The fifty-two-inch wide plank on the third-floor landing remains an eye-catching example of the proprietors' disregard for the rule reserving trees over 24 inches in diameter for the Royal Navy (see fig. 4.4).[94] This latest clash with Wentworth resolved in the proprietors' favor on May 3, 1773, after John Adams, an acquaintance of Gardiner's and the other proprietors as well as their lawyer, argued their case to the Admiralty Court. For a few years, at least, the proprietors were able to more directly extract trees from the Kennebec.[95]

Despite the strategic importance of mast trees, it was usually more profitable to cut lumber than masts in the Maine woods. Pines felled in violation of Crown policy accounted for much of the lumber produced in eighteenth-century New England. This lumber, as boards and staves, was then profitably exported to the Caribbean and other Atlantic destinations.[96] Parts of local trees not typically processed into lumber also possessed medicinal uses. The bark, needles, and twigs of white pine could treat kidney problems, for example. In his *American Herbal* of 1801, Samuel Stearns noted the stimulant, diuretic, detergent, and antiseptic properties of white pine turpentine. Balsam of turpentine (*terebinthina*), the thick material remaining in the still after the essential oil has been removed, could be found throughout period medicine recipes. Turpentine, tar, and pitch from a variety of pine species were often used to give unguents and plasters their sticky consistencies and were useful in other industrial processes. The hackmatack or tamarack tree, better known as the Eastern larch (*Larix laricina*), became prized in wooden shipbuilding in the eighteenth century while its bark was valued for its therapeutic properties. The Micmac, for instance, used hackmatack bark to treat colds, fevers, and infections.[97] Gardiner used various turpentine mixtures and natural balsams in his medicines, though one can only surmise whether these ingredients arrived from Maine as part of his lumber trade. Other New England practitioners also recognized the

Figure 4.4 Floorboard at Pownalborough Courthouse, ca. 1761. The visibility of wide floorboards, particularly this one measuring more than 52 inches across, throughout the Kennebeck proprietors' Pownalborough courthouse flouted the Crown's forest policy. The proprietors did not even bother to have this board cut down to a width of 22 or 23 inches (just below the legal limit) as was often the custom to circumvent the restrictions. Image from the author's collection, courtesy of the Lincoln County (ME) Historical Association.

medicinal potential of local pitch pine, larch, and fir.[98] In some areas of Maine, lumbering became the equivalent of a monoculture so that settlers depended for subsistence almost entirely on imported provisions, according to some reports, such as those provided by the Kennebeck proprietors. Taxes or rents could be paid in lumber, but people could not survive on wood alone. In 1779, Timothy Parsons of Pownalborough offered any price in silver, paper, or lumber for a bushel or two of seed barley for the town. Many inhabitants had been without bread for a month, living only off what they could get from the river such that "Harty Men Are brought to Meare Skeletons," he wrote, while "A number of others Lay helpless for want of proper Sustenance."[99] Intensifying trade in a more connected Kennebec Valley meant returns for the owners of land, transportation, and supplies, though often at the expense of those who rented and worked the land.

While Gardiner reaped the rewards of his provisioning and lumber trades, Jepson's potash works neared ruin. The salt had sold well enough in Britain in 1765 and 1766 that Jepson bought more of it from other producers in Connecticut to increase the amount he could send Gardiner for the next season's export. Gardiner thought it could fetch £40 to £42 per ton in the summer of 1765 when he discharged people's debts to him or the Kennebeck Company by shipping their potash to London free of commission. This boon, however, lasted only a few years. In 1767, the price of potash dropped by more than 20 percent to under £30 per ton, throwing Jepson's plans into disarray. Exporting potash provided a convenient way to move value across the Atlantic often regardless of profit margins, but Jepson had spent too much amassing the salt in anticipation of high prices in London. Generating credit overseas was only worth so much. This miscalculation left him with a significant deficit in his personal finances and in the partnership's. Gardiner tried to wring some last profit from the venture, but when it failed, Jepson lost much of his available money. Potash manufacturing nevertheless persisted throughout the region despite the lean times.[100] Gardiner did not anticipate Jepson's struggles since to his eyes Jepson "had more advantages all things considered than a Customary young man." He had provided medicines for their store at no markup, arranged overseas trade, and advised Jepson on mercantile matters for nearly a decade.[101]

Though Jepson's losses took Gardiner by surprise, they should not have been wholly unexpected because of the risks associated with transatlantic trade. Nor were they particularly uncommon as Gardiner had witnessed over the years. When bad debts arose, he relied on legal action; forced the sale of land; and seized goods in efforts to recover money owed him. Gardiner assured tardy debtors that "[they] may depend upon being Sued."[102]

He honored promises to pay but rarely granted anyone "one moment more then [*sic*] you Promised." Suits appear fairly commonplace during medical disputes in eighteenth-century New England. Despite his recourse to legal means of debt recovery, Gardiner felt antipathy toward the colonial courts. He often spoke disparagingly of lawyers and Connecticut's General Court, which, he protested, has "no more right to take cognisance of Judicial matters more then [*sic*] the Devil has."[103] When Gardiner entered into bond, even for small amounts, he ensured that the debtor first had supplied security in the form of mortgages on land or attachments to goods.[104] Due to the scarcity of money, he also took a litigious approach to notes of hand. Gardiner demanded Jepson sue everyone who did not fulfill the notes he collected on Gardiner's behalf. Gardiner remained distrustful, if not outright disdainful, of colonial paper money following the currency manipulations of the 1750s and 1760s. He sought hard money in his transactions and accepted discounts to see sterling rather than local bills or notes.[105]

Such precautions, however, did not prevent Gardiner from receiving little of his share of the seven-year partnership with Jepson. He was owed more than £2,800 and an additional share of the remaining £1,421 4d of good debts in 1765 after supplying more than £4,000 of medicines over the course of the partnership.[106] By 1771, Jepson not only had failed to remit those proceeds but had also taken for himself more than £1,000 of Gardiner's money from private trade, by Gardiner's count, instead of conveying it to Boston as provisions. These actions reverberated across Gardiner's networks all the way to London. Throughout the later 1760s he endeavored to mollify merchants who tired of his late or incomplete remittances. An act of Parliament removing the drawback on certain drugs had also increased the prices of medicinal ingredients Gardiner imported, which only worsened the state of his accounts.[107] Many signs pointed to Jepson's responsibility for the mismanagement of the potash works and Gardiner's private trade. Yet, in a decision unusual for Gardiner, he could not bring himself to sue Jepson, his former apprentice and partner, whom he saw as a son.[108] Gardiner followed his maxim that "nothing is worth more then [*sic*] it will Fetch" in many facets of his life, business and personal, it seems.[109]

In responding to such challenges, Gardiner, like others in the medicine trade, enmeshed himself in the fabric of colonial commerce, which appears at once particular to New England but also connected to broader imperial systems of extraction. Similar circumstances provided opportunities for traders to diversify their businesses during the eighteenth century but also could result in losses or failure. Gardiner's position within the Kennebeck Company enabled him to convert goods and promises to pay into more desirable and less risky stores of value, namely land. As a proprietor, Gardiner extracted rents, sold provisions, and felled trees. That these ventures relied on credit and connections from his medicine trade further entangled British medicines and colonial natural resources. Not only were medicines made from colonial plants, but they also helped fund their extraction. Despite several challenges to his trade in the 1760s, Gardiner continued his transatlantic and regional ventures until rebellion shook the foundations of New England landownership and the North American medicine trade.

Conclusion

In the early 1770s, accusations of Toryism echoed across the Kennebec Valley fed by long-standing disputes between the settlers who worked the land and the proprietors who owned it. As the political storm clouds gathered, nascent cries for liberty merged with latent antimonopoly and anti-Anglican sentiments to spark several outbursts of violence in the autumn of 1774.[110] On September 23, Silvester Gardiner fled his house along the Kennebec River, narrowly eluding a mob hunting tea and Tories. Jacob Bailey, the reverend at Gardiner's church in Pownalborough, recounted the activity of those fraught days and nights when he was variously harangued, assaulted, and shot at. Around midnight on the twenty-third, an estimated 150 men brandishing axes and clubs surrounded Gardiner's house. Some forced their way inside where they ransacked Gardiner's belongings and drank several gallons of his rum. They rifled through his personal papers in hopes of finding clues to the whereabouts of the tea and other goods they presumed he had hidden nearby. The mob also flung Gardiner's most reliable surveyor, John Jones, headfirst into the river and "dragged him about till he was almost torn to pieces," according to Bailey.[111] As one of the principal landowners in the Kennebec who controlled much of the local trade, as well as a vocal Anglican and loyalist, Gardiner made an obvious target for the northern backcountry's frustration.

Gardiner's career illustrates the interstices of New England commerce in a connected Anglo-American world and bears the mark of patterns reshaping long-distance trade and healthcare. John Singleton Copley's 1772 portrait of Gardiner's second wife Abigail depicts her wearing a fashionable Turkish-inspired dress in a pose reflecting the family's social standing after nearly forty years of business (see fig. 4.5).[112] By the mid-eighteenth century, medicines connected medical practitioners in New England to the slave trade, plantation complex, armed forces, and land speculation. Medicine was hardly the only thread connecting these extractive enterprises,

Figure 4.5 Portrait of Abigail Pickman Gardiner, ca. 1772. Both Silvester Gardiner and his second wife Abigail Pickman, appear in portraits by Copley in contemporary fashions indicative of the family's wealth and stature in Boston after nearly forty years of business. John Singleton Copley (American, 1738–1815), *Mrs. Sylvester (Abigail Pickman) Gardiner*, ca. 1772, oil on canvas, 50 3/8 × 40 in. Brooklyn Museum, Dick S. Ramsay Fund, 65.60.

but they shared a reliance on certain kinds of medicines provided through an emergent infrastructure that made possible their consumption across a range of geographies. The pragmatic approach to healthcare witnessed in New England reflected particular colonial exigencies on the one hand and common concerns of quality and efficacy on the other. While Gardiner prepared many items in Boston, his inventory also relied on transatlantic trade. Imported medicines supported the New England medical community and generated credit that could be applied to other pursuits, such as land speculation or coastal trade, offering the possibility of more stable wealth. For the land and those who worked it, by contrast, these medicines offered another vector for the extension of power over their economic lives and even over their bodies.

Resumption of colonial warfare in the 1770s threatened to upend the markets that had structured economic life and political allegiance in the Atlantic colonies. The provenances of the portraits Copley painted of Silvester Gardiner and Abigail Pickman Gardiner epitomize the divergent paths of New England's medical community. After its confiscation along with the rest of his estate in 1776, Gardiner's portrait was reappropriated by the Massachusetts Committee of Sequestration to his daughter Abigail Whipple in 1778 and then descended in the Gardiner family of Gardiner, Maine, before arriving at the Seattle Art Museum. Abigail's portrait, by contrast, avoided seizure and traveled with the Gardiners to Poole (Dorset, England), where the family lived for a time after they left Boston in 1776, before passing through multiple hands to the Brooklyn Museum.[113] Like these two paintings, medical practitioners in New England found themselves caught between imperial prospects and local politics. The medicine trade underwent several reorganizations during the ensuing years given its reliance on a set of monetary, legal, and societal conditions, though ultimately remained founded on many of the hierarchies, infrastructures, and expectations that had taken root over the course of the century. During the imperial crisis of the 1760s and 1770s, dependence on British medicines led to shortages that in some cases turned into opportunities for the merchants of medicines who had begun to appear in the colonies. The expense of medicines also prompted discussions about alternatives to imported remedies across the British empire. Nevertheless, the trade's long-distance organization and underlying principles proved resilient even if it began to look different at the local level.

FIVE

Self-Sufficiency in a Bottle

By the early 1770s, the medicine trade was the fastest growing of Britain's manufactured goods trades, stretching across the Atlantic and Indian Oceans. It supported colonial labor regimes and depended on the markets they provided to return substantial profits. As some accrued wealth and power from them, others began to question a reliance on medicine imports. Customers from India to Pennsylvania considered medicinal self-sufficiency with new urgency, but also continued to rely on the kinds of remedies that moved across long distances. If they had looked closely, officials may have seen that local medicine production was hardly a recent development given the healthways that persisted amid imperial healthcare infrastructures. Crises in the colonies brought these concerns to a head. As it had before, war increased demand for medicines but now severed many commercial ties, while a credit crisis limited the ability of traders at the transatlantic and regional levels to distribute British medicines. Merchants of medicines found themselves on opposing sides of the American Revolutionary War, during which some fared better than others. John Greenleaf thrived while Silvester Gardiner found himself living in exile; and, in London, Thomas Corbyn and Joseph Gurney Bevan hurried to recoup overseas debts that suddenly seemed more inaccessible than ever.

Similar challenges to the medicine trade's organization arose across the empire, commercially intertwined as it had become. British medicines offered the prospect of convenient treatments but remained an expensive option considering the amount of material that had to move and the work that such activity required. As the East India Company (EIC) faced declining profits in the second half of the eighteenth century, its directors looked to cut costs. Company officials debated whether imported medicines were worth the expense and considered building a laboratory to produce

British-style medicines locally in Madras. Meanwhile, half a world away, a similar debate was occurring at the Pennsylvania Hospital. Philadelphia, the most populous city in British North America by the mid-eighteenth century, boasted lively mercantile and medical communities in which several pharmacies had begun to expand their manufacturing capacity, resembling their counterparts across the Atlantic.[1] That these businesses possessed transatlantic ties, chemical equipment, and trading expertise enabled the hospital's managers to reduce their imports. To that end, they built a laboratory and contracted nearby vendors, two decisions that would prove useful during the war.

In terms of organization, the imperial crisis represented a turning point for the trade, but in terms of products, markets, and access, it marked the endurance of a system established over the preceding decades by institutions that largely stayed in place: empire, large-scale agriculture, and military fiscalism. Medicines were already entrenched across imperial labor regimes and would remain so. After legal exports of British medicines to the colonies in revolt ceased in 1775, American practitioners initially struggled to meet the demand for medicines that followed the sickness and injuries of war. They turned to goods seized from prize ships, loyalists, and prisoners of war to alleviate shortages in the Continental Army and Navy. Soon, medicines made in Atlantic port cities, namely Boston and Philadelphia, filled some of the void left in the absence of shipments from London. The war years saw significant state funding to encourage American medicine manufacturing built on the foundation established under the colonial system. Following the Peace of Paris in 1783, transatlantic trade largely returned to its prewar activity, but the medicine trade did not entirely follow the trend. American manufacturers began to build their own export trades and import less even as medicines continued to flow from Britain to other colonial spaces. Medicines had become inextricable from formulations of state power and labor across the Anglo-American world though some of the pathways they followed had shifted.

Supplying the Madras Army

While previous decades saw the entrenchment of British medicines in the EIC's settlements across South Asia, the infrastructure for importing medicines remained a topic of debate during the 1760s. The debate concerned whether the Company ordering system could bear India's growing military population and the related healthcare expenses. Expectations alone could not make medicines arrive on time, unspoiled, or in the quantities desired, necessitating adaptations by the surgeons on the ground. British medicines themselves also retained only tenuous supremacy over more local supplies within the EIC's medical service as warfare and scarcity took their toll on inventories. Faced with declining profits, rising costs, uncertain provisions, and an increasing population of patients, officials sought new routes to medical self-sufficiency in the Indian settlements. A plan in 1765 to establish a laboratory at Fort St. George, Madras, reflected just how muddled the supply of medicines within the EIC had become. While British medicines remained a favored mode of treatment among troops and employees, observers were unconvinced they were worth the price of importation. The medical bureaucracy instituted by the EIC's directors over the next several decades at once deepened reliance on European medicines while also supporting a search for local substitutes and manufacturing capacity in India. The urgency of finding effective medicines in the face of alarming mortality had spurred the turn toward imported remedies earlier in the eighteenth century; now it provoked a moment of doubt about the long-term viability of that system.

The demands of treating groups of patients supported by the useful concept of specifics had institutionalized a dependence on imported medicines within the EIC's medical service but also sparked discussion about the costs of such an arrangement. Relying on shipments from London meant Company surgeons often suffered inadequate supplies, long delays, and careless packaging that caused spoilage.[2] Spending on medicines had increased over the years (see fig. 5.1). As the EIC's territorial authority expanded and profits waned, the directors considered alternatives to their dependence on imported medicines.[3] Transporting large quantities of medicine from Europe was an expensive undertaking, particularly in light of the interstate competition occurring in South Asia characteristic of the dominant mercantilist political economy. Insufficient healthcare, however, could endanger recent territorial gains or tarnish the company's reputation if too many people got sick. Self-sufficiency, one of the pillars of this political economy, offered a potential solution to the directors' problem of medical supply.[4] For more than a century, the goal of medicinal self-sufficiency had spurred a search for useful plants to encourage a positive balance of trade and military advantage, but now this idea was applied to the manufacture of medicines at the imperial rather than national scale. Instead of flows of bullion, officials thought of budgets and manpower. If medicines could be found or produced locally, the EIC would put itself in a better position vis-à-vis its rivals, such as the Dutch, Spanish, Portuguese, and French, who had also established colonies in South Asia.[5] The EIC's attempts to achieve medicinal

Figure 5.1 The East India Company's spending on medicines, 1740–90. Over the course of the eighteenth century, Company officials spent more on medicines for settlements across South Asia, despite the hazards of long-distance shipping and the availability of apparently more local sources of medicines. Fort St. George (34 percent), Bengal (28 percent), and Bombay (24 percent) together received the majority of these supplies (nearly 90 percent), though smaller settlements also saw rising imports. Spending is calculated by value in pounds sterling and adjusted for inflation (1730 = 1). Commerce Journals, 1735–90, IOR/L/AG/1/6/11–20, India Office Records.

self-sufficiency took two forms after the Seven Years' War: building colonial laboratories and encouraging the purchase of local remedies. These may seem like contradictory pursuits since they drew on healthways often described separately, but they were both motivated by similar concerns of imperial self-sufficiency.

The Seven Years' War had seen the EIC increase its territorial sovereignty in South Asia as well as its human footprint in the region, which strained its medical service. In general, Company medical establishments operated with greater military discipline as the century wore on, despite the lingering fiscal issues.[6] One way to import fewer medicines was to make them at scale in India. Following a spike in spending on medicines for the settlement, Gilbert Pasley (d. 1781), the surgeon at Madras, sent to the Council at Fort St. George a list of "Materials necessary to form an Elaboratory" in October 1765. The council, supportive of the proposal, ordered the papers transmitted to the Court of Directors in London as quickly as possible.[7] Earlier that year, the directors had written to the council announcing their "Inclination for preparing Medicines [at Fort St. George]." Pasley had responded with skepticism at first, finding little infrastructure to support production of the desired medicines. Establishing a laboratory in Madras would require the appointments of an apothecary "who has been bred up in an Elaboratory" in Britain and an assistant "in Case of Accidents." The prominence of medicine manufacturing had indeed come a long way since earlier in the century. Not only were people with enough experience to produce quality medicines at sufficient scale lacking at the settlement, but so were "every Utensil and Necessary for the Business," including "Chemical Vessells [for] preparing compound Medicines," he observed. The apothecary employed at Madras at the time was only a "Monthly Servant" and therefore could leave at any time for more lucrative employment at another settlement or in private trade. Despite the limitations he saw, Pasley assured the council that with the appointments of a full-time apothecary and an assistant, as well as the provision of proper equipment, "the Branch of Pharmacy may be conducted so as to be usefull" in India.[8]

Pasley's choice of the word "usefull" situates the laboratory scheme in the wider dialogue about improvement that began to inform British imperial political economy during the eighteenth century.[9] Efforts by Company surgeons to increase the effectiveness of healthcare in Madras occurred alongside efforts to install economically significant plants in the settlement. James Anderson (1739–1809), a surgeon and amateur botanist, helped found the Madras botanical garden and set its initial goals. Thanks to Company grants following Robert Kyd's 1786 proposals demonstrating the economic utility of botanical gardens in Asia, gardens were established at Company settlements in Calcutta, Madras, Bombay, and St. Helena by 1790. They housed a range of commercially and medically useful plants. Central to Anderson's plan for the Madras garden was introducing the nopal plant and the cochineal insects that lived on it to erode the Spanish monopoly of cochineal dye production.[10] He also touted the virtues of the "many plants us'd here in Medicine by the Natives." Another goal of Anderson's was to grow "such useful plants as are to be found here for Medicinal purposes as well as objects of Commerce," including camphor, benjamin, indigo, cotton, sugar, and dates, which would simultaneously lessen the dependence on imported European medicines and generate profit for the EIC.[11] Reflecting earlier attempts to establish a source of medicinal plants in the Atlantic world, the garden combined in one institution a variety of valuable plants intended to improve Britain's standing amid the interstate competition of the day.[12]

The desire to establish a laboratory in Madras came at a time when the directors sought to revamp what they saw as an unstable and expensive system of healthcare, reliant on a shifting group of British suppliers and requiring precise labeling of items before export. An Indian laboratory would, on the one hand, lessen the EIC's dependence on merchants of medicines in London and, on the other, institutionalize a higher degree of standardization in the preparation of the medicines consumed by Company employees stationed abroad. For a time, it seemed that the project would move forward quickly faced with increasing demand for medicines and the threat of armed conflict. The Company engineer at the settlement, a Mr. Call, produced plans for the laboratory to be built "in the interior part of the Ravelin before the South West Curtain" inside Fort St. George. Call noted that the number of walls and arches in this part of the fort were conducive to a "Bomb Proof Laboratory," which would prove advantageous in case of a siege. For safety concerns, the laboratory would also be detached from, rather than connected to, other structures should a fire occur. Yet the directors refuted Call's plans, noting the "seeming impropriety of Building a solid strong Work any where for such a purpose as a Laboratory," since common practice had been to build laboratories as slight buildings so that in case of an explosion (from an experiment gone wrong, one presumes) the structural resistance would be minimal. With a bombproof laboratory, they reasoned, "such an Accident might be attended with dangerous effects on the other Works and Buildings" nearby.[13] The needs of Madras as a militarized settlement influenced Call's initial design for the laboratory, which conflicted with the contemporary view of laboratories as prone to explosions and dangerous to the structures around them.

While building a laboratory in Madras seemed possible for a time, the directors eventually settled on a different answer to the question of supply that maintained their business with manufacturers in London. In November 1766, the Court of Directors announced that it had granted a monopoly to the Society of Apothecaries for supplying medicines to the Indian settlements. The apothecaries' laboratory offered the appearance of regulation and quality they had desired in the previous year's laboratory plan.[14] The directors wrote to the Council at Fort St. George in early 1767 that they found it unnecessary to "comply with your Request of an Assistant Apothecary, or the Indent for Materials to form an Elaboratory" given their decision to contract the apothecaries. They saw the yearly exports, now supplied by the apothecaries, as sufficient for the settlement's needs. Spending on medicines exported to Fort St. George stayed level for several years before substan-

Figure 5.2 Sea view of Fort St. George on the Coromandel Coast, ca. 1781. Seally et al., *Complete Geographical Dictionary* . . . vol. 1. Reproduced by kind permission of the Syndics of Cambridge University Library.

tial increases in the 1770s and 1780s.[15] The directors' reversal of opinion from suggesting a laboratory in October 1765 to abandoning the plan in March 1767 underscored the allure of British medicines and overshadowed the consideration that had been given the project in Madras by those who dealt with the limitations of long-distance healthcare every day.

The initial optimism that the Society of Apothecaries would lessen the expense and improve the efficiency of medicine supply turned to frustration within several years. Medicines would often arrive rotten or damaged—useless, in other words—after the long voyage from Britain. Upon inspecting a shipment in 1773, Pasley reported that "those Medicines we have received are much damaged on account of the careless and injudicious Package of them." Making matters worse, moisture draining from boxes of Glauber's salts had corroded the surgical instruments included in that shipment, exacerbating the loss.[16] The Fort St. George Military Department cautioned the directors to end the "careless mode, now observed" of packing the mineral acids in the same cases as the "most combustible materials" to avoid any further explosions in the cargoes. In the meantime, spending had risen to new heights.[17] Additional changes to the distribution methods were seen as necessary to ameliorate the problems of price and quality that worsened as medicine exports soared.

The extension of inland trade along the Coromandel Coast and into Bengal during the eighteenth century had opened new avenues for the dissemination of European medicines. It also brought Company surgeons into contact with medicines from the local "bazaars" where they encountered South Asian medicinal knowledge, including the procuring and preparing of ingredients that offered the possibility of cheaper, fresher remedies. What comprised this "country medicine" ran the gamut of medicinal items acquired locally rather than shipped from Europe. The category of "country medicine" was also employed by Europeans in the West Indies to separate remedies encountered locally from imported ones, though this label did not preclude their use. The 1780s saw a number of proposals for the bulk purchase of "country medicine" for Company hospitals, troops, and ships. In particular, these medicines were often directed toward the increasing numbers of South Asian men serving in the Native Infantry for whom indigenous remedies were believed at first to be more effective than European ones. European descriptions of the indigenous residents of settlements often suggested an inert rootedness to that place distinct from the agentic mobility assigned to the Europeans in binary portrayals of encounter. Something similar can be said for the distinctions drawn between country and European medicines in the company-state as surgeons, employees, and officials looked to Indian remedies as localized novelties that were also capable of delivering relief.[18]

Simply because country medicines appeared more local to European eyes did not mean they were necessarily more convenient, cost-effective, or even proximate than imported ones. Many arrived in India via long-distance trade routes of their own: aloes from China and Borneo, mercury from Tibet, benzoin from Sumatra and Java, and gamboge from Cambodia and Thailand.[19] Medicines bought at markets or from Indian practitioners embodied their own mobility and complexity, just not in ways that necessarily fit Company officials' or surgeons' expectations. As they were used more frequently on Britons in the company-state, these medicines began to be included in well-known catalogs of foreign plants and *materia medica* lists, such as the *Edinburgh New Dispensatory*. Reciprocally, Tamil physicians experimented with chemical medicines, such as nitrous, muriatic, and vitriolic acids, which contributed to the syncretic training found in some Indian hospitals by the early nineteenth century.[20] This level of practical hybridity should come as no surprise since the very stuff of British compound medicines required ingredients extracted from colonial landscapes and reconstituted in a European intellectual framework stripped of many of their previous meanings. Such connectedness, and such power, was acted out across the pages of contemporary recipe books and the apparatuses of countless laboratories.

The presence of a mixed European and Indian military in South Asia encouraged the use of a variety of medicines since practicality and accessibility shaped Company medical practice rather than any strict adherence to a single doctrine. European soldiers in the Madras army were granted an allocation of imported and country medicines every four months. Common items included Peruvian bark, bitter cathartic salts, and opium. The quantities sent into the field fluctuated with the number of the sick and wounded; and, if supplies were delayed, regimental surgeons could procure medicines at the "Army Bazars" that followed the regiments.[21] When fevers arose among the troops, surgeons requested immediate stocks of imported opium, salts, and bark. Often these did not arrive in time, whether from Fort St. George or from London.[22] Though country medicines could theoretically assuage some of these shortfalls, officials in London reproached the Company apothecaries working in Indian hospitals for buying too many local medicines. The frequency of these censures in the 1770s illustrates the durable popularity of country medicines within the Company medical service, especially in the case of perishable items or when European medicines were scarce.[23] A provision allocated 400 rupees a month for the acquisition of country medicines for the Bombay marine to supplement European medicines, for instance.[24]

Reliance on European medicines coexisted with attempts to introduce more local medicines into the Company service. The Madras Military Department considered many of the country medicines provided for troops "the most important Medicines in use[,] essential to the proper treatment of the Sick and wounded." Because they were often difficult to acquire in the field, and at what were seen as "exorbitant rates," maintaining a regular supply of medicines remained paramount to the commander in chief of the Madras army.[25] The settlement's medical storekeeper made 15 percent commission on the packages of European medicines and deliveries of country medicines. He supplied tamarinds, camphor, jaggery, taddy, and other items acquired outside the EIC's infrastructure, as well as items more familiar to European pharmacopoeias, such as aloes, china root, dragon's blood, opium, and sago. Company spending on these remedies varied with "Market Prices," sometimes dramatically, and therefore led to new charges of overspending. Nevertheless, regimental surgeons depended on medicines available on the spot because many could not be preserved during a long journey away from a Company hospital.[26] Opportunities for significant markups also encouraged surgeons to resell European medicines earmarked for other uses. For example, extract of thebaica (a preparation of opium) or *sal succinum* (a salt of amber, or succinic acid) could yield 100 percent profit based on its invoiced price in some cases.[27] In 1792, the Hospital Board mandated regular supplies of country medicines to Indian hospitals, thereby institutionalizing the local sourcing of at least a portion of Company medicine stocks. Medicines received in this manner included widely used local plants and regional substitutes for European remedies, such as "Hindustanee Fever Pill," a replacement for imported Peruvian bark.[28]

Three decades after the directors had considered building a European-style laboratory in Madras, bulk medicine manufacturing arose in colonial South Asia during the 1790s from the exigencies of the EIC's health infrastructure and made possible by the instrumentalization of specifics more broadly across the British empire by this time. Company surgeons in Bombay began to manufacture some medicines alongside those they received from Europe. Especially for private practice, surgeons prepared their own volatile drops, *elixir proprietatis*, gummosa pills, and mercurial plasters to treat venereal disease rather than import them like other articles.[29] In 1795, mills for the "purpose of Powdering Bark" and other drugs were sent to each settlement.[30] They enabled the local preparation of Peruvian bark to treat endemic fevers as well as of other compound medicines that required powdered drugs. Soon after, the medical board observed "with much satisfaction" that the demand for some medicines had lessened in the recent orders from India. The directors attributed the change, which "afford[ed] [them] much Pleasure," to the board "having adopted the improvements made of late years in the Pharmacopoeias" and Company surgeons increasingly asking for "the materials for Medical preparations instead of the Preparations themselves."[31]

European pharmacy remained a standard within the company-state despite the expediency of local medicines in many circumstances. Widespread medicine production threatened to reduce the variation in medical practice across colonial spaces where a range of treatment options had prevailed due to day-to-day needs. By 1800, British medicines, either imported or made locally, displaced country medicines in many once-hybrid spaces, and the roles of Indian medical practitioners in Company hospitals were reduced as part of a wider rejection of indigenous methods of health and hygiene. At the same time, British medicines were more frequently given to Native troops, even though colonial bodies were increasingly seen to respond to medicines differently than their British counterparts. Medicine ordering also came under greater metropolitan regulation as spending on imported medicines continued to rise during the Mysore Wars of the 1790s. Shipping became more routine over time as well. The Society of Apothecaries supplied the EIC's army of around 300,000 men through the 1820s. Indigenous healthways nevertheless endured as Indian practitioners incorporated European medicines into techniques that persisted into the twentieth century.[32] Even though European medicines emerged once again ascendant in the Company medical service, the second half of the eighteenth century had been a moment of uncertainty when officials and surgeons experimented with alternative ways of achieving the healthcare considered essential for the EIC's definition of territorial authority in South Asia. The decades-long search for medicinal self-sufficiency shaped South Asian medical institutions and demonstrated the potential of British medicines to project power abroad even if that self-sufficiency was never fully achieved.

Supplying the Pennsylvania Hospital

Meanwhile, questions about the necessity of receiving medicines from London also arose in Philadelphia. Similar conversations could occur in South Asia and North America because both locations featured colonial institutions reliant on imported medicines, regional trading hubs, and hybrid medical communities by the second half of the eighteenth century. Pennsylvania surpassed New England as the leading North American destination of British medicines outside the West Indies until Virginia and Maryland together outstripped it in the 1770s. Anglo-American men and women with varying degrees of medical training (formal or otherwise) worked alongside practitioners from the African American and Lenape populations in the area. Chemical processes undertaken in laboratories, shops, kitchens, and other domestic spaces allowed people in a range of societal positions to access the regional medical marketplace.[33] In 1765, John Morgan returned to Philadelphia from his studies in Europe carrying medicines made at Plough Court and an urge to separate pharmacy from other medical practice. Morgan's vision became reality as businesses devoted specifically to manufacturing medicines arose in the city.[34] Philadelphia also boasted the Pennsylvania Hospital, the first general hospital in the colonies and soon the site of another debate about importing British medicines. Its managers maintained close ties to merchants of medicines in London, but also forged links to local manufacturers who began to supply the hospital in the 1770s. These decisions fell short of achieving full self-sufficiency, but they established connections among local merchants of medicines that would prove vital in the coming years.

The Pennsylvania Hospital blended carceral, charitable, and commercial goals under one roof. In 1751, several Philadelphians, including Benjamin Franklin, proposed a hospital to the Pennsylvania Assembly, which enacted it into law shortly thereafter "for the Reception and Relief of Lunaticks, and other distemper'd and sick Poor within this Province." The planners evoked humanitarian rhetoric of medical care for those who could not afford it, but also designed to remove the poor and mentally ill from the city's streets. Those initially involved in the institution, erected in 1755 at Eighth and Pine Streets, included men with connections across Europe and the colonies. Joshua Crosby, a merchant, was elected the first president of the hospital board, a post Franklin filled after Crosby's death in 1756. The medical staff consisted of Thomas Bond, his brother Phineas Bond, and Lloyd Zachary; and in 1759 added Cadwalader Evans, who had studied in England, worked in Jamaica, and received medicines from London.[35] Of the hospital's initial capital stock, more than £2,000 came from private donations and an additional £2,000 arrived as matching funds from the colonial assembly. Entreaties were made to Quakers in London—notably Silvanus Bevan and Thomas Hyam—to secure land and funding for the hospital from the Pennsylvania proprietors. The proprietors, however, proved reluctant to support the hospital, and, as a result, much of the fundraising fell to its managers in Philadelphia. Operating revenue came from a variety of charitable and commercial efforts, including multiple working farms on the property that supplied the hospital with milk and produce, as well as surplus agricultural products that were sold locally.[36]

Money from these endeavors sustained the hospital's medicine stock, much of which arrived from Plough Court. The invoices, however, were often more than the managers expected and more than they could afford. They instructed the hospital's staff to sell any extra medicines on "more Advantageous Terms" to recoup some of the costs. Samuel Rhoads built a storeroom and shop in the back room of the hospital where imported medicines could be kept clean and dry before sale or use.[37] The managers also received gifts of medicines from nearby residents, such as Deborah Morris in 1755, and in other cases purchased items from estate sales.[38] Silvanus and Timothy Bevan continued to send medicines throughout the 1750s and 1760s, but multiple years could elapse between shipments and orders, delays that frustrated both the hospital managers and the Bevan brothers. Nevertheless, it was not yet in question that the medicines should come from London. Products from Plough Court made up a significant portion of the hospital's inventory. An order from January 1766, for example, included 100 pounds of potassium nitrate (saltpeter, almost 2,000 doses), 100 pounds of Epsom salts (about 1,600 doses), and 50 pounds of *lapis calaminaris* (a zinc silicate used in skin ointments).[39] For a time, the managers repaid their debts within a year, and the Bevans readily extended them credit.[40] Yet the managers' ability to pay promptly did not last. Timothy Bevan wrote in 1768 that the hospital seemed in "immediate want of a Supply of Medicines, and that the Lowness of [its] funds ha[d] prevented [the managers] from ordering any." In response, he offered "any Credit that may be necessary" to ensure that the hospital had sufficient stocks of British medicines. Continuing to import medicines, Bevan noted, would be "more advantageous to the Charity than to purchase them in your City," given his opinion of Plough Court's remedies. Using rhetoric of charity, he sold the managers medicines before they necessarily needed or were able to pay for them, which indebted the managers to Plough Court for decades.[41] This cycle could happen because both sides agreed that British medicines should be the foundation of the hospital's healthcare.

Negotiation of the hospital's medicine supply came down to details of long-distance trade. The managers protested that Bevan overcharged them, "charge[ing] [some medicines] at a higher price than we could have bought them at here [in Philadelphia]," an accusation that merchants of medicines, such as Timothy Bevan or Thomas Corbyn, were accustomed to hearing by this time. The managers found Bevan's margins excessive, blaming them for an inability to pay. They also appealed to Bevan's "Noble Charity" and "long connections in trade to these Colonies" for better prices, arguing that since the hospital served many European patients, its benefits extended beyond the North American colonies.[42] Apparently, Bevan's charity stretched only so far. He saw no moral obligation to provide medicines at lower prices, he countered, thanks to the purportedly generous terms of credit and quality of goods he already offered. In what had become a fairly standard back and forth for the medicine trade, Bevan admonished the managers that products from Plough Court should not be compared to those from "common Vendors of Druggs and Medicines" found elsewhere. Nonetheless, in 1765, he agreed to grant the managers a 5 percent discount for prompt payment (in lieu of one year's credit) on their next orders. Bevan, however, did not give the managers the 10 percent discount they desired, apologizing that his "reputation as Tradesmen seems to require it, & hope therefore you will not think it impertinent." As a conciliatory gesture, he promised a donation of £50 to the hospital.[43] These conditions did not help the managers pay their bills.

With their debts mounting, the hospital managers—like the directors of the East India Company—considered building a laboratory to produce more of their own medicines. That the managers saw the cost of imported goods as unnecessarily high, dependent on both prices set by merchants and duties levied by the government, spurred this turn toward self-sufficiency. In August 1768, they appointed Samuel Rhoads, Jacob Lewis, and Isaac Greenleaf to purchase materials and employ tradesmen to erect a two-story laboratory on the premises. By all accounts the project was successful, although the structure remained unfinished almost two years later in April 1770.[44] Bevan, meanwhile, continued to offer long terms of credit in hopes that the managers would "not hesitate at sending us further Commissions on account of not being able to furnish immediate Remittances." Though the managers rarely sent timely remittances, the size of their orders did not shrink for several more years. Popular imported items remained Peruvian bark, cream of tartar (powdered sodium potassium tartrate, used for the treatment of fevers), oil of vitriol, Epsom salts, Glauber's salts, and *ferrum vitriolatum* (ferrous sulfate, an astringent, tonic, and diuretic also used by chemists in the distillation of the spirit of vitriol, another common remedy).[45] An order sent by the managers to London in October 1773 requesting materials typically used for manufacturing medicines suggests that the laboratory was operational by this time. The order included ingredients for treating venereal disease and fevers, as well as vials, bottles, corks, and a marble mortar and pestle necessary in a laboratory.[46] The list of medicines on hand at the hospital in the spring of 1773 was extensive based on a contemporary inventory. It included drugs from across the British and Spanish empires as well as chemicals, compound medicines likely made on site, and patent medicines prepared in Europe.[47] In 1775 when transatlantic imports ceased, the managers were able to rely more heavily on local sources of medicines, including several nearby pharmacies.

Medicine manufacturers in eighteenth-century Philadelphia served local customers as well as an expanding regional trade. Christopher Marshall (1709–97) founded his druggist and chemist shop at the corner of Front and Chestnut Streets in 1729. His sons, Christopher Jr. (1740–1806), Charles (1744–1825), and Benjamin (1737–78), all gravitated toward trade, with Christopher Jr. and Charles joining their father's business. During the 1760s, Christopher and his sons distributed medicines to New York, North Carolina, Nova Scotia, Rhode Island, and Virginia by sloop or schooner, typically about ten to twenty voyages per year. The brothers also turned their resources to the provisioning trade. In 1764, they sent seventeen barrels of fish to Jamaica, followed by ten barrels of pork to Antigua in 1766. Such shipments omitted medicines since estates in the British West Indies still imported much of their supply from London. When they sent provisions further afield in larger ships, such as to Paramaribo (Suriname), Christopher Jr. sought assistance from their brother Benjamin, now a general merchant. About once a year the brothers, similar to Silvester Gardiner, balanced their London accounts with North American goods. Christopher Jr. and Charles, for example, sent beeswax and snakeroot (both used in medicine manufacture) aboard the *Minerva* to Plough Court.[48] The elder Marshall had taken on Christopher Jr. and Charles as partners in 1765, and they took over the business after their father's retirement in 1772. Charles was well regarded for his skill as an apothecary, botanist, chemist, and druggist, by many accounts. They began to produce more of their own medicines but continued to rely on London manufacturers for botanical simples, metallic compounds, and some salts and saline preparations, which they sold wholesale to country doctors across the Mid-Atlantic colonies.[49]

Other important pharmacies in the city included those of Isaac (d. 1801) and Moses (d. 1810) Bartram. Isaac and Moses came from a large family of botanists, apothecaries, and medical practitioners; both were sons of John Bartram Sr. (1699–1777), who had achieved fame exchanging plants and seeds from his botanical garden on the western bank of the Schuylkill River. Colonial gardens once again buttressed a project of medicinal self-sufficiency.[50] Moses had tried to establish a pharmacy on his uncle's North Carolina plantation during the 1750s in an effort to supply the upper South with locally made medicines. He had also been master of the *Corsley* in 1756 before returning to Philadelphia and entering into partnership with his brother Isaac. The brothers, alternatively described as pharmacists, druggists, apothecaries, and merchants by contemporaries, kept shops close to one another near the Delaware riverfront. Isaac's was located at 39 North Third Street, and Moses's could be found nearby at 58 North Second Street. They also maintained a laboratory for chemical preparations separate from their shops, customary at the time for reasons of safety and secrecy. Many of the drugs Isaac sold came from Bartram Gardens, his father's botanical garden. Other items arrived from Thomas Corbyn, who supplied several contacts in Philadelphia, including the Marshalls, Esther White, and Townsend Speakman. Speakman advertised using those imported medicines to prepare medicine chests for "shipping, plantations, or iron works." Bartram sold rosewater, one of his major products, to a range of wholesale customers, many of whom were women participating in the city's medical marketplace.[51]

Women healers could assume a variety of commercial roles in the British North American colonies, unregulated as they were by the state or preprofessional organizations that structured some European medical communities.

In 1779, Margaret Hill Morris made and sold medicines, diagnosed illnesses, and provided nursing care in Burlington, New Jersey, not far from Philadelphia. Morris, alongside other free and unfree women of various classes and affiliations across the Delaware Valley, participated in the region's medical marketplace. They established transatlantic connections to rival their male counterparts. Morris, for example, purchased imported medicines to supplement herbs she found locally and remedies she made herself. One of Morris's friends, Grace Buchannan, had trained as an apothecary in London before setting up shop in Burlington.[52] Though it remained difficult for most men and women involved in the medicine trade to secure contracts or establish transatlantic ties, and few had the family resources of the Bartrams or Marshalls, Philadelphia could soon boast an array of nascent merchants of medicines.

By the 1770s, some had begun to question the prevailing organization of transoceanic medicine trades. The managers of the Pennsylvania Hospital and directors of the East India Company considered building laboratories to lessen their institutions' dependence on imported medicines. While only the laboratory in Philadelphia was completed as planned, discussions about the cost of medicines prompted the search for local alternatives while simultaneously reinforcing British medicines as the accepted norm in many colonial spaces. Medicine manufacturing gradually developed in the North American colonies and Indian settlements where indigenous and British practices blended for a time due to the scarcity and cost of imported medicines. Even after several decades of growth, the long-distance medicine trade carried risk and assured debt for all involved. The hospital managers fell behind in their payments to Timothy Bevan, who soon struggled to recoup Plough Court's overseas debts for reasons both within and beyond his control.

The Center Cannot Hold

Decades of credit extensions encouraged exports but also increased the risks of the transatlantic medicine trade, as the hospital managers saw. The total commercial debt owed to Britain from the thirteen colonies rose during the eighteenth century to a peak of £6 million sterling in 1774.[53] This figure reflected thriving long-distance exchange from one perspective, but from another it suggested the extent to which such exchange hinged on promises to pay guaranteed by the future value of colonial natural resources. By some estimates, less than a third of amounts owed was returned annually from the colonies.[54] Across many transatlantic trades, including the medicine trade, exporters waited for sums held overseas. This did not stop them from sending out more goods and extending credit, though it depleted their capital stocks and left many merchants of medicines vulnerable to shifts in state policy affecting their access to overseas markets or raw materials.

Despite general economic growth, the 1760s and 1770s experienced dramatic fluctuations in British overseas trade. Medicine exports weathered a credit crisis in 1763 and benefited from improvements in manufacturing and infrastructure, as well as relatively cheap capital. Abundant credit led to increased borrowing by colonists, pushing exports to new heights in the early 1770s. Exports to the North American and Caribbean colonies had entered a prolonged phase of growth since about 1756, a period that also saw rising military and enslaved populations in many of the places where markets developed. The sugar complex transformed energy into capital and was itself sustained by the investment of mercantile profits, much of which came from commission agents in London. Colonial traders and planters thus became debtors to London merchants.[55] This credit supported countless trades, enabling outsized profits for a time. It also meant that the fortunes of export trades, such as in medicines, ultimately depended on the booms and busts of sugar or tobacco. All sorts of property became security for these debts as a result of easy credit. Many things would be taken or sold if mortgages and bonds were collected. Nothing was safe in a crisis.[56]

Even as their laboratories, warehouses, and counting houses bustled with activity, merchants of medicines fretted about the future of the overseas medicine trade under the Hanoverian state. During the reign of George II, smuggling had persisted at alarming rates, and government revenue tables rendered the system of taxes more perplexing than ever. Then, in the early 1760s, new duties were added to articles exported to the American colonies. Pervasive opposition to the customs regime and a spate of industrial disputes of unprecedented severity followed across Britain. The Townshend duties of 1767 provoked uproar in the colonies on the topic of trade policy. Such measures designed to assuage the growing national debt proved disappointments after the Seven Years' War. Political economists faced the conundrum of how to control trade without stifling economic growth or pushing consumers to seek illicit goods during what was widely considered a critical moment for the empire.[57] Tax burden was generally increasing as well, peacetime taxation having risen from below £2 million in 1688 to £15 million in 1786. Many of these revenue policies inflated the cost of raw materials, which infringed on medicine manufacturers' profit margins, especially in the export trades. In general, drugs saw steeper import duties over the course of the eighteenth century, even as medicine exports remained nominally untaxed following the Walpole Reforms of the 1720s. Foreign drugs had been taxed in previous centuries because they were judged to be luxuries. Yet in the eighteenth century, they had become widely used as remedies and essential to a range of manufacturing pursuits—a disparity between policy and commercial practice that did not go unnoticed by merchants of medicines.[58]

A lack of consensus on the medicine trade's political economy prompted criticism of state revenue policy from merchants of medicines. They found customs duties, particularly as applied to drugs, outdated to the point of encouraging smuggling and limiting the availability of raw materials. The official rates used to calculate the duties paid on certain imports had stayed constant since the 1660 *Book of Rates* in some cases, meaning that the charges on many drugs no longer reflected their current values in a system of global commerce and bulk manufacturing.[59] Several merchants of medicines petitioned the treasury in 1769 that the duties on imported drugs were compiled from "excessive high" rates tabulated when drugs "were of much greater value and less known than they are now both at home and abroad," which overvalued imports and led to exorbitant duties. So egregious were the overvaluations, they argued, that those who legally imported some key drugs could easily be undersold by smugglers, who paid no duties, which in turn diminished the state's tax revenue. The market prices of several drugs, including jalap and ipecac (ipecacuanha), were indeed lower than their rates at times during the 1760s and 1770s. With their high value and low bulk, drugs were a popular choice for illicit trade.[60] The "greatest part, if not the whole Quantity" of many drugs, the petitioners estimated, "are Smuggled [into England] and pay[] no Duty at all," demanding immediate attention to the customs.[61] Merchants of medicines faced the unfavorable proposition of buying these drugs in bulk with cash while selling them on credit, making payments often slow and irregular.

This was a policy issue, because smuggling hurt government revenue collection, but it was also a medical one. The petitioners warned of illegal cargoes containing impure goods that would threaten "the Health of his Majestys Subjects," especially if they were made into medicines. Merchants of medicines thus linked economic concerns about their overseas trade to the public good, tapping into broader concerns about the health of the armed forces at the time.[62] State officials, by contrast, sought to generate revenue from the import and export of raw materials given the trade's volume. Policy intended to fill state coffers could theoretically enable someone living in another country with lower duties to purchase raw materials reexported from England more cheaply than someone living in England who desired those goods for domestic consumption or industry. Observers worried that foreigners would gain improved access to colonial goods they could not import directly, particularly goods from the Americas and Asia for which Britain acted as Europe's primary entrepôt. Solutions from the petitioners included a reduction in the rates, duties, and drawbacks (effectively tax rebates) on drugs that were imported from their places of growth in accordance with the Navigation Act and higher duties on drugs arriving from someplace other than their place of growth. Such a proposal incentivized medicinal self-sufficiency projects across the empire since it would privilege raw materials grown or collected in British colonies, yet leave the acquisition of important drugs grown in Spanish or Portuguese colonies indirect, expensive, and susceptible to smuggling. Merchants of medicines recognized that "encouragement from the government" was necessary to support domestic manufacturing.[63] They were not alone in lobbying for favorable trade policy. The drawbacks to manufacturers from exports of refined borax and camphor—two noteworthy drugs used in both medicinal and industrial processes—had already been adjusted after petitions earlier in the century.[64] Nevertheless, the petitioners' concerns largely went unattended. In their report, customs officials W. Musgrave, Edward Hoges, H. Pelham, and J. Jeffreys agreed that it was "absolutely necessary" to determine "how far [duties] are too high at present in proportion to such Prices [of drugs]," but concluded that the proposals were in "no respect agreeable to the present System of Custom Duties."[65] The customs appeared too entrenched, particularly in light of the pressing revenue needs of the 1760s, and medicine exports continued to flourish for several years after 1769 in spite of these complaints.

The credit crisis that gripped the Atlantic world in 1772 was preceded by the prosperous years of 1770 and 1771, following the end of the embargo on exports of some British goods to the American colonies in 1769 and 1770. Contemporary accounts describe wharves piled high with British goods, generous credit, and rising prices during these halcyon years. "There is a great flood of goods [in Boston] than ever was known," recounted a friend of John Adams in June 1771.[66] Medicine exports continued to rise dramatically during these years, unaffected by the nonimportation movement. In 1772, however, a credit crisis shook Scotland and soon spread to the British colonies in Asia and the Americas, where it disrupted the balance of payments. Bills of exchange became scarce, and in response, specie began to flow from the colonies to Britain. Deleterious effects of the crisis were most acutely felt in tobacco-producing regions, notably Virginia. Many London mercantile houses curtailed their involvement in the tobacco trade, which, in turn, precipitated a permanent decline in the transatlantic commission system. While commission trading largely dried up in the tobacco trade after the crisis, many merchants trading on their own accounts with access to indirect trade managed to obtain specie, bills of exchange, and commodities to pay their debts and continue business. At first, the medicine trade seemed to emerge unscathed as exports reached a new peak from 1772 to 1774. That the medicine trade was an amalgam of the commission and commercial systems allowed it to escape much of the credit shortage that followed the crisis in other trades. Participants often traded both on commission and on their own accounts with merchants of medicines in London, who also acted as agents for colonial produce.[67] Nevertheless, the transatlantic medicine trade would soon face another test.

Things Fall Apart

Across the British North American colonies, signs of a shift in the transatlantic medicine trade appeared during the Stamp Act crisis. The act, approved by Parliament in early 1765, exposed fissures in colonial medical communities that would widen preexisting divides over religion, business, and politics in the coming decades. Opposition to the new taxes on printed materials drove seven physicians and fourteen apothecaries to join many of Philadelphia's principal merchants in signing a nonimportation petition in October of that year. Among the signers were Isaac and Moses Bartram, Christopher Marshall and his sons, Phineas and Thomas Bond, and others with ties to the transatlantic medicine trade, such as Thomas Lightfoot. The signers contended that the act would increase the cost of imports, thus deepening colonial debt, but also decrease the availability of specie, making it more difficult for local merchants to pay the duties or conduct their trade.[68] Colonial nonimportation movements politicized transatlantic trade and the local manufacture of various goods, including medicines. They added new urgency to the medicine production that already existed in the North American colonies and was beginning to occur at larger scale and greater frequency in the 1760s.

Other colonial merchants of medicines opposed nonimportation because of their ties to British manufacturers and political bent. Silvester Gardiner, for example, did not sign a nonimportation agreement circulated around Boston in 1767. He would later join other merchants in an address to Governor Hutchinson on May 30, 1774, publicly signaling his Tory leanings. The addressers, as they were subsequently labeled, became targets of public ire, such as by tarring and feathering, though Gardiner eluded such ignominy. John Greenleaf, in contrast, signed the 1767 nonimportation agreement and counted among his regular customers men like Joseph Warren, who would prominently take up the revolutionary cause.[69] In spite of another pact signed in Boston on August 1, 1768, many traders continued to import British goods. Gardiner's name appeared among those accused of violating the agreement in the August 17, 1769, edition of *the Boston Chronicle*. The publisher John Mein likely printed the list in retaliation for being accused of importing banned goods himself. The next week, Gardiner's name appeared again, this time on the *Chronicle*'s front page alongside such notable offenders as John Rowe, Elbridge Gerry, and John Hancock.[70] While Gardiner once again avoided significant consequences, others accused of importing British goods did not fare as well. Jane Eustis left Boston after the notoriety of her alleged commerce in banned goods. She had supported nonimportation in 1768, one of seven women to sign the subscription; yet, after the year of nonimportation passed, she resumed importing British goods. When another pact appeared in 1769, she did not add her name, leading her to be called an importer. After a series of attacks on local merchants perceived to be loyalists, Eustis posted a front-page notice in *the Boston Gazette* that she would soon be sailing for England and desired to sell the remainder of her wares, which would have included medicines. She had maintained a thriving business in British goods throughout the 1760s, despite the boycotts, but felt it no longer safe to continue by 1770.[71]

Meanwhile, Martin Brimmer took up with the Sons of Liberty while continuing to sell medicines from his shop in Boston. He was one of the 150 members at the Liberty Tree Tavern in Dorchester for the 1769 annual commemoration of the Stamp Act riots that had shaken Boston in August 1765. According to witnesses, more than 300 revelers flew flags in "a Variety of Colours," played music, fired cannons, and presented forty-five toasts, including one wishing "Success to the Manufactures of America." A list of the Sons of Liberty prepared for an English newspaper in 1775 listed Brimmer among sixty-one notable members.[72] Like Greenleaf, Brimmer found himself immersed in the city's popular politics. He nevertheless continued to prepare medicines for people from across the political spectrum, including men later condemned as loyalists, such as Benjamin Church, Silvester Gardiner, John Jeffries, William Paine, and Charles Russell.[73]

Gardiner's local standing, by contrast, had begun to erode due to his politics and reputation for litigious land speculation. Gardiner eventually lost his investments in Maine as religious, political, and personal disputes divided the Kennebeck Company. The rift between Gardiner, a Tory and Anglican, and John Hancock, a Whig and Congregationalist, exemplified the divisions among the proprietors.[74] Gardiner was publicly labeled a "rank Jacobite" and "slaunch [staunch] Courtier in the present reign" due to his actions of the previous years.[75] In 1776, he departed Boston with other loyalists before the Continental Army confiscated his inventory of medicines and instruments. The Sons of Liberty and the Committee of Correspondence in Pownalborough happily received the news that Gardiner had "fled from Boston with our enemies," and they seized much of his property in the Kennebec. The committee began leasing out Gardiner's lands, for example, charging a rent of £5 for his farm on Swan Island. After leaving Boston, Gardiner resided in New York City, Halifax (Nova Scotia), and Poole (Dorset, England) before returning to Rhode Island, the place of his birth, and passing away in August 1786 at the age of seventy-nine. The ships in Newport harbor were noted to have flown their colors at half-mast out of respect, and in Maine a marble cenotaph in Christ Church (Gardiner, Maine) today recalls Gardiner as "a learned physician . . . in transacting business, indefatigable, sagacious, and vigilant."[76] Another practitioner with loyalist sympathies, John Halliburton, left his Newport medical practice in 1775 before settling in 1782 with his family in Halifax, Nova Scotia, where Halliburton's brother-in-law, James Brenton, served as an assistant judge on the Supreme Court. There he resumed private practice and became head of the Royal Navy's medical department.[77]

The Boston medical community faced a regional marketplace in flux during these years. With the outbreak of formal hostilities between the thirteen colonies and Britain in 1775, transatlantic trade between them ground to a halt. The same held true for the medicine trade, even though British medicines remained the standard of care for the Continental Army and Navy. Medicine exports to the North American colonies plunged by more than 85 percent from 1774 to 1777. Shipments to the West Indies fared better, but total medicine exports from London fell by more than 60 percent (see fig. 2.1). In 1774, more than 65,000 pounds of medicines went to New England, 80,000 pounds to Pennsylvania, and 90,000 pounds to Carolina by official counts. They received none in 1777.[78] On December 14, 1775, shortly before King George III's proclamation closing the colonies to all trade, apothecaries for His Majesty's General Hospital in Boston took medicines valued more than £500 from John Greenleaf's shop by order of General Howe. Confiscations like this further depleted already thin colonial inventories.[79] Brimmer's medicine trade largely ceased as he turned to selling New England and Caribbean goods in the absence of transatlantic supplies.[80] Shortages of medicines, glass vials, laboratory equipment, and surgical instruments were acutely felt in the early war years.[81] Some practitioners collected herbs to replace imported medicines. Supplies also arrived from captured British stores or via French and Dutch traders who provided arms, ammunition, and medicines to the rebellious colonists under a form of state-sponsoring smuggling.[82] Drawing on decades of experience, others turned to local production supported by state funding. The temporary breakdown of the medicine trade's dominant infrastructures accelerated the emergence of a more autonomous American trade as people found alternative ways of acquiring medicines detached from the long-distance ties and hierarchies that had shaped its development.

Prevalent disease made medical supplies a vital concern in both the British and American armed forces. Fevers, particularly malaria (ague) and yellow fever, had a long history of inflicting severe losses on military expeditions and settlement schemes in the Americas. Now they assailed the military and civilian populations unsettled by war and lacking experienced practitioners and medical supplies. Most of the newly arriving British and German soldiers had never before encountered malaria and thus were very susceptible to the disease, whereas American colonists in endemic areas had developed a measure of differential resistance. The Continental Army, however, was not spared the ravages of sickness. By some estimates, disease deaths outnumbered battlefield deaths by a factor of ten to one during the American Revolutionary War.[83] Healers on either side could not expect consistent supplies of European medicines, nor would inoculation or immunity provide much protection. The British Army employed doctors, surgeons, and apothecaries who received supplies from London and could compound their own remedies.[84] Soldiers, nevertheless, looked warily on doctors in the armed forces. Apothecaries, druggists, and chemists provided some medicines for the Continental Army, but their private stocks proved inadequate for the scale of the conflict. In the absence of transatlantic trade, supplies remained unpredictable, especially when some merchants of medicines with large inventories, such as Silvester Gardiner, proved unwilling to aid the revolutionary cause. Those who did deliver medicines often saw only sporadic compensation from a Continental Congress with little money in the 1770s. In the field, the exigencies of military service brought together learned and popular medical practices. American healers treated malaria with powders of barks or roots, mixtures of brimstone and sugar, and water from iron mines. Cinchona, however, remained the only reliable treatment, but it was expensive and scarce for the British Army, more so after Spanish officials prohibited export of the bark in 1778 to keep it from enemy hands.

The Americans managed to maintain more consistent supplies of the bark throughout the war, which practitioners used in conjunction with locally sourced drugs, such as snakeroot, to alleviate fevers. Cinchona was also desperately needed in the West Indies and India during these years, which strained British supplies even more.[85]

Confiscations provided some relief for the shortages that faced the American medical community. In the autumn of 1778, John Greenleaf still sold fresh aloes and guaiacum, meaning he could acquire drugs of Asian and South American origin in spite of the restrictions on trade.[86] American privateers preying on British shipping delivered medical supplies taken from prize ships. Massachusetts officials sent medicines recovered from the sloop *Julius Caesar* to Greenleaf in December 1776, for example. Three months later, he received another package of medicines worth more than £116 from the *Three Friends*, recently taken by the brigantine *Tyrannicide*.[87] Greenleaf maintained his retail and wholesale business even though shipments from Hughes & Whitlock in London had ceased after November 1774. He supplied medicine chests to army regiments and navy vessels, a task once fulfilled by Gardiner, and received some cash in return. Appropriation of Gardiner's inventory in 1776 was also a boon to the Continental Army, reflecting the urgency with which the new government addressed the replenishment of medicine supplies. Each chest Greenleaf provided was valued around £20, which, by comparison, was only 1 percent of the medicines (valued at £2,000) confiscated from Gardiner.[88] Greenleaf also received medicines from French sources in 1777, possibly from the medical staff of the French auxiliary force in North America.[89]

A similar arrangement developed in Philadelphia where Christopher Jr. and Charles Marshall, perhaps inspired by their father, a "fighting Quaker," provided medicines and surgical supplies to the Continental Army.[90] With many of their transatlantic ties severed, the brothers turned to a constellation of colonial merchants, ship captains, and captured stores to acquire European goods. They usually put up multiple orders per month for the Continental Congress, medicines that then went to battalions from New Jersey, North Carolina, Pennsylvania, and Virginia. The most popular items were cathartics (for fevers), analgesics (for pain), and antiseptics (for wounds). Between April and June 1776, for example, the Marshalls supplied more than 94 pounds of cream of tartar, 30 pounds of Peruvian bark, 20 pounds of *sal nitri*, and 270 pounds of Epsom salts. Their medicine chests typically contained many chemical remedies. They also sold medicines to the Pennsylvania Hospital and received confiscated items, including Epsom salts, camphor, and opium, which were all medicines provided to the British troops.[91] The Marshalls, like Greenleaf, managed to stock Asian and South American drugs alongside European medicines during the war, indicating that they maintained some overseas connections. They used their older brother Benjamin's connections to draw bills in Amsterdam and Suriname for trade with Dutch merchants. With the treaty of French alliance signed on February 6, 1778, medical supplies also began to arrive from France, and the worst of the medical shortages abated.[92] During the war, long-distance supply chains had not entirely halted but shifted due to the new circumstances.

Despite its chronic lack of cash, the Continental Congress also supported the local manufacture of medicines during the war. By many accounts, the Revolutionary War accelerated the scope, scale, and necessity of American pharmacy. In 1775, a resolution established a medical department based on the British Army model. John Morgan, who knew well London's merchants of medicines and had advocated the manufacture of medicines in Philadelphia before the war, was appointed its director-general and chief physician. He emphasized the post of apothecary for procuring and dispensing medicine to the armed forces. Another resolution in 1777 officially laid out the duties of four apothecaries-general who would supervise collection, preparation, and delivery of medicines for the hospitals and army. Collectively, these decisions affirmed a commitment to the local production and distribution of medicines. New domestic sources of essential items, including purging salts, nitre, and glassware, were soon established. The apothecaries-general kept a storehouse in Yellow Springs, Pennsylvania, and Andrew Craigie, an apothecary-general, prepared a variety of medicines at his shop in Carlisle. Craigie had been medical commissary and apothecary to the Massachusetts forces since May 1775 and was noted for his skill in medicine, although little evidence survives of his training. Congress had begun to direct larger sums toward the cash purchase of medical supplies, instead of relying on credit or demanding middlemen use personal funds and then wait for reimbursement. Congress allocated more than $940,000 for the hospital department in 1778 alone, in contrast to the roughly $27,000 budget for medical supplies in 1776.[93]

By funding supply and demand, state institutions encouraged a more self-sufficient American medicine trade that, nevertheless, owed much to the preexisting imperial infrastructures. Greenleaf's sales to military customers persisted even after much of his other business ceased during the war years. From 1776 to the beginning of 1778, he put up more than £1,800 worth of medicines for the State of Massachusetts Bay, Continental Navy, and Board of War. Payment arrived via a combination of cash, medicines captured as prizes, and cash advances, which were not necessarily worth much at the time.[94] Many embraced the opportunities offered by spikes in illness and injury, as well as shortages of medicines and physicians in war-torn communities. Women's healing work and knowledge of local medicinal plants proved lucrative, while other women gained employment as military nurses.[95] Spending on medicines for the Pennsylvania Hospital declined because more were being made on the premises or purchased locally. An inventory of the hospital's stock from 1782 showed a range of medicines, as well as items indicative of its fully operational laboratory, such as two large marble mortars, scales, apothecaries' weights, one large brass mortar, one leaden cistern, one copper digester, and a variety of glassware.[96] The Continental Congress rented the laboratory to make medicines for the army beginning in 1781 and paid in specie.[97] After the war's end, much of the hospital's medicinal stock continued to arrive from local druggists and chemists, such as Moses and Isaac Bartram and Christopher Jr. and Charles Marshall, unlike in previous decades when much of the stock had arrived from London at higher cost.[98]

There is perhaps no better example of the war's effect on American medicine manufacturing than the *Lititz Pharmacopoeia*, first printed in Philadelphia in 1778. Its full Latin title roughly translated promised simple and efficacious medicines to those who used the text. Likely compiled by William Brown using whatever sources he had at hand, the pharmacopoeia illustrates the convergence of European medical tradition (namely humoral medicines) and colonial experience (an emphasis on specifics) in the recipes used at military hospitals. It represented an early glimpse of bulk medicine manufacturing in the United States and an attempt at achieving self-sufficiency, like other attempts in laboratories and gardens across the European empires. Recipes used native drugs and locally produced chemicals in addition to what could be imported in spite of the British blockade. Brown's text drew on common British works, including the *Pharmacopoeia Edinburgensis* (1756) and the *Pharmacopoeia Londinensis* (1746), which contained recipes popular with merchants of medicines in London. Soon after the pharmacopoeia appeared in print there followed a compendium of recipes for the French military hospitals in North America compiled by Jean-François Coste, chief physician to the French army. Whereas the *Lititz Pharmacopoeia* relied on North American plants, including sassafras and snakeroot, Coste's did not. The former colonists used native drugs extensively, but struggled to find substitutes for essential foreign ones, such as Peruvian bark (cinchona), which remained in short supply.[99] Together these

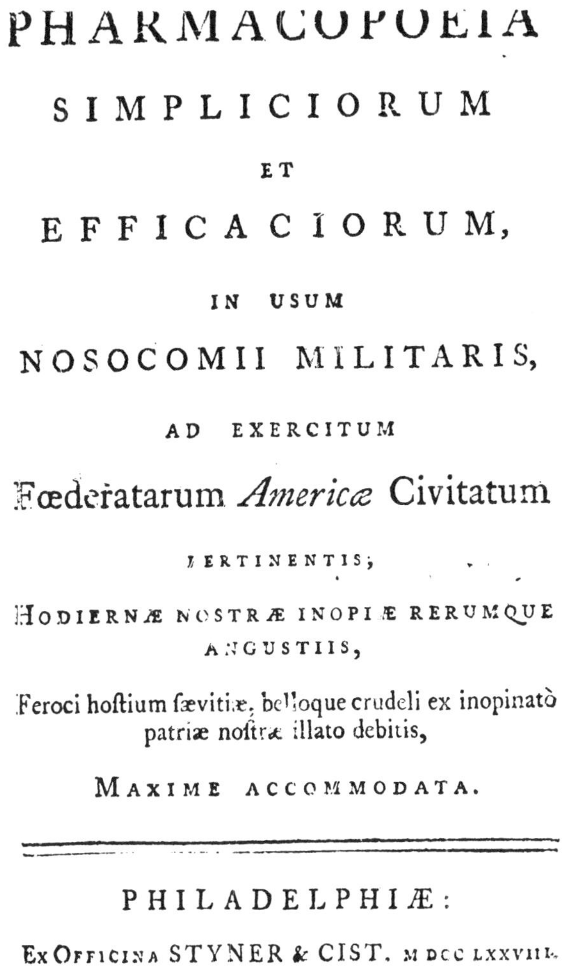

PHARMACOPOEIA

SIMPLICIORUM

ET

EFFICACIORUM,

IN USUM

NOSOCOMII MILITARIS,

AD EXERCITUM

Fœderatarum *Americæ* Civitatum

PERTINENTIS;

HODIERNÆ NOSTRÆ INOPIÆ RERUMQUE ANGUSTIIS,

Feroci hoſtium ſævitiæ, belloque crudeli ex inopinatò patriæ noſtræ illato debitis,

MAXIME ACCOMMODATA.

PHILADELPHIÆ:

Ex OFFICINA STYNER & CIST. M DCC LXXVIII.

Figure 5.3 Title page of the *Lititz Pharmacopoeia*, 1778. Note the emphasis on simple ("simpliciorum") and efficacious ("efficaciorum") remedies in an American context. Brown, *Pharmacopoeia simpliciorum et efficaciorum*.

experiences diversified what counted as acceptable medicines and enshrined simplicity and efficacy as benchmarks for medical practice moving forward, as made visible in the pages of the *Lititz Pharmacopoeia* compared to contemporaneous European texts. During these years, merchants of medicines across North America filled much of the void left by the abrupt closure of the legal transatlantic trade and, in the process, shifted the balance of the medicine trade.

A Revolution in Trade?

The transatlantic medicine trade's disruption affected its participants differently. Silvester Gardiner abandoned his shop in Boston, whereas John Greenleaf found new business with the Continental Navy. Martin Brimmer's doors remained open, but he increasingly sold goods other than medicines. In Philadelphia, activity increased at the Marshall brothers' laboratory. Merchants of medicines in London, meanwhile, struggled to settle accounts with their American customers. From a continental perspective, an American medicine trade emerged from the war with improved manufacturing capacity enabling a greater degree of self-sufficiency. A similar trend occurred in other sectors where goods and services that had been purchased from abroad were temporarily or more permanently produced in the new nation. Many of the ties severed by the war healed after it ended, returning the transatlantic medicine trade to a familiar shape, though with different people occupying producer, distributor, and consumer roles in some cases. American merchants of medicines continued to rely on British drugs, glass, and credit even as they cultivated their own long-distance networks. For example, Edward Stabler in Virginia purchased glass bottles and vials in bulk from suppliers in Philadelphia, Baltimore, and London.[100] The medicine trade embodied the ambivalence of the word *revolution*, which connotes a sudden or complete change in modern parlance compared to its older meaning of returning to an initial position in terms of celestial motion.[101] Regarding this moment primarily as a rupture in the medicine trade obscures that the characteristics central to medicines becoming global commodities had already developed throughout the eighteenth century.

Repercussions of the imperial crisis rippled across the British medicine trade's overseas markets. In practice, they took the form of mounting debt and greater caution on the part of exporters. Joseph Gurney Bevan had more than £2,000 tied up in Caribbean debts from 1776 to 1780. The majority of Plough Court's overseas customers could be found in the Caribbean at this time, which made the disruptions to transatlantic trade and the sugar economy significant to the pharmacy's bottom line.[102] Old debts left unsettled during the hostilities were difficult to resolve afterwards. Britain struggled to rebuild its overseas trade in an Atlantic system no longer unified under its imperial dictates.[103] Sometimes Bevan withheld goods as a penalty for late payments, but he also continued shipments in hopes of maintaining a good relationship, which would lead, in his thinking, to eventual returns. Obtaining security often did not stop him from resorting to suits or mortgages.[104] In 1781, his accounts had barely improved. Bevan complained that

Table 5.1 Duties on selected drugs, 1769–86

Drugs	1769	1786
Ambergris	11/7 $^{10}/_{20}$	16/6
Balsam	1/11 $^{4}/_{20}$	11d
Borax	3/3	3/8
Rhubarb	4/10 $^{10}/_{20}$	5/6
Sago	31.14.2 $^{75}/_{100}$	37.16.1 $^{77}/_{100}$
Scammony	3/10 $^{10}/_{20}$	5/6

Notes: All values are pounds sterling (s/d) per pound weight, except ambergris (per troy ounce) and sago (per £100 gross price).
Source: Report on Two Memorials, 21 March 1769, T 1/470/194–99, Treasury Board Papers; Commutation for Duties on Sundry Drugs, 10 February 1786, Add MS 38409, vol. 220, Liverpool Papers.

his business was "really straitened by the detention of money in Jamaica" and that he did not want to venture much more in overseas trade.[105] With most of his North American business blocked and deprived of remittances from the Caribbean, Bevan accrued more debt at home to extend credit abroad in hope of future repayment.

Trade policy remained a thorn in the side of London's merchants of medicines. The duties on imported drugs had not declined after 1769 despite war, lobbying, and the financial struggles of the chartered joint-stock companies. Rather, import duties had increased by almost a shilling in some cases (see table 5.1). For several medicinal drugs—ambergris, bezoar stones, coral, musk, and rhubarb, among others—the duty "may be said to amount to a prohibition" on legal importation, according to witnesses. Other drugs popular in medicine manufacturing, including cantharides, cassia, and galbanum, were noted as having excessive but not yet prohibitive duties levied on their importation. Smugglers along the English coast at Deal and Dover sold rhubarb for less than its duty of five shillings and six pence per pound. Smuggled goods may have comprised as much as one-fifth of total imports during this period.[106] High duties on drugs, sometimes even more than their wholesale prices in London, raised concerns among merchants of medicines who struggled through an economic recession in the 1780s. These taxes also affected other manufacturing industries since drugs, such as borax, were essential to ceramics and metalworking.[107] In 1783, the first stamp tax was applied to proprietary medicines in Britain to help the treasury pay the interest on loans from the war.[108] Such policies were designed to replenish state coffers rather than to aid manufacturers or practitioners and represented another step in the commodification of medicines.

Customs policy helped separate the compound medicine and drug trades by erecting impediments to importing some items rather than others. Manufacturers found ways to acquire raw materials through multifaceted supply chains that had grown and reacted to the customs system for almost a century. Some purchased contraband ingredients, while others, such as Joseph Gurney Bevan, claimed that they, on principle, did not deal in smuggled goods. Principles, however, did not stop Bevan from sending contraband musk to Barbados for Gera Alton, an apothecary and surgeon, in 1788.[109] Other drug importers circumvented the customs in spite of the losses and penalties they risked if caught dealing in contraband goods because "many of the very Articles were and are now sold in the Shops for less than their Dutys," George Webster protested.[110] Price data from merchant correspondence suggests that some drugs were indeed selling for nearly the amount of their duties in the early 1770s.[111] Shipments arrived for the partnership of Primatt & Maud from a network of suppliers connecting Rouen, Rotterdam, Hamburg, Messina (on the island of Sicily), and Amsterdam. Such transactions required competency, if not fluency, in a variety of foreign monies, exchange rates, and languages. Likewise, Corbyn & Company received shipments from across continental Europe. Neither the East India Company nor the British government could stem the flow of drugs from these sources.[112] Other merchants of medicines gradually moved away from drug importation to focus on manufacturing or scaled back their businesses entirely in response to the difficulties they faced in the 1770s and 1780s.

When Bevan's accounts did not improve after the Treaty of Paris in 1783, he resolved to no longer allow such generous terms of credit in overseas trade. That he had been "so grievously disappointed of remittances" from the West Indies during the war years made it difficult for Bevan to make payments to his own creditors "with the punctuality [he] desire[d]."[113] He more often demanded immediate settlement of debts and was increasingly "willing to forbear at present any coercive steps . . . proper and necessary" to achieve it. In some cases, Bevan threatened to cut off all credit to correspondents with accounts in arrears, which would have been especially potent if doing so blocked their access to the Quaker banks Bevan used to conduct trade.[114] Mortgages were another tactic of medicine exporters to recoup debts that by this time were often secured by land or, more frequently, human collateral. Then, when mortgages produced insufficient returns, Bevan terminated credit and sold assets. Generally, Bevan desired payment by bills of exchange drawn on British banks or by Caribbean agricultural products (ginseng, ginger, rum, or sugar) since he did not have the luxury of waiting for capital tied up in bonds or mortgages abroad.[115]

Imported medicines remained integral to plantation health regimes in response to continuing patterns of mortality and the pro-natal policies of the abolition era. Enslaved women's health received greater attention under the rhetoric of amelioration that did little to change conditions on most sugar estates in the 1780s and 1790s. Campaigns to promote childbirth emerged from concerns for the future of the plantation system as planters and abolitionists faced the consequences of demographic decline for the British imperial economy during debates about the slave trade. After 1807, Britons sought economically efficient means to encourage reproduction within the institution of Caribbean slavery, the bodily costs of which fell to mothers rather than plantation owners. Pro-natal policies implemented across the region included the attempted medicalization of midwifery and the marginalization of Afro-Caribbean healing practices. British medicines meanwhile featured prominently among imported provisions as they had for some time. The plantation complex thereby continued to offer lucrative markets for British medicines into the next century.[116]

Bevan's difficulties recouping overseas debt, however, indicated a broader shift in the organization of Plough Court's export trade in the 1780s. He transferred some of his more troublesome customers to West India merchant houses, so that correspondents could draw bills on the houses in London as an easier way to send returns. Bevan more frequently relied on Birkbeck, Blakes & Company to ship goods and collect debts.[117] The partnerships of Lascelles & Daling and Long, Drake & Long also worked with Plough Court. As sugar factors and commission agents, they purchased goods from Bevan and shipped them to planters for whom the houses served as agents. The Lascelles operated perhaps the principal West India merchant house of the seventeenth and eighteenth centuries. Family members participated in various transatlantic commodity trades; invested in shipping, the slave trade, and plantations; and benefited from their positions within the state bureaucracy.[118] These merchant houses acted as middlemen for Bevan. He still dealt with many customers directly and provided financial services for them, but he no longer had to trouble with some aspects of shipping and insurance.

Control of the trade had thus begun to slip from the grasp of the merchants of medicines who had shaped its infrastructure. Christopher Jr. and Charles Marshall's account with Bevan remained a topic of contention for years after 1783. Their lack of interest in his products concerned Bevan. The Marshalls received some items from Plough Court in the 1780s, but their reliance on British exporters waned compared to before the war. Learning from their experiences during the war, the brothers imported some medicines from Amsterdam often for less than from London.[119] By 1786, they were also producing more chemicals in their laboratory that they sold locally, used in manufacturing, and even exported to Britain. The Marshalls shipped chemicals and drugs—snakeroot, sarsaparilla, and sassafras, for example—to London in the 1790s.[120] Isaac and Moses Bartram had also increased the scale of their regional trade over the intervening years. The Bartrams' shops stocked the produce of British settlements in Asia and the Caribbean alongside that of North America, including camphor, Carolina pink root, and Barbadian aloes. Isaac also carried out an extensive trade with medical practitioners in the Pennsylvania interior, such as Samuel Fahnestock, who was one of Lancaster County's foremost doctors.[121] Medicines not sourced or made locally continued to arrive from London. Clarke, Jacam & Clarke, druggists at 8 Barbican, sent goods to Isaac (including pink flint popper bottles) and paid Joseph Gurney Bevan cash on his account after receiving bills of exchange. Following the war, Isaac still moved value around the Atlantic world to pay his bills and wrote to many of his former correspondents, but he increasingly did so through his own infrastructure on his own terms. In 1789, Bevan wrote to Isaac promising "thou canst import from hence with advantage" even though it had been "upwards of five years since I sent thee the last parcel" and an unsettled account remained between them.[122]

Since expectations about efficacy had already overcome the problem of distance, as captured in the more capacious definitions of specifics used by the close of the century, these postwar infrastructures enabled the production of such medicines to occur beyond its initial locus in London. Plentiful imports of foreign ingredients enabled American apothecaries, chemists, and druggists to manufacture an expanding variety of products illustrated by a profusion of printed stock lists and advertisements. In many ways, the catalogs resembled those that had circulated in London earlier in the century. Professional and governmental oversight of American pharmacies would not arrive until later in first half of the nineteenth century.[123] As American medicine came into its own, observers evaluated the viability of the new nation by the health of its body politic. Just as adulterated medicines or moldy drugs threatened British notions of state power and hierarchy, so too did medicine and public health take on political significance within the early republic beyond simply their commercial benefits.[124]

Constructing a medicine trade in the United States required both substitutes for imported drugs and chemical manufacturing capacity. In the 1780s, people scoured the countryside, particularly the area around the James River in Virginia, for supplies of snakeroot, an aromatic herb used

Figure 5.4 Snakeroot vine with a dead American robin in the foreground, 1731. Depictions of Virginia snakeroot began to appear in natural history texts alongside other examples of American flora and fauna during the eighteenth century, such as in Catesby's popular *Natural History of Carolina* (shown here). "*Turdus pilaris, migratorius*. The Fieldfare. *Aristolochia &c*. The Snake-Root," plate 29 in Catesby, *Natural History of Carolina*. Courtesy of the John Carter Brown Library at Brown University.

to treat fevers and as a stimulant. Though it had been traded for decades, snakeroot experienced a resurgence in this period. Advertisements from Pennsylvania and Virginia newspapers practically overflowed with the drug. Pink root, an anthelmintic (anti-worms) found across the upper South was also fast becoming a fixture of period herbals and a popular export. British mercantile houses desired many of these American plants for their medicinal properties, especially against the fevers and parasites that continued to assail populations across the empire, and practitioners across Europe integrated them into their practices.[125] Businesses soon arose that added value to the plants before their export. John Heard (1744–1834), for example, produced medicinal waters from snakeroot at a distillery in Ipswich, Massachusetts. The roots had to be boiled in water before ingestion, so distillation offered another mode of preservation and preparation.[126] The 1780s and 1790s witnessed continued efforts in the nascent United States to achieve greater self-sufficiency from European imports by blending European and local medicinal techniques. Printed books, notably Johann Schoepf's *Materia Medica Americana* (1787) and Samuel Stearns's *American Herbal* (1801), publicized the country's natural resources and indigenous *materia medica*. In Pennsylvania, the botanist, clergyman, and teacher, Gotthilf Heinrich Ernst Mühlenberg (1753–1815) classified regional plants in an effort to find a more nativist and syncretic kind of medicinal self-sufficiency.[127]

American partnerships soon traded American, Asian, and European medicinal products to regional and global customers. British trade with the United States had reopened on July 2, 1783, by an order in council. Though it fell short of the free trade some wanted, this development did not stifle American exports to Britain and its remaining colonies as well as to new European and Asian trading partners. After an economic downturn in the 1780s that hurt merchant shipping, the next decade saw US economic activity, including commodity exports, surpass prerevolutionary levels.[128] An apothecary by trade, William Stearns exported pink root, sassafras, snakeroot, and ginseng from Salem, Massachusetts, to European markets via the mercantile house of David Taylor & Sons in London, who handled more than 3,000 cases of medicinal bark in 1805.[129] Though exports of American medicinal plants had occurred in some form for centuries, going back to the introduction of sassafras in the late sixteenth century, these exports occurred at greater scope, scale, and capitalization, as well as with a different relationship to the state.[130] A shortage of cash also pushed Bevan to accept a larger portion of remittances as American produce. Unlike Corbyn, who had accepted sugar, cotton, and tar, Bevan received mainly drugs, such as snakeroot, ginger, and sarsaparilla. These he sold to well-known London drug merchants Vezey & Company and George Webster, with whom he often did business and from whom he was able to get good prices for them. Though Bevan shied away from acting as a commission agent, selling American plants helped recoup some of Plough Court's debts more punctually than could often be accomplished through other money forms.[131] Profits from the drug trade reshaped the physical landscape of the early republic. For example, Stearns served as director of the Salem Turnpike and Chelsea Bridge Corporation, and constructed buildings in the port of Salem that was experiencing a period of dramatic growth thanks to the China and Caribbean trades.[132]

The arrangements that emerged from the revolutionary decades reflected the extent that manufactured medicines had become a fixture of the imperial system during a moment when many questioned the trade's prevailing organization. American merchants of medicines could lessen their dependence on London because of the infrastructure they had built over several decades and that had been activated by necessity and state funding during the war. The balance of transatlantic credit relations shifted in response to political-economic conditions and efforts to achieve medicinal self-sufficiency. Despite his best efforts, many of Joseph Gurney Bevan's customers across the United States and Caribbean failed to fully repay their debts after the war. When Bevan left Plough Court in 1794, his accounts remained unsettled, and he was "more intent than ever in collecting the remains of [his] injured property in the West Indies."[133] American merchants of medicines increasingly navigated transatlantic commerce less beholden to those in London who had steered the medicine trade for nearly half a century, though its global reach and reliance on imperial institutions remained.

Conclusion

The last quarter of the eighteenth century saw a credit crisis in the British Caribbean leave many properties worth less than their mortgages, profits decline for the East India Company, the customs regime fall further out of date, and a global war elevate the cost of freight and insurance, interrupt supply lines, shake territorial control, and take thousands of lives. Within this broader context, the debt accumulated during the growth of the medicine trade threatened a system caught between long-distance supply and local manufacturing. Much overseas trade was swept up in a wave of uncertainty at this time, but pharmacies typically faced greater instability than established mercantile houses due to their smaller capitalization and tighter margins. The American Revolutionary War temporarily disrupted the transatlantic medicine trade, providing an opportunity for the bulk production of medicines in the United States funded, in part, by the new state and national governments. By the start of the 1780s, some merchants of medicines in London had grown weary of overseas trade, though it was not for some years until American-made medicines supplanted British ones in domestic markets. Elsewhere, medicines from Britain continued to arrive at plantations, trading settlements, and military hospitals in bulk quantities and at great expense.

Reliance on a certain kind of medicine, however, was not in question. The war years encouraged the self-sufficiency only previously discussed in Madras or Philadelphia and brought about several visible changes to the directionality of the long-distance medicine trade, but the relationship of medicine to state power and labor remained fundamentally unchanged. Characteristics associated with modern medicine (portability, standardization, commensurability, economies of scale)—and thus, medicine as a global commodity—had developed throughout the eighteenth century, rather than as a particular outcome of the imperial crisis. When transatlantic trade resumed in the 1780s, a new crop of merchants of medicines asserted their claims to its future. In London, ownership of Plough Court would leave the Bevan family, while Thomas Corbyn would retire to the suburbs, leaving his son in charge of exports. British-manufactured medicines would remain central to colonial health regimes across Asia, Australia, Africa, and the Caribbean. A more robust and widespread medicine industry soon influenced the institutionalization of healing in the American Republic that would debate the place of manufacturing, agriculture, and enslavement within its society.[134] Patterns of commerce, warfare, and unfree labor—of empire, in other words—established over the course of the eighteenth century would continue to dictate healthcare's terms.

Conclusion

British medicine and empire had become entangled, but the relationship hardly remained static. As merchants of medicines reevaluated their priorities after the wars of the late eighteenth century, state and East India Company officials pondered the future of overseas settlement and the place of imported medicines in it. Planters and merchants also reaffirmed medicines' place in the plantation health regimes of the abolition period. By some claims, the British medical marketplace boasted more than 30,000 druggists and apothecaries in 1786, which did not even include the at least 700 surgeons and apothecaries documented across the armed forces and plantations. Such a crowded field signaled the ascendancy of a commercialized healthcare for many by this time, but it also made turning a profit more dependent on a steady supply of cheap ingredients and markets for manufactured medicines.[1] New duties levied in 1799 brought higher charges on drugs that could leave little room for those profits. The EIC absorbed annual losses importing drugs because yearly sales of some, such as rhubarb, paid more in duties than they returned to the company. Rhubarb was an extreme case, but other medicinal drugs also saw a large proportion of their sales revenue go to the customs.[2] Robert Wigram (1744–1830), a notable drug merchant and eventual member of Parliament, summed up the feelings of many involved in the medicine trade by calling for a "general change of Commerce that must take place," or else, he warned, Britain would lose the American as well as the European drug trades.[3]

Though Wigram spoke of the commerce in raw materials, his perspective also applied to the trade in medicines at the turn of the nineteenth century. Participants understood the trade's connection to policy decisions, capital markets, interstate competition, and commodity prices but neglected its reliance on the sailors, servants, slaves, and soldiers who ingested medicines, an abstraction that not only shaped the trade's scale but also conceptions of the products involved in it. Wigram's warning did come to pass, to a degree, as other European states and the United States built up their own medicine trades over the coming decades. Not only had American pharmacists, such as the Bartrams and Marshalls in Philadelphia, begun to make their own medicines, but drug merchants, such as William Stearns in Massachusetts, had also started to encroach on other aspects of the British trade. By the first quarter of the century, some American-made medicines had replaced British ones in domestic markets. Once their British patents expired, popular medicines, such as Hooper's Female Pills and Turlington's Balsam, were produced and sold across the country under US patents.[4] Yet as Americans forged new commercial ties westward and overseas, so too did London-based merchants of medicines find new markets in India, Africa, and Australia as the British empire continued to underwrite the production, distribution, and consumption of their products.[5]

Tremendous effort and expense were put forth to deploy British medicines because they supported a political economy based on military fiscalism, territorial acquisition, and institutionalized human enslavement. These characteristics of early modern empire encouraged the globalization of the medicine trade, which subsequently threatened those very same political-economic developments at several points.[6] The imperial framework that the medicine trade had helped construct during the eighteenth century brought a renewed commitment to overseas trade, manufacturing, and ties to the state as industry donned new corporate and institutional guises in the century ahead. Large shipments went to plantations and hospitals overseas while the state provided lucrative contracts and protected long-distance trade. The institutions, such as the army, navy, and trading companies, carrying out this work far from Britain's shores received manufactured medicines in increasing quantities. These products had thus become a permanent, and quotidian, means of expressing British power in faraway places. Their global distribution linked places and people by the expectations of profit, health, and work they embodied.

Meanwhile, medicine manufacturing in London experienced a generational transition as members of the cohort who had guided the trade through its transformation into a globalized enterprise were now either dying or retiring. Its successor would still look overseas but in different ways and to different places. Thomas Corbyn passed away on February 25, 1791, after retiring some years earlier to tend a garden and contemplate religion in Bartholomew Close near Smithfield. Though considered a "pious" and "sincere Christian" by his peers, Corbyn was also seen to possess "a considerable portion of sectarian bigotry" leading to several nicknames, including "hat regulator" and "Pope Corbyn." The former sobriquet he gained for often disapproving of the size and shape of hats worn by other Friends, and the latter for his "rigid precepts" while holding a high position in the London Meeting. Given the incongruity of comparing a Quaker to the pontiff, it might also have been a reference to the wealth of Rome on account of the sizeable estate Corbyn had amassed after decades spent in the medicine trade. Corbyn owned several properties and many government annuities at the time of his death, enabling him to leave significant sums to his children, former employees, and charity.[7] His son John Corbyn carried on the pharmacy but struggled to settle accounts related to his father's estate into the 1820s. Nonetheless, Corbyn & Company persisted as overseas sales to Canada and Australia grew and those to the Caribbean rebounded in the first half of the nineteenth century. Customers from all over wrote to the partnership without prior contact requesting medicines in quite a departure from the lengths Thomas Corbyn first went in the 1740s to establish reliable connections across the Atlantic world.[8]

The 1790s also saw the storied Bevan family depart the medicine trade. After Joseph Gurney Bevan left in 1794, the Plough Court pharmacy passed from the family that had started it almost a century earlier to the partnership of Samuel Mildred and William Allen (1770–1843), also known as the "Spitalfields Genius" for his exploits in the realms of business and science. This arrangement was short lived, and in 1798 Allen partnered with Luke Howard (1772–1864) to form the partnership of Allen & Howard. While both men are perhaps better known individually for other activities, Allen as a prominent abolitionist and Howard as an amateur meteorologist who developed a nomenclature system for clouds, together the two Fellows of the Royal Society became successful medicine manufacturers.[9] Allen and Howard purchased goods in larger quantities than their predecessors. They continued Plough Court's overseas trade, and local sales also nearly quadrupled from 1781 to 1808. To do so, Allen and Howard used debt more frequently to pay their bills, while receiving fewer long-term promises to pay and keeping labor costs low. Under their care, Plough Court remained a pillar of the British medicine trade, and of its expanding empire, into its second century.[10]

Soon emerged the large corporations that marked the medicine trade's next phase in the mid-nineteenth century whereby Plough Court would eventually become part of a name more recognizable today, GlaxoSmithKline. Through a series of mergers and acquisitions, Plough Court endures as a fragment within GlaxoSmithKline, itself the result of a 2000 merger of Glaxo Wellcome and SmithKline Beecham. Henry Wellcome established Burroughs Wellcome & Company in 1880 to sell medicines across the British empire, the proceeds of which were used to establish the Wellcome Trust, one of the world's largest medical charities, and the Wellcome Collection, where much of the research for this book occurred.[11] That the records of an eighteenth-century pharmacy lurk in the overlooked corners of a twenty-first-century pharmaceutical giant should no longer seem contradictory, nor should the incorporation of Plough Court into GlaxoSmithKline. Little separated the worlds of medicine, commerce, and state power in the early modern period; and the routines that developed from that proximity have proved resilient and generative. Eighteenth-century medicine manufacturing partnerships were highly capitalized with close ties to the state and overseas markets provided by colonial settlements, the slave trade, and the armed forces. They relied on credit to make remedies in bulk and benefited from ubiquitous mortality as well as the desperation of those hoping to forestall it. Such hopes, in turn, crowded local healthways with an imperial one enabled by long-distance medicines and universal subjects. The global spread of manufactured medicines convenient for bulk application supported first an idea of human sameness that was then bent into a justification of physiological difference once it threatened to undermine the hierarchies that had come to organize early modern empire and capitalism.

The events described in the preceding chapters took place during a period of far-reaching changes in European commerce, governance, and healthcare that were imposed across distances by the twinned violence of merchant capitalism and empire. It is difficult to disentangle these developments, especially from the perspective of those who ingested medicines or were subjected to the hierarchies they informed. The rise of a global medicine trade upset many health practices but could not fully replace local ones. A mass market, nevertheless, emerged where customers could expect that the medicines they paid for would live up to new models of efficacy, established not in a laboratory or medical text but through the processes of exchange. Yet these claims remained largely aspirational. The gains in public health and sanitation that spurred real declines in mortality occurred only later in the nineteenth century and, then, mainly in metropolitan locations rather than colonial ones.[12] Out of these conditions did emerge the long-distance delivery of medicines useful to individuals and institutions alike. Druggists, chemists, and apothecaries around the Atlantic world turned to novel financial instruments and established modes of trust to yoke the manufacture of medicines to expanding systems of overseas trade and resource extraction. Medicines also greased the wheels of numerous transatlantic trades and lined the pockets of those fortunate to attain contracts to supply them. Like so many other trades, the one in medicines rewarded some and excluded others from the generational wealth derived from overseas empire and healthcare within it.

Out of this moment arose infrastructures of healthcare built on methods of production and distribution that drew on credit and the slave trade, alongside other forms of trust and power. At a more epistemic level, such infrastructures also relied on commensurable remedies and interchangeable bodies, two forms of abstraction that arose from the particular conditions of medicine, empire, and commerce in the long eighteenth century. Constitutional medicines that considered individual characteristics were impractical at the scales and distances demanded by the British empire. Faced with astonishing mortality and tenuous power dynamics, Britons abroad looked to medicines that instead offered the prospect of simple treatment for many. In this way, mobility transformed what medicines were expected to do, how they were valued, and for whom they were intended. Long distances required medicines understood to work on disparate patients, free and unfree, often situated far from both the individuals' and the medicines' places of origin. Mobile people who had been detached from familiar cultures and forced into groups seemed to require a certain kind of bulk treatment, which elevated new logics of healing while raising questions of expense, self-sufficiency, and bodily difference. Colonial medical work did not just influence the emergence of what has been called European "scientific medicine"—a story others have told—but reshaped the very expectations of health and value that undergirded global exchange.[13] The medicine trade abetted the abstraction of the natural world so central to capitalism's emergence as an ideological system whereby imaginary or aspirational value displaced the real value of bodies, medicines, and wellness.[14] These material pressures could have led in other directions, as illustrated by the theoretical potential of the universal body or medicinal self-sufficiency, but did not in this case. Instead, these alternate paths were slowly foreclosed as the relevant ontological frameworks lost much of their analytical power when integrated into the day-to-day fabric of long-distance exchange. Not only did medicines support the intellectual and physical work of empire, but its commercial form shaped the practice of medicine and the fundamental understandings underpinning it.

Though the nineteenth century would see the arrival of several things that made the medicine trade look and work differently around the Atlantic world—such as germ theory, chemical testing, and professional regulation—how people understood medicines as global commodities tied to the fortunes of empire and capitalism carried over from this early modern moment. Rarely perhaps did these eighteenth-century medicines demonstrate curative properties by today's pharmacological standards, but they proved effective vehicles for extending commercial ties, valorizing particular ideas or practices, providing succor, and exerting power. As a result, British medicines spread hopes of health and productivity throughout a system that used them to support the exploitation of the very people who were often the intended consumers of the products. These medicines embodied the promise and violence of a nascent globalization, leading to a depersonalization of care in line with exigencies of imperial commerce. Looking beyond the structural forces of commerce and empire, there was also something essentially embodied about the spread of these commodities. The acts of holding, smelling, and tasting a pill or elixir; hoping it will alleviate pain, provide a living, or save a life, all serve as reminders that medicines were not always impersonal, standardized, or convenient despite how they were marketed or understood. While a bitter tincture of snakeroot today may seem unconventional, and perhaps even outlandish, the idea that there exists a medical solution appears very contemporary and hardly out of place among the one-size-fits-all approaches characteristic of institutionalized healthcare. Anxieties about efficacy, quality, or expense hardly seem old-fashioned, while the tension between ontological and physiological understandings of disease and treatment has returned in the form of precision medicine and other targeted therapeutic options for those who can afford them. What remains difficult, then as now, is recognizing the systems of power and exclusion behind those options before they become so ingrained in the commerce and culture of healing that their coercive histories fade into inevitability.

ACKNOWLEDGMENTS

More assistance than I can hope to adequately acknowledge in these pages made the publication of this book possible. My parents, Lydia and Andrew, set an example of empathy, generosity, and inquisitiveness about the world that I try to follow every day. I am fortunate to have the support of my sister Grace and a wider circle of Dorners, Feduses, Follansbees, and Dunnings. Several people who saw the beginning of this project did not get to see the end of it. In particular, my grandmother instilled in me the value of reading and curiosity. She would have appreciated this. Mom, that I may have to include you in this list is not something I ever expected to write. Friends and teammates have nourished my spirit at every turn, all the more so given the relocations that come from a peripatetic academic life. Claire Dunning's capacity as a scholar and partner remain unmatched since I wrote similar words in the dedication to my dissertation. I would not have made it here without you, Claire. The determination with which you carry so much that you did not choose to pick up reminds me of what together we can realize.

Financial backing from Brown University, the International Seminar on the History of the Atlantic World at Harvard University, the Institute of Historical Research, the William L. Clements Library, and the Program in Early American Economy and Society at the Library Company of Philadelphia made possible the archival research for the book. This research would not have been productive, however, without assistance from the archivists and librarians at the American Philosophical Society, Baker Library, Barbados Department of Archives, Boston Athenaeum, British Library's APAC Room, Caird Library, Cambridge University Library, Clements Library, Countway Library of Medicine, Hispanic Society of America, Historical Society of Pennsylvania, London Metropolitan Archives, Maine Historical Society, Massachusetts Historical Society, Massachusetts State Archives, National

Archives of the UK, New England Historic Genealogical Society, Newport Historical Society, Philips Library (PEM), Rhode Island Historical Society, Royal College of Physicians, Royal College of Surgeons, Swem Library Special Collections, and Wellcome Library. Special thanks are due Janet Payne at the Worshipful Society of Apothecaries, Jill Veitch at GlaxoSmithKline, James Green at the Library Company of Philadelphia, and the volunteers at the Lincoln County (ME) Historical Society for their contributions at formative stages. Given the changing structure of the academic job market, I have been fortunate with employment, though outside the tenure track. I am grateful for the time and space provided by a lectureship at Stanford University and the Patrick Henry Postdoctoral Fellowship at Johns Hopkins University that have allowed me to finish this project.

My ledger reveals intellectual debts to many. Conversations with Patrick Wallis and Richard Drayton helped me construct a methodology in the project's early days. Since then, seminars and writing groups at Brown University, Stanford, the Omohundro Institute of Early American History & Culture, and Johns Hopkins have helped hone my prose and refine my arguments, as have colleagues at sundry conferences. The participants in the Omohundro Institute's 2017 Scholars' Workshop encouraged the interventions at the heart of chapter 3. I fondly recall my time in Williamsburg amid such an affirming and productive intellectual community. Conversations with Paula Findlen, Jonathan Gienapp, and Caroline Winterer at Stanford came at pivotal moments in the process of turning a dissertation into a book. The Stanford Spatial History Project provided a space where the book's maps could come into being with the assistance of my research assistant at the time, Natty Jumreornvong. Thanks are due as well to Farid Azfar for sharing several of his sources relating to the South Sea Company. A manuscript conference hosted by François Furstenberg and Philip Morgan at Johns Hopkins in the fall of 2018 delivered a necessary final push. My thanks to all the participants—Phil, François, Amanda Herbert, Casey Lurtz, Mary Fissell, and the graduate students—for their insightful engagement with my work and the useful conversations that followed.

The book bears the imprint of my advisors at Brown University: Tim Harris, Harold Cook, and Seth Rockman. In addition to offering mentorship and training, they fostered an environment where I could explore a constellation of interests. They made me a better researcher, writer, and colleague. A portion of chapter 4 appeared under a different guise in the *William & Mary Quarterly* in 2015. The guiding hands of Josh Piker, Meg Musselwhite, and the anonymous reviewers and former editors helped transform a graduate research paper into an article worthy of the *Quarterly*, an early step on a path that has led here. Nadine Zimmerli also receives my gratitude for seeing promise in this project and for her contributions to the manuscript.

The editors and other staff at the University of Chicago Press have proved careful stewards for the manuscript through the publication process. They commissioned thoughtful reviewers whose detailed comments pushed my thinking on content, structure, and argument. Tim Mennel has offered sage advice from before he had any reason to do so. The production and marketing teams ensured this book actually achieved physical form.

Kate Blackmer greatly improved the book's maps. Several friends took the time to read portions of the manuscript—thanks Luke and Stuart! Claire has read more versions of this book than anyone besides myself and has rescued it on more than one occasion.

Revisions occurred during a period of illness and loss in my family, and the book's final form emerged sharper from it. After all, one of the book's central premises is that what is often felt as personal is, in fact, constructed to be impersonal—entries in a ledger. This disconnect has been as jarring to read in an eighteenth-century archive as it has been to encounter in twenty-first-century life. Sickness and commerce are ubiquitous presences in our lives, intimately tied to our perceptions of ourselves and the world around us, yet they remain subject to material power outside our control. It feels urgent now, as it always has been, to identify the extent to which such abstraction has shaped the ways some profit from the bodies of others.

Hyattsville, Maryland
September 8, 2019

NOTES

INTRODUCTION

1. Griffenhagen and Bogard, *Drug Containers*, 75–76; Watters, "Turlington Balsam Phail."
2. On wood and gunpowder, see Roberts, "Pines, Profits, and Popular Politics"; Cressy, "Saltpetre, State Security." For the importance of medicines to interimperial competition, see Dorner, "Medicines and Mercantilism."
3. It is fortunate that some have survived from this period. In one notable case, the journals and ledgers of the English East India Company's drug trade from 1730 to 1784 were destroyed alongside other "useless records" in 1860 to clear space in India House. Papers Relating to the Destruction of Useless Records (1858–1881), IOR/H/722, pp. 253, 257–59, India Office Records and Private Papers, Archives and Manuscripts, British Library (hereafter cited as India Office Records). For details of Plough Court's history, see Chapman-Huston and Cripps, *Through a City Archway*.
4. Commodity studies have become a staple of early modern studies, especially for those who work in a transnational frame. Historiographically, commodities played an important role in organizing early approaches to commodity chains and a world economy. They have been used to write increasingly material histories of empire that link distant places and people, and blend questions of what things meant and what things did across broad geographies and chronologies. The category of commodity has remained capacious, effectively incorporating the human, the environmental, and the manufactured. Commodities have also offered a concrete approach for reevaluating early modern developments, such as the emergence of industrial capitalism or long-distance modernity. For some examples, see Bair, *Commodity Chain Research*; Pomeranz and Topik, *World that Trade Created*; Hancock, *Oceans of Wine*; Norton, *Sacred Gifts*; Schwartzkopf and Sampeck, *Substance and Seduction*; Smallwood, *Saltwater Slavery*; Rappaport, *Thirst for Empire*; Satia, *Empire of Guns*; Beckert, *Empire of Cotton*; DuPlessis, *Material Atlantic*; Anishanslin, *Portrait of a Woman in Silk*; Smith, "Amidst Things."
5. For a useful overview of this historiography, see Porter, *Disease, Medicine, and Society*, 40–41. For another perspective on scientific progress, see the essays in Daston and Park, *Cambridge History of Science, vol. 3*.
6. For example: Winterbottom, "China Root," 42; Cook and Walker, "Circulation of Medicine"; and especially in the study of Caribbean slavery, such as Paugh, *Politics*

of Reproduction, as discussed in chapter 3. In a twentieth-century context, see Tomes, *Remaking the American Patient*.

7. How medical practices looked and evolved in a variety of local contexts across this geographical span are stories told by other historians, for example: Gómez, *Experiential Caribbean*; Sweet, *Domingo Álvares*.
8. James Delbourgo offers one example of this codependence being made in London through collecting, classifying, and categorizing specimens arriving there, see Delbourgo, *Collecting the World*.
9. For characterizations of the cotton mill or the iron manufactory as crucibles of industrialization, see Beckert, *Empire of Cotton*; Satia, *Empire of Guns*. The term *laboratory* also reflects period parlance, for example: Inventory and Valuation of Stock, 25 Dec. 1761, MS 5452/2, Corbyn & Co. Papers, Wellcome Library (hereafter cited as Corbyn & Co. MSS). On the topic of laboratories as modern, see Kohler, "Lab History."
10. Walker, "Remedies from the Carreira da India," 177–78; Cook and Walker, "Circulation of Medicine," 339. Eighteenth-century exporters separated drugs from other medicines in the lists they sent to overseas correspondents, see examples in Foreign Letter Book (1742–1755), MS 5442, Corbyn & Co. MSS. The term *pharmaceutical* has been used as an adjective since the early seventeenth century, but as a noun only since the nineteenth. *OED Online*, s.v. "pharmaceutical," March 2018.
11. Wallis, "Exotic Drugs," 20–22, 26, 34, 36–38; Cook and Walker, "Circulation of Medicine," 339–40; Rutten, *Blue Ships*, 11–12; Porter and Porter, "English Drugs Industry," 279; Roberts, "Import of Drugs into Britain."
12. Porter, *Health for Sale*, 40; Berg, *Age of Manufactures*; Weatherill, *Consumer Behaviour*.
13. Weisser, *Ill Composed*, 2–4; Seth, *Difference and Disease*; Wear, *Knowledge and Practice*, 67; Gómez, *Experiential Caribbean*, 129; Schiebinger, *Secret Cures of Slaves*, 12, 123–26. Though scientists today can struggle to chemically isolate an active ingredient from a long-used plant, that does not mean that the medicines made from it were rubbish. Despite the later stigma that premodern medicines often did more harm than good, they were only harmful sometimes; more often, they were palliative and mostly harmless, though rarely curative in our terms. Yet for patients and practitioners, merchants and vendors, many substances had strong bodily effects that "worked" from their perspectives. Winterbottom, "China Root," 40–41; Vaughan, *Curing Their Ills*, 7.
14. On biological versus cultural approaches to historicizing taste, see Norton, *Sacred Gifts*, 7–9; Mintz, *Sweetness and Power*, esp. chap. 1.
15. For a useful overview of the new history of capitalism, see Beckert and Rockman, *Slavery's Capitalism*; Beckert and Desan, *American Capitalism*; Matson, special issue, *Journal of the Early Republic*; "Forum: Paper Technologies of Capitalism." Historians have identified the thick intellectual traffic between the realms of commerce and science in the early modern world: Cook, *Matters of Exchange*; Margócy, *Commercial Visions*; Delbourgo, *Collecting the World*. Less attention, however, has been paid to the economic implications of the confluence of infrastructures of capitalism, empire, and medicine in this period.
16. For these changes, see Brewer, *Sinews of Power*; Wennerlind, *Casualties of Credit*; Hancock, *Oceans of Wine*; Stern, *Company-State*; Drayton, *Nature's Government*. Medicines, however, are typically left to their own literature, for example: Curth, *Physick to Pharmacology*; Poynter, *Evolution of Pharmacy*; Anderson, *Making Medicines*; Higby and Stroud, *Apothecaries and the Drug Trade*. Newer work has sought to connect the fields, such as that of Cook, Walker, Winterbottom, and Breen cited elsewhere in this introduction.

17. Wallis, "Exotic Drugs," 37; Mortimer, *Dying and the Doctors*; French and Wear, *Medical Revolution of the Seventeenth Century*.
18. Shaw and Welch, *Making and Marketing Medicine*, 17–20, 22–23. For the Egyptian, Greek, Roman, and Arabian antecedents, see Court, "Pharmacy from the Ancient World."
19. Wallis, "Exotic Drugs," 26–27, 29, 37; Cook, "Good Advice," 22; Cook, *Decline of the Old Medical Regime*; Jenner and Wallis, *Medicine and the Market*; Burnby, "Printer's Ink"; Watson, "Trading Accounts"; Porter, *Health for Sale*. For the European context, see Gentilcore, *Healers and Healing in Early Modern Italy*; Lindemann, *Health and Healing in Eighteenth-Century Germany*; Brockliss and Jones, *Medical World of Early Modern France*; Cavallo, *Artisans of the Body*.
20. Harrison, "Treatment of Fevers"; Porter and Porter, "English Drugs Industry," 279; Simmons, "Medicines, Monopolies and Mortars," 226; Golinski, "Chemistry." Chemical manufacturing as an export industry had existed in the Netherlands since at least 1700, with branches of the trade stretching back further, see Davids, *Dutch Technological Leadership*, 195–96, 198.
21. McKendrick, Brewer, and Plumb, *Birth of a Consumer Society*, 2; Brewer and Porter, *World of Goods*; Breen, *Marketplace of Revolution*, 57–58; De Vries, *Industrious Revolution*; King, "Accessing Drugs," 49–50.
22. Cook, "Rose Case Reconsidered," 527–28; "Attempted Legislation," 198–200; Barrett, *History of the Society of Apothecaries*, 136–37; Porter, *Health for Sale*, 27–29; Loudon, *Medical Care*; Pelling, *Medical Conflicts*; Cook, *Decline of the Old Medical Regime*.
23. Deneweth and Wallis, "Development of Medical Care," 543–44, 551; Wallis, "Early Modern Medical Economy," 477; Wallis, "Exotic Drugs," 35–36; Bamji, "Medical Care in Early Modern Venice"; Wallis and Pirohakul, "Medical Revolutions." On domestic medicine in this period, see Leong, "Making Medicines"; Pennell, "Perfecting Practice."
24. For example, Wear, "Continuity and Union," 319.
25. Chakrabarti, *Materials and Medicine*, 4–5, 9, 13; Dorner, "Medicines and Mercantilism," 32, 50, 52–53. On the interplay of botany, commerce, and politics that motivated certain medical practices and the search for valuable plants, see also Schiebinger, *Plants and Empire*; Bleichmar, "Atlantic Competitions"; De Vos, "Natural History"; Drayton, *Nature's Government*; Raj, *Relocating Modern Science*; Harris, "Long-Distance Corporations."
26. This gives new emphasis to what Pratik Chakrabarti has called the "imperialism of goods" (Chakrabarti, *Materials and Medicine*, 6). On the emergent logics of enslavement, see such recent works as Turner, *Contested Bodies*; Burnard, *Planters, Merchants, and Slaves*; Roberts, "Labor and Agriculture."
27. John Brewer suggested a strong state, whereas J. H. Elliott has proposed a weaker version. Brewer, *Sinews of Power*; Elliott, *Empires of the Atlantic World*. Subsequent work reinforces the idea of a weaker state that nonetheless engineered conditions suitable for the growth of empire and trade: Desan, *Making Money*; Beckert, *Empire of Cotton*; Satia, *Empire of Guns*.
28. Szechi and Holmes, *Age of Oligarchy*, 76–77, Compendium G.2(ii). On the British state's role regulating the economy, see O'Brien, "Central Government"; Davis, "Rise of Protection." Britain lacked a stamp tax on proprietary medicines until 1783, for example. Griffenhagen, *Medicine Tax Stamps*; Alpe, *Medicine Stamp Duty*.
29. Szechi and Holmes, *Age of Oligarchy*, 149–51; Berg, *Age of Manufactures*, 117, 121–22; Price, "What Did Merchants Do," 269–70; Davis, "English Foreign Trade, 1700–1774."

30. Ledgers of Imports and Exports, England and Wales (1696–1780), CUST 3/1–82, National Archives of the UK (hereafter cited as CUST 3).
31. Turner, *Contested Bodies*; Paugh, *Politics of Reproduction*.
32. For such comparisons: Kuriyama, *Expressiveness of the Body*; Beukers et al., *Red-Hair Medicine*; Berger, *Ayurveda Made Modern*; Mukharji, *Nationalizing the Body*; Osseo-Asare, *Bitter Roots*; Gómez, *Experiential Caribbean*.
33. Brown, *Reaper's Garden*, 1–6, 15–19, 54–55, 58–59; Gómez, *Experiential Caribbean*, 15.
34. On these other factors' contributions to colonial demand, see Berg, *Age of Manufactures*, 128–30. For a discussion of labor exploitation for production, see Ashworth, *Industrial Revolution*, 129. Recent commodity studies have done much to unite producer and consumer views of the eighteenth-century British empire, for example Eacott, *Selling Empire*; Anishanslin, *Portrait of a Woman in Silk*.
35. The earliest example of this consumerist literature remains McKendrick, Brewer, and Plumb, *Birth of a Consumer Society*; whereas more recent studies have expanded the idea in an eighteenth-century context, including Breen, *Marketplace of Revolution*; Breen, "Baubles of Britain"; Van Horn, *Power of Objects*; Haulman, *Politics of Fashion*. Some historians have attributed many of the advances in the medical marketplace to an increasingly choosy clientele that had more to spend on health: Porter, *Disease, Medicine, and Society*, 44; Digby, *Making a Medical Living*; Jenner and Wallis, *Medicine and the Market*. In contrast, Chris Evans foregrounds the importance of individuals who had less say in what objects they received, such as the enslaved laborers who did not use iron hoes by choice; see Evans, "Plantation Hoe."
36. For example, Styles, "Involuntary Consumers," 9; Evans, "Plantation Hoe," 77. For further discussion of involuntary consumers and how to write about them, see Hartigan-O'Connor, *Ties That Buy*, 5; Finn, *Character of Credit*, 9–10; Martin, *World of Goods*.
37. Cook, "Markets and Cultures," 142.
38. More recent scholarship on the contribution of medical discourse to hardening categories of human difference builds on the earlier work of Richard B. Sheridan and includes: Gómez, *Experiential Caribbean*; Schiebinger, *Secret Cures of Slaves*; Hogarth, *Medicalizing Blackness*; Seth, *Difference and Disease*; Turner, *Contested Bodies*; Curran, *Anatomy of Blackness*, chap. 4.
39. Morgan, *Laboring Women*, 154.
40. Schiebinger, *Mind Has No Sex*, chap. 8; Ulrich, "Midwifery and Mortality," 28–29; Wear, *Knowledge and Practice*, 48–55, 65–67; Gómez, *Experiential Caribbean*, 58–64; Bashford, "Medicine, Gender, and Empire," 117; Sweet, *Domingo Álvares*, 143–44.
41. This is also not to say that British medicine was innovative or dynamic while non-European medical cultures were the opposite. Every system of medical practice in the early modern period could appear either progressive or traditional depending on the context. For example: Crawford, *Andean Wonder Drug*; Mukharji, "Embedded Traditions"; Bhattarcharya, "From *Materia Medica* to the Pharmacopoeia"; Gómez, "Circulation of Bodily Knowledge," 395–96; Weaver, *Medical Revolutionaries*.
42. Such a combination of the local and the global with the personal and the impersonal has been identified in the emergence of a long-distance modernity in the mid-eighteenth century. Vernon, *Distant Strangers*, xi–xii, 15–16. Though this was hardly an inexorable march toward modernity. Breen, "Drugs and Early Modernity," 4; Gómez, *Experiential Caribbean*, 5–6; Latour, *We Have Never Been Modern*.
43. The shift from seeing any person as naturally unique to naturally common has been proposed as an indicator of medical modernity. Cook, "Markets and Cultures,"

128–30; Winterbottom, "China Root," 34. Additionally, colonial experiences with medicines not only informed learned medical practice in Europe, as Mark Harrison has shown, but also shaped the very conceptions of bodies, disease, and exchange underneath that. Harrison, *Commerce and Empire*, 1–4.

CHAPTER ONE

1. On these eighteenth-century commercial trends, see Hoppit, *Risk and Failure*, 8.
2. On competition, see Shaw and Welch, *Making and Marketing Medicine*, 17–20, 22–23; and on collaboration, see Wallis, "Competition and Cooperation," 48, 50, 54, 60–61.
3. Porter and Porter, "English Drugs Industry," 282; Shaw and Welch, *Making and Marketing Medicine*, 22; Campbell, *London Tradesman*, 63 (quotation).
4. Kett, "Provincial Medical Practice," 19–20; Sonnedecker, *History of Pharmacy*, 104; Worling, "Pharmacy in the Early Modern World," 67; Loudon, *Medical Care*, 28; Burnby, *Study of the English Apothecary*, 12; Digby, *Making a Medical Living*, 30; Wallis, "Consumption, Retailing, and Medicine," 27.
5. The medicine trade now offered an opportunity to enter the commercial classes. For trends among the eighteenth-century professions, see Szechi and Holmes, *Age of Oligarchy*, 154–56.
6. "Eighteenth-Century London Chemists," 178–79; "Retail Pharmacy," 130–63; Haycock and Wallis, *Quackery and Commerce*, 35, fn. 139.
7. Loudon, *Medical Care*, 133–38; Crawford, "Patients' Rights," 395–96.
8. Schiebinger, *Plants and Empire*, 96, 99–100; Leong, "Making Medicines," 147; Rankin, *Panaceia's Daughters*; Wear, *Knowledge and Practice*, 48–55, 65–67; Pollock, *With Faith and Physic*, 26, 102–4; Mortimer, "Rural Medical Marketplace," 74–76.
9. Though in other cases, in Italy, for example, unconventional healers were often simply licensed, see Gentilcore, *Healers and Healing in Early Modern Italy*.
10. Bell and Redwood, *Progress of Pharmacy*, 33–34; Porter and Porter, "English Drugs Industry," 281–83; Cook, "Medicine," 418–20.
11. Cook, *Decline of the Old Medical Regime*, 20–27.
12. Dopson, "London Chemists' Shops," 718–19; "Visitation of Apothecarys, Chemists & Druggists' Shopps in London," 17 May 1737, MS 2152, Royal College of Physicians (hereafter cited as RCP Visitation Books).
13. Pelling, *Medical Conflicts*; Cook, "Rose Case Reconsidered," 527–28.
14. For greater detail on the Society of Apothecaries' activities during this period, see Hunting, *Society of Apothecaries*; *Apothecary Display'd*, 13 (quotation), 36–39.
15. Barrett, *History of the Society of Apothecaries*, 136–37. A number of London's livery companies similarly sought to force practitioners into their membership again in this period, many via acts of Common Council (city, not Parliament). For more discussion on how the livery companies were not as inimical to economic and technological development as many historians have thought, see Berlin, "Guilds in Decline."
16. CJ XXV, 508: 11 Feb. 1748; "Attempted Legislation," 198–200. A second petition soon arrived from apothecaries not free of the society corroborating the allegations of the first (CJ XXV, 536: 4 Mar. 1748). On the outcome of this petition, see CJ XXV, 650: 2 May 1748; CJ XXV, 651: 3 May 1748; LJ XXVII, 230: 6 May 1748; LJ XXVII, 231: 9 May 1748. Until September 1752, the English calendar lagged behind the Gregorian calendar by eleven days and marked the beginning of the new year on March 25, rather than on January 1. Years have been shifted to begin on January 1, rather than March 25, though no shift has been made to account for the eleven days.

Therefore, what was denoted in period writing as 4 March 1747 reads here as 4 March 1748. After September 1752, dates are in the new style as became the convention.

17. Royal and Sun Alliance Insurance Group, CLC/B/192, London Metropolitan Archives (hereafter cited as Sun Insurance). On fire insurance, see Cockerell and Green, *British Insurance Business*. The Sun Insurance Office (established in 1710) maintained a near monopoly of insuring industrial property during the eighteenth century.
18. The datasets for 1730, 1760, and 1790 were generated from MS 11936/30–33, 130–35, 364–75, Sun Insurance, and are treated as a constant value series. Adjusting for inflation amplifies these trends. After 1760, assets increased by nearly 60 percent. Though the medicine trade seems to have been growing by this metric of insurance policies, the precise loci of that growth remain difficult to pinpoint. For further discussion of these datasets, see Dorner, "Manufacturing Pharmaceuticals," app. A. On contemporaneous developments, see Wallis and Pirohakul, "Medical Revolutions"; Wallis, Wallis, Burnby, and Whittet, *Eighteenth Century Medics*, xii.
19. For comparison with the brewing industry, see Sumner, "London Porter," 292.
20. Policy #174178, MS 11936/130, Sun Insurance; Policy #177810, MS 11936/135, Sun Insurance.
21. Nagy, *Popular Medicine in Seventeenth-Century England*, 73–76, 79.
22. This represented a quickening of seventeenth-century trends; see Porter, *Health for Sale*, 36–38; Haycock and Wallis, *Quackery and Commerce*; Wulf, *Brother Gardeners*, 134.
23. Spilsbury, *Friendly Physician*.
24. Alsop, "British Imperial Medicine," 33, 24.
25. Burnby, "Printer's Ink," 162–63; Crellin, "Dr. James's Fever Powder," 136–43; Crellin and Scott, "Lionel Lockyer."
26. Daffy's Elixir Tracts, William H. Helfand Popular Medicine Collection, Library Company of Philadelphia (hereafter cited as Helfand Collection).
27. RCP Visitation Books (1732–1747), MS 2152 (quotation); Whittet and Burnby, "Corbyn and Stacey," 42, 44–45; Epstein and Prak, *Guilds, Innovation, and the European Economy*. On this idea of transition in the eighteenth century, see Jacob and Kadane, "Weber's Protestant Capitalist," 23.
28. Thomas Corbyn to Dr. Rutty, 12 Jan. 1743, MS 5442, Corbyn & Co. MSS.
29. Advice and opinion on differences between Mary Clutton, Morris Clutton and Thomas Corbyn, 3 Jan. 1745, MS 5436/3, Corbyn & Co. MSS. For background on the obstacles facing a young merchant at this time, see Grassby, *Business Community*, 88–89.
30. Thomas Corbyn and Morris Clutton to Mary Clutton, 14 Apr. 1747, MS 5436/6/2, Corbyn & Co. MSS. The Quaker community blended kinship, friendship, and religion to sustain commercial networks; see Davidoff and Hall, *Family Fortunes*, 215–16.
31. Bond: Thomas Corbyn to John Hanbury, 25 Dec. 1753, MS 5436/17/10, Corbyn & Co. MSS; Bonds, promissory notes, and receipts relating to loans made to Clutton and Corbyn, 1745–53, MS 5436/4/1–25, Corbyn & Co. MSS. For further discussion of family credit, see Hunt, *Middling Sort*; and on previous forms of borrowing within the medicine trade, see Shaw and Welch, *Making and Marketing Medicine*, 125.
32. Muldrew, *Economy of Obligation*, 4–5; Wennerlind, *Casualties of Credit*, 8, 1–3; Murphy, *Origins of English Financial Markets*. On bankruptcy, see Hoppit, *Risk and Failure*, 51, 55, 57, 96, 180–81.
33. Abstract of accounts between Morris Clutton and Thomas Corbyn up to 16 Mar. 1747, MS 5436/8, Corbyn & Co. MSS; Account book of the trade in copartnership between Morris Clutton & Thomas Corbyn, 1747–54, MS 5437/1/1, Corbyn & Co. MSS.

34. RCP Visitation Books (1748–1754), MS 2153.
35. Though the bark had been around for more than a century, it gained popularity with the expansion of overseas colonization, military populations, and plantation agriculture in zones of endemic malaria. Schiebinger, *Plants and Empire*, 10; Chakrabarti, *Materials and Medicine*, 187–88; McNeill, *Mosquito Empires*, 63, 74–75.
36. Thomas Corbyn to Elijah Collins, 13 Oct. 1746, MS 5442, Corbyn & Co. MSS.
37. Account book of the trade in copartnership between Morris Clutton & Thomas Corbyn, 1747–54, MS 5437/1/1, Corbyn & Co. MSS.
38. Account book of the trade in copartnership between Morris Clutton & Thomas Corbyn, 1747–54, MS 5437/1/1, Corbyn & Co. MSS; Porter and Porter, "English Drugs Industry," 289–90; Thomas Corbyn to Robert James, 17 Mar. 1753, MS 5442, Corbyn & Co. MSS (quotation); Thomas Corbyn to Peter Sonmans, 15 Jan. 1755, MS 5442, Corbyn & Co. MSS. For comparison to other London merchants, see Price, *Capital and Credit*, 22–25.
39. Davidoff and Hall, *Family Fortunes*, 208.
40. Bonds, promissory notes, and receipts relating to money borrowed by Thomas Corbyn, MS 5439/1/1–14, Corbyn & Co. MSS; "The Property of Morris Clutton in the Joynt Trade or Copartnership with Thomas Corbyn," 1754, MS 5437/1/2, Corbyn & Co. MSS.
41. Price, *Capital and Credit*, 50–53.
42. Bonds, promissory notes, and receipts relating to money borrowed by Thomas Corbyn, MS 5439/1/1–14, Corbyn & Co. MSS. The Quaker community in London provided support in business matters but did not prevent members from interacting with nonmembers (Davidoff and Hall, *Family Fortunes*, 56–57). Muldrew, *Economy of Obligation*, 315, 328–31.
43. Bonds and related papers concerning money lent by Thomas Corbyn, 1760–80, MS 5439/2/1–5, Corbyn & Co. MSS. Sumner, "London Porter," 291; Stone and Stone, *An Open Elite*; Price, *Capital and Credit*, 48–49.
44. Accounts and papers relating to money lent by Thomas Corbyn to John Hunt, 1752–62, MS 5439/3/1–7, Corbyn & Co. MSS.
45. See figure 2.1.
46. Peter Renaudet medical apprentice notebook, ca. 1741, Library Company of Philadelphia (hereafter cited as Renaudet Notebook); Manufacturing recipe book (1748–1847), MS 5446, Corbyn & Co. MSS; Thomas Corbyn to Peter Renaudet, 12 Sept. 1743, MS 5442, Corbyn & Co. MSS.
47. Thomas Richardson to Thomas Corbyn, 26 Jan. 1759, Letter Book (1751–1761), Vault A #1596, Thomas Richardson Papers, Newport Historical Society (hereafter cited as Richardson MSS); McCusker, *Money and Exchange*, 135–36, 154.
48. Thomas Corbyn to Josiah Beale, 10 Dec. 1742, MS 5442, Corbyn & Co. MSS.
49. Thomas Corbyn to Thomas Lightfoot, 18 Sept. 1742, MS 5442, Corbyn & Co. MSS; Thomas Corbyn to Thomas Lightfoot, 10 Sept. 1743, MS 5442, Corbyn & Co. MSS.
50. Burnby, "English Apothecary," 192.
51. Invoice book, 1749–1806, MS 5878, f. 25, Primatt & Maud Papers, Wellcome Library (hereafter cited as Primatt & Maud MSS). For the Atlantic perspective, see Hartigan-O'Connor, *Ties That Buy*, 104, 107–8.
52. Thomas Corbyn to Elijah Collins, 20 Jan. 1746, MS 5442, Corbyn & Co. MSS.
53. Thomas Corbyn to Isaac Greenleaf, 22 Sept. 1746, MS 5442, Corbyn & Co. MSS.
54. Copy of a Letter Sent with Each Chest, 15 Mar. 1745, MS 5442, Corbyn & Co. MSS.

55. Thomas Corbyn to John & Esther White, 7 Mar. 1747, MS 5442, Corbyn & Co. MSS.
56. Gauci, *Emporium of the World*, 82–83; Davidoff and Hall, *Family Fortunes*, 250, 200, 202; Grassby, *Business Community*, 89. The high risks of transatlantic trade led to more and larger partnerships as well (Morgan, "Business Networks").
57. MS 11936/30–33, 130–35, 364–75, Sun Insurance.
58. Hancock, *Citizens of the World*, 52; Bengal General Letter, 1 Apr. 1760, Bombay Despatches, IOR/E/4/617, p. 41, India Office Records; Draught Company's General Letter to Fort St. George, 23 Jan. 1759, Madras Despatches, IOR/E/4/861, p. 1006, India Office Records; Policy #49534, MS 11936/30, Sun Insurance.
59. Morris, "Silvanus Bevan," 2; Wood, "Time of Transition," 30–39; Cantor, "Quakers in the Royal Society," 181–82, 185–87; Chapman-Huston and Cripps, *Through a City Archway*.
60. Timothy Bevan to the Court, 22 Oct. 1773, IOR/E/1/57, f. 385r, India Office Records.
61. See the Plough Court Ledgers, AH181, AH207, AH228, AH226, Allen & Hanburys Manuscripts, GlaxoSmithKline (hereafter cited as Allen & Hanburys MSS); Plough Court Practical Journal, 1782–90, AH201, Allen & Hanburys MSS; Plough Court Order Book, 1776–77, AH140, Allen & Hanburys MSS; Chapman-Huston and Cripps, *Through a City Archway*, 25.
62. Morris, "Silvanus Bevan," 3; Plough Court Practical Journal, 1782–90, AH201, Allen & Hanburys MSS; Ackrill and Hannah, *Business of Banking*, 18–21, 41; Holmes and Szechi, *Age of Oligarchy*, 153; Mokyr, *Enlightened Economy*, 220–24. Such relationships were not exclusive to Plough Court or to the Quaker community; see Note: Thomas Corbyn to Mary Reynolds, 14 Apr. 1749, MS 5436/6/6, Corbyn & Co. MSS; Account Book of Robert Scott, ca. 1790s, MS 3181, Gilbert Laurie & Co. Papers, Wellcome Library (hereafter cited as Gilbert Laurie & Co. MSS). Plough Court's connection to private banking mirrored a trend in the gunmaking industry, see Satia, *Empire of Guns*, chap. 2.
63. Madge, "Cookworthy," 3.
64. Receipt for the value of Nicholas Marshall's one-third share in the partnership of Corbyn, Brown, & Marshall, 1767, MS 5439/15/3, Corbyn & Co. MSS. For these comparisons, see Grassby, *Business Community*, 248, 255; Clark, "British Earnings."
65. Wages Book (1762–1770), MS 5444, Corbyn & Co. MSS (quotations); Wages Book (1770–1825), MS 5445, Corbyn & Co. MSS; Articles of Copartnership between Thomas Corbyn, John Brown and Nicholas Marshall, 1 Jan. 1762, MS 5438/1, Corbyn & Co. MSS; Articles of Copartnership between Thomas Corbyn and John Brown, 1 Jan. 1767, MS 5438/2, Corbyn & Co. MSS.
66. Davids, *Dutch Technological Leadership*, 194, 425.
67. Balance Sheet, 1770, MS 5439/18, Corbyn & Co. MSS.
68. Articles of Copartnership between Thomas Corbyn, John Brown, John Beaumont and George Stacey, 1 Jan. 1781, MS 5438/3, Corbyn & Co. MSS.
69. Davidoff and Hall, *Family Fortunes*, 250, 200, 202; Grassby, *Business Community*, 93; Inventories and Valuations of Stock, 1761–70, MS 5452/1–9, Corbyn & Co. MSS.
70. Porter and Porter, "English Drugs Industry," 289; Watson, "Trading Accounts," 58, 66.
71. Hancock, "'Domestic bubbling'," 679; Froide, *Silent Partners*.
72. Szechi and Holmes, *Age of Oligarchy*, 159, 163, 388; Berg, *Age of Manufactures*, 34–36, 53–56, 132–33.
73. Manufacturing Recipe Book (ca. 1782), pp. 149, 175, MS 5447, Corbyn & Co. MSS; Balance Sheet, 1770, MS 5439/18, Corbyn & Co. MSS.

74. On the experience of such spaces, see Wallis, "Consumption, Retailing, and Medicine." For examples of these containers, see Hudson, *English Delftware Drug Jars*; Griffenhagen and Bogard, *Drug Containers*.
75. Inventory and Valuation of Stock at Coldbath Fields and Holborn, 25 Dec. 1761, MS 5452/1–4, Corbyn & Co. MSS.
76. Winterbottom, "China Root," 22; Wallis, "Exotic Drugs," 33.
77. Inventories and Valuations of Stock, 1761–70, MS 5452/1–9, Corbyn & Co. MSS; Inventories and Valuations of Stock, 1754–73, MS 5451, Corbyn & Co. MSS; Wallis, "Consumption, Retailing, and Medicine," 36.
78. For example: Policy #581619, MS 11936/376, Sun Insurance; Policy #571187, MS 11936/370, Sun Insurance. Sometimes the policies stipulated that there was no laboratory on the premises, such as Policies #626435, #626734, MS 11936/398, Sun Insurance. On the history of laboratories in Europe, see Crosland, "Early Laboratories"; Kohler, "Lab History."
79. Alchemy, under the guise of chemistry, influenced a number of early industrial professions, including distillers, perfumers, metallurgists, and apothecaries. Smith, "Laboratories," 302–3, 305; Newman, "Alchemy to 'Chymistry,'" 502, 516–17.
80. Simmons, "Stocks and Science," 144–45; Allen, "Distilling Household Medicine," 103, 110.
81. Wages Books (1762–1770, 1770–1825), MS 5444 & 5445, Corbyn & Co. MSS.
82. Fumerton, *Unsettled*, 26, 13; Mokyr, *Enlightened Economy*, 97; Allen, "Great Divergence in European Wages," 414–15, 427; Berlin, "Guilds in Decline."
83. Inventories and Valuations of Stock, 1761–70, MS 5452/1–9, Corbyn & Co. MSS. For comparison, see Shaw and Welch, *Making and Marketing Medicine*, 55–57, 63–66.
84. Invoice book, 1749–1806, MS 5878, Primatt & Maud MSS; Policy #584817, MS 11936/376, Sun Insurance; Policy #624279, MS 11936/398, Sun Insurance; Policy #548656, MS 11936/356, Sun Insurance; Policy #571162, MS 11936/370, Sun Insurance.
85. Warren, "Oldest Pharmacy in London," 187, 189; Policy #584817, MS 11936/376, Sun Insurance.
86. Allen, "Distilling Household Medicine," 106, 113.
87. Account Book of Robert Scott, ca. 1790s, MS 3181, Gilbert Laurie & Co. MSS.
88. Flor. Benzoin Costing, May 1760, MS 5448/2, Corbyn & Co. MSS. Quincy, *Complete English Dispensatory*, 323 ("Specific against"). Trained as an apothecary, Quincy published numerous tracts on medical topics, including pharmacy. Quincy's *Dispensatory* went through twelve editions before William Lewis revised it in 1749 into his *New Dispensatory* (1753). Howard-Jones, "John Quincy."
89. Manufacturing Recipe Book (1748–1847), pp. 165, 167, 379, MS 5446, Corbyn & Co. MSS; Recipe book fragments, MS 5450, Corbyn & Co. MSS; Invoice book, 1749–1806, MS 5878, Primatt & Maud MSS.
90. Leong and Rankin, *Secrets and Knowledge*; Smith, *Body of the Artisan*; Long, *Technical Arts*.
91. On the gift exchange of recipes, see Leong and Pennell, "Recipe Collections," 139; Pennell, "Perfecting Practice," 238, 250–51. On industrial espionage in chemical manufacturing, see Harris, *Industrial Espionage*, 114–27.
92. Manufacturing Recipe Book (ca. 1782), pp. 175, 225, 251, 263, and 267, MS 5447, Corbyn & Co. MSS.
93. Shaw and Welch, *Making and Marketing Medicine*, 256–57; Spilsbury, *Friendly Physician*, 2 ("efficacious"), 5.

94. Manufacturing Recipe Book (ca. 1782), MS 5447, Corbyn & Co. MSS; Manufacturing Recipe Book (1748–1847), p. 361, MS 5446, Corbyn & Co. MSS; Amber Costings, July 1765, MS 5448/6, Corbyn & Co. MSS; Lewis, *Edinburgh New Dispensatory*, 484; Lewis, *New Dispensatory*, 1st ed., 290–91 (p. 291: "very elegant," "almost useless," "wholesale dealers"). On Lewis, see Gibbs, "William Lewis;" Sivin, "William Lewis." On Boerhaave's influence in Britain, see Davids, *Dutch Technological Leadership*, 511.
95. See the introduction, note 28; and chapter 2, notes 8, 10.
96. Report on Two Memorials, 21 Mar. 1769, T 1/469/157–58, Treasury Board Papers, National Archives of the UK (hereafter cited as Treasury Board Papers); Report on Two Memorials, 21 Mar. 1769, T 1/470/194–99, Treasury Board Papers. Petitioning can be an indicator of merchant association: Gauci, *Politics of Trade*, 127; Gauci, *Regulating the British Economy*, 110; Olson, *Making the Empire Work*; Hoppit, *Britain's Political Economies*, 153, 156–59, 160–62. On this wave of petitioning, see Holmes and Szechi, *Age of Oligarchy*, 49–50.
97. (Copy) Articles of Copartnership of the Laboratory Stock, 31 Dec. 1774, box 64/15, Laboratory Stock Records, Worshipful Society of Apothecaries (hereafter cited as Laboratory Stock Records); Hunting, *Society of Apothecaries*, 153; Simmons, "Medicines, Monopolies and Mortars," 222, 231; *Origin, Progress and Present State*, 15.
98. Joseph Cruttenden to William Arbuckle, 16 Aug. 1716, p. 99, in Steele, *Atlantic Merchant-Apothecary* ("goodnesse"). The production of medicines stood at the heart of the Society's rivalries with the Society of Chemical Physicians and College of Physicians in the second half of the seventeenth century: Cook, *Decline of the Old Medical Regime*; Cook, "Rose Case Reconsidered," 527–28, 533–34; Cook, "Society of Chemical Physicians."
99. Barrett, *History of the Society of Apothecaries*, 129; Court Minute Book (1745–1767), ff. 21, 23, MS 8200/7, Worshipful Society of Apothecaries; Hunting, *Society of Apothecaries*, 160; Laboratory Stock Agreement, 1767, MS 8215, Laboratory Stock Records.
100. Laboratory Stock Minutes (1741–1751), 22 Dec. 1743, MS 8220, Laboratory Stock Records; Laboratory Stock Rough Minutes (1760–1765), 22 Dec. 1761, MS 8221/2, box 92, Laboratory Stock Records.
101. Members of the Laboratory Stock Committee inspected these books to ensure proper bookkeeping. "Rules & Orders for the better management of the Elaboratory Stock," 9 May 1702, box 64/15, Laboratory Stock Records; Laboratory Stock Minutes (1741–1751), 3 Mar. 1742, 2 Jan. 1745, 6 Feb. 1745, 2 May 1744, 5 June 1745, 16 Feb. 1749, MS 8220, Laboratory Stock Records; Laboratory Stock Rough Minutes (1741–1746), ff. 29–30, MS 8221/1, box 92, Laboratory Stock Records (quotation).
102. "Rules & Orders for the better management of the Elaboratory Stock," 9 May 1702, box 64/15, Laboratory Stock Records; Laboratory Stock Minutes (1741–1751), 2 Dec. 1741, MS 8220, Laboratory Stock Records.
103. Laboratory Stock Minutes (1741–1751), 3 Oct. 1744, 16 Feb. 1749, MS 8220, Laboratory Stock Records. These charges echoed earlier criticisms leveled by the College of Physicians at the apothecaries in the 1690s (Cook, "Rose Case Reconsidered," 533–34). Joseph Cruttenden, an apothecary, remarked in 1716 that he could "buy [medicines] cheaper of the Towne Chymist but then they are not to be depended on" (Joseph Cruttenden to William Arbuckle, 16 Aug. 1716, p. 99, in Steele, *Atlantic Merchant-Apothecary*).
104. Court Minute Book (1745–1767), 16 Mar. 1758, f. 127v, MS 8200/7, Worshipful Society of Apothecaries; Laboratory Stock Subscriber Share Account (1728–1817), MS 8226, box 123, Laboratory Stock Records; Laboratory Stock Agreement, 1767,

MS 8215, Laboratory Stock Records; Lists of Proprietors of Laboratory Stock (1767–1797), M6, Laboratory Stock Records.

105. Joseph Gurney Bevan to John Kett, 31 Dec. 1787, AH035, Allen & Hanburys MSS; Estes, *Dictionary of Protopharmacology*, 8.
106. Plough Court Ledger (1781–1784), ff. 150–70, AH207, Allen & Hanburys MSS; Plough Court Ledger (1789–1796), ff. 95–104, 229, 233, AH228, Allen & Hanburys MSS.
107. Plough Court Order Book, 1776–77, AH140, Allen & Hanburys MSS. On connections between medicines and other types of manufacturing, see Davids, *Dutch Technological Leadership*, 501–2.
108. Plough Court Ledgers, AH207, AH228, AH226, Allen & Hanburys MSS.
109. Shaw and Welch, *Making and Marketing Medicine*, 81–83; Lists of Customers/Debtors of Thomas Corbyn & Co., 1762–70, 1779, MS 5439/12/1–6, Corbyn & Co. MSS.
110. Plough Court Ledgers, AH228, AH226, Allen & Hanburys MSS.
111. Plough Court Practical Journal, 1782–90, AH201, Allen & Hanburys MSS. On glass: Thomas Corbyn to Elijah Collins, 14 Oct. 1748, MS 5442, Corbyn & Co. MSS; Copy of Certificate for Glass, 24 Jan. 1750, MS 5442, Corbyn & Co. MSS; Certificate for Glass, 1 May 1752, MS 5442, Corbyn & Co. MSS; Crellin and Scott, *Glass and British Pharmacy*.
112. Plough Court Ledger (1789–1796), AH228, Allen & Hanburys MSS; Plough Court Ledger (1797–1808), AH226, Allen & Hanburys MSS; Schiebinger, *Plants and Empire*, 96.
113. Manufacturers in the United Provinces, for instance, were largely dependent upon merchants and their patronage, see Davids, *Dutch Technological Leadership*, 524–25.
114. Newson, *Making Medicines in Early Colonial Lima*, 2, 5–7; Numbers, *Medicine in the New World*.

CHAPTER TWO

1. Thomas Corbyn to Elijah Collins, 21 Dec. 1747, MS 5442, Corbyn & Co. MSS; Joseph Gurney Bevan to Armstrong & King, 14 Mar. 1780, AH033, Allen & Hanburys MSS (quotation).
2. Hartigan-O'Connor, *Ties that Buy*, 98; Wood, *Radicalism*, 140–41; Muldrew, *Economy of Obligation*, 315; Finn, *Character of Credit*, 3–4. The eighteenth century has been portrayed as a moment when a premodern world of deferred payment and barter gave way to a modern world of risk taking based on credit as contract rather than credit as relationship. As Wood describes, by the mid-eighteenth century more colonists could borrow or trade using impersonal notes or paper money, but, as others like Hartigan-O'Connor and Finn have shown, economic activity for many remained deeply personal. Nevertheless, these examples demonstrate that the streamlined, impersonal world found by some historians was far from the only reality, even in the major transatlantic trades. On kin and obligation, see O'Neill, *Opened Letter*, 87; Ben-Amos, *Culture of Giving*, 47.
3. Brown, *Knowledge is Power*, 112; Pettegree, *Invention of News*, 296; Haggerty, "Trade and the Transhipment," 172; Zahedieh, "Making Mercantilism Work," 152–55; Mathias, "Risk, Credit and Kinship."
4. Smoak, "Counterfeit Coins"; Hartigan-O'Connor, *Ties that Buy*, chaps. 3–4.
5. Thomas Mayleigh Account Book, Society of Apothecaries; Voyages 16491, 16581, Transatlantic Slave Trade Database (hereafter cited as Slave Voyages).
6. On the slave trade's knowledge production: Smallwood, *Saltwater Slavery*; Murphy, "Collecting Slave Traders"; Delbourgo, *Collecting the World*.

7. The medicine trade thus portrays the ongoing replacement of embedded, personal trust with mobile, impersonal facts during the eighteenth century. In contrast to other accounts, the abstraction of monetary value via money forms does not seem to be solely responsible for commensurability among objects. For background on the idea of commensurability, see Cook, "Commensurable Materials," 99, 101, 104, 110; see also Baucom, *Specters of the Atlantic*; 16–17, 111. For an example of period attempts to evaluate drugs abroad, see Steel, *Portable Instructions*.
8. Crouch, *Complete View*, 267; 8 Geo. I, c. 15, c. 16, pp. 389–405, in Pickering, *Statutes at Large*, vol. 14; Szechi and Holmes, *Age of Oligarchy*, 76–77, Compendium G.2(ii); Kennedy, *English Taxation*, 96–99; CUST 3/1–82.
9. Joseph Cruttenden to Habijah Savage, 16 Apr. 1713, pp. 53–54, in Steele, *Atlantic Merchant-Apothecary*; Joseph Cruttenden to Habijah Savage, 25 Aug. 1712, pp. 42–44, in Steele, *Atlantic Merchant-Apothecary* ("our wise Parliament"); Joseph Cruttenden to Thomas Barton, 22 Aug. 1715, pp. 83–84, in Steele, *Atlantic Merchant-Apothecary*; Joseph Cruttenden to John Nicolls, 17 Mar. 1715, pp. 74–76, in Steele, *Atlantic Merchant-Apothecary* ("custome ruines all").
10. Holmes and Szechi, *Age of Oligarchy*, 77; CJ XIX, 646: 19 Oct. 1721 (quotation).
11. Wallis, "Exotic Drugs," 29; 17 Geo. II, c. 31, pp. 236–39, in Pickering, *Statutes at Large*, vol. 18; Crouch, *Complete View*, 38–40, 112–45; Kennedy, *English Taxation*, 135–36; CJ XIX, 710: 13 Jan. 1722.
12. CUST 3/1–82. As a constant value series, data from the customs ledgers allow for comparison over time. Trade statistics were compiled and duties collected using English measures, specifically the English pound in these ledgers. For further discussion of the ledgers and constant versus current value series, see McCusker, "Current Value," 613, 615, 618; McCusker, "Weights and Measures," 604, 612. On demography, see Hancock, *Oceans of Wine*, 166.
13. Davis, "English Foreign Trade, 1700–1774," 290–92, 295–96; Szechi and Holmes, *Age of Oligarchy*, 153, 380. For comparisons to other trades, see Wallis, "Exotic Drugs," 29–30. On the question of change in the eighteenth century, see Veluwenkamp and Veenstra, "English Merchant Colonies," 20–21.
14. Hoppit, *Risk and Failure*, 98, 101–3, 122, 127, 130, 139. For the heavy losses sustained by merchant shipping during England's seventeenth-century wars, see Grassby, *Business Community*, 92.
15. For example, Price, "What Did Merchants Do?"; Hancock, *Oceans of Wine*; Jarvis, *Eye of All Trade*.
16. O'Neill, *Opened Letter*, 140–41, 143, 145; Trivellato, *Familiarity of Strangers*, 178–87; Steele, *English Atlantic*, 213–15; Brown, *Knowledge is Power*, 130–31, 270–71.
17. Brown, *Knowledge is Power*, 115–17; O'Neill, *Opened Letter*, 155, 168.
18. Joint Laboratory Stock Articles of Agreement, MS 8214, box 225, Laboratory Stock Records. Joseph Cruttenden to James Henderson, 13 Mar. 1714, p. 63, in Steele, *Atlantic Merchant-Apothecary*.
19. Joseph Cruttenden to Thomas Barton, 12 Mar. 1717, pp. 109–11, in Steele, *Atlantic Merchant-Apothecary*; "Copy of What Sent in Capt Bond," 27 June 1710, p. 13, in Steele, *Atlantic Merchant-Apothecary*; Joseph Cruttenden to Conrade Adams, 13 Mar. 1710, pp. 6–7, in Steele, *Atlantic Merchant-Apothecary*; Steele, *Atlantic Merchant-Apothecary*, xiii.
20. Corbyn to John Pickering, 1 Sept. 1746, MS 5442, Corbyn & Co. MSS (quotation). Trivellato, *Familiarity of Strangers*, 3–4; Muldrew, *Economy of Obligation*, 148; Granovetter, "Problem of Embeddedness;" Greif, "Institutions and Impersonal Exchange."

21. Thomas Corbyn to Cadwalader Evans, 18 Apr. 1750, MS 5442, Corbyn & Co. MSS; Thomas Corbyn to John Pleasants, 12 Feb. 1742, MS 5442, Corbyn & Co. MSS.
22. Landes, *London Quakers*, 104; Granovetter, "Impact of Social Structure"; Wilson, *Pious Traders;* Zahedieh, "Making Mercantilism Work," 154–56. For example: Thomas Corbyn to Christopher Gadsden, 10 Feb. 1746, MS 5442, Corbyn & Co. MSS; Thomas Corbyn to Robert James, 29 Aug. 1746, MS 5442, Corbyn & Co. MSS.
23. Joseph Gurney Bevan to Francis Richards, 19 June 1778, AH246, Allen & Hanburys MSS.
24. Inventory of the personal estate of the late William Hunter, 7 Feb. 1777, Vault A, Box 90, Folder 9, William Hunter Papers, Newport Historical Society (hereafter cited as Hunter MSS); "Joseph Gamble Jackman Inventory," 6 Oct. 1786, Inventories of Property and Personal Effects (1764–1888), Barbados Department of Archives (hereafter cited as BDA Inventories).
25. Letter to new correspondents, 23 Dec. 1749, MS 5442, Corbyn & Co. MSS; Thomas Corbyn to Christopher Marshall, 7 Mar. 1747, MS 5442, Corbyn & Co. MSS; Thomas Corbyn to Webb, ca. 1752, MS 5442, Corbyn & Co. MSS.
26. Thomas Richardson to Thomas Corbyn, 30 Oct. 1751, Letter Book (1751–1761), Vault A #1596, Richardson MSS; Thomas Richardson to Thomas Corbyn, 20 Jan. 1759, Letter Book (1751–1761), Vault A #1596, Richardson MSS.
27. Elliott, *Empires of the Atlantic World*, 50; Steele, *English Atlantic*, figs. 2, 3, pp. 57–58.
28. On glass, see Griffenhagen and Bogard, *Drug Containers*, 11–12, 19–22; MacLeod, "Accident or Design," 780–83; Wallis, "Consumption, Retailing, and Medicine," 10–13; Certificate for Glass, 1 May 1752, MS 5442, Corbyn & Co. MSS; Thomas Corbyn to John Worrell, 24 Feb. 1752, MS 5442, Corbyn & Co. MSS; "Copy of a Letter sent with each Chest," 15 Mar. 1745, MS 5442, Corbyn & Co. MSS; Thomas Corbyn to Robert James, 29 Aug. 1746, MS 5442, Corbyn & Co. MSS; Thomas Corbyn to Christopher Gadsden, 10 Feb. 1746, MS 5442, Corbyn & Co. MSS (quotation). On the similar day-to-day functions of locally-circulating coins of low denomination, see Peterson, "World in a Shilling."
29. Griffenhagen and Bogard, *Drug Containers*, 14; Thomas Corbyn to John Easton, 7 Sept. 1742, MS 5442, Corbyn & Co. MSS.
30. Thomas Corbyn to John Pleasants, 12 Feb. 1742, MS 5442, Corbyn & Co. MSS; Directions for the Use of a Box of Medicines per Hunt & Greenleaf, Apr. 1750, MS 5442, Corbyn & Co. MSS; Thomas Corbyn to Othniel Beale, 10 Feb. 1746, MS 5442, Corbyn & Co. MSS.
31. Chakrabarti, *Materials and Medicine*, 210.
32. Thomas Corbyn to John Wills, 4 Sept. 1751, MS 5442, Corbyn & Co. MSS; "Copy of a Letter sent with each Chest," 15 Mar. 1745, MS 5442, Corbyn & Co. MSS (quotation); Thomas Corbyn to John Easton, 7 Sept. 1742, MS 5442, Corbyn & Co. MSS.
33. Shaw and Welch, *Making and Marketing Medicine*, 126–29, 143.
34. Joseph Gurney Bevan to Benjamin Rush, 6 Apr. 1786, AH035, Allen & Hanburys MSS.
35. Joseph Gurney Bevan to John Thomson, 18 May 1784, AH031, Allen & Hanburys MSS.
36. Pelling, *Medical Conflicts*; Cook, *The Decline of the Old Medical Regime*. See also chapter 1 for further discussion of medical regulation.
37. Steel, *Portable Instructions;* Teigen, "Taste and Quality."
38. For example: Thomas Corbyn to Christopher Marshall, 5 Dec. 1748, MS 5442, Corbyn & Co. MSS.

39. Joseph Gurney Bevan to William Fyfe, 19 Apr. 1783, AH031, Allen & Hanburys MSS. See also the dispersal of care among a range of caregivers, healers, and practitioners in colonial New England: Mutschler, "'Social Credit.'" On identifying quality in textiles, see Reddy, *Rise of Market Culture*.
40. See the early eighteenth-century debate over garbling (the close inspection of raw materials for purity and quality): "The Garbling-Act: With Short Remarks relating to the East-India Company," 1708, CUP645.b.11 (1), Petitions, General Reference Collection, British Library; "Reasons for Passing the Bill for Regulating the Abuses of the Office of Garbling," ca. 1708, 816.m.12 (111), Tracts Relating to Trade Etc., General Reference Collections, British Library; "The Case of the Act for Garbling of Spices," ca. 1708, CUP645.b.11 (1), Petitions, General Reference Collection, British Library; "Reasons Humbly Offered for continuing the Law made for Garbling of Spices and Drugs," ca. 1708, 816.m.12 (113), Tracts Relating to Trade Etc., General Reference Collections, British Library. For a definition of garbling, see Roberts, *Map of Commerce*, 42. On impure remedies, see Wear, *Knowledge and Practice*, 90–91, 94.
41. Thomas Corbyn to Christopher Marshall, 7 Mar. 1747, MS 5442, Corbyn & Co. MSS.
42. For the repeal of the Garbling Act, see CJ XV, 588: 4 Mar. 1708; CJ XV, 601: 11 Mar. 1708; LJ XVIII, 505–506: 11 Mar. 1708. On patients' legal rights, see Porter, *Health for Sale*; Crawford, "Patients' Rights," 381–82, 383–85, 387–88, 393. Medical litigation remained basically the enforcement of contracts. Conversely, charges for medicines were rarely denied.
43. Thomas Corbyn to Christopher Marshall, 7 Mar. 1747, MS 5442, Corbyn & Co. MSS.
44. Thomas Corbyn to Robert James, 1744, MS 5442, Corbyn & Co. MSS.
45. Thomas Corbyn to Isaac Greenleaf, 24 Jan. 1747, MS 5442, Corbyn & Co. MSS.
46. For example, Cook, *Matters of Exchange*; Grassby, *Business Community*, 298.
47. Thomas Corbyn to Robert James, 1744, MS 5442, Corbyn & Co. MSS; Thomas Corbyn to Isaac Greenleaf, 22 Sept. 1746, MS 5442, Corbyn & Co. MSS.
48. Thomas Corbyn to Robert James, 1744, MS 5442, Corbyn & Co. MSS ("speak for themselves"); Thomas Corbyn to Elijah Collins, 30 Apr. 1754, MS 5442, Corbyn & Co. MSS ("good Repute").
49. Joseph Gurney Bevan to William Fyfe, 19 Apr. 1783, AH031, Allen & Hanburys MSS.
50. [Philadelphia Yearly Meeting], *Collection of Memorials*, 374–76; Thomas Corbyn to John & Esther White, 18 Apr. 1750, MS 5442, Corbyn & Co. MSS. For background on commission trading, see Steele, *English Atlantic*, 216; Zahedieh, *Capital and the Colonies*, 101–3; Hancock, "World of Business to Do."
51. Thomas Corbyn to Christopher Marshall, 12 Sept. 1747, MS 5442, Corbyn & Co. MSS; Thomas Corbyn to John & Esther White, 30 Jan. 1748, 25 Apr. 1748, MS 5442, Corbyn & Co. MSS.
52. Thomas Corbyn to Thomas Lightfoot, 7 Mar. 1747, MS 5442, Corbyn & Co. MSS; Thomas Corbyn to Esther White, 10 June 1752, MS 5442, Corbyn & Co. MSS.
53. Thomas Corbyn to John & Esther White, 19 Sept. 1758, 22 Feb. 1752, MS 5442, Corbyn & Co. MSS.
54. Thomas Corbyn to Robert Foster, 8 Nov. 1753, MS 5442, Corbyn & Co. MSS; Thomas Corbyn to John & Esther White, 8 Nov. 1753, 31 Jan. 1754, MS 5442, Corbyn & Co. MSS.
55. Thomas Corbyn to Esther White, 10 June 1752, MS 5442, Corbyn & Co. MSS (quotation). On the gendering of debt, see Hartigan-O'Connor, *Ties That Buy*, 110.
56. Thomas Corbyn to Elijah Collins, 30 Apr. 1754, MS 5442, Corbyn & Co. MSS.

57. Davidoff and Hall, *Family Fortunes*, 202–3. For an example of this trend in the Madeira trade, see Hancock, *Oceans of Wine*, 179–84.
58. Soll, "Information Management," 356, 369; see also Littleton, "Evolution of the Journal Entry."
59. Thomas Corbyn to Robert Foster, 8 Nov. 1753, MS 5442, Corbyn & Co. MSS; Thomas Corbyn to John & Esther White, 8 Nov. 1753, MS 5442, Corbyn & Co. MSS (quotation).
60. Thomas Corbyn to Robert Penny, 4 Oct. 1750, MS 5442, Corbyn & Co. MSS. On Quaker trading scruples, see Kent, "Quaker Ethic"; Fox, *Warning to All the Merchants*; Fox, *Line of Righteousness*. Theological principles often gave way before the realities of commerce, for example Grassby, *Business Community*, 295.
61. Thomas Corbyn to Esther White, 10 June 1752, MS 5442, Corbyn & Co. MSS.
62. Steele, *English Atlantic*, 222; Zahedieh, "Making Mercantilism Work," 158; Shammas, "Revolutionary Impact," 183; Hartigan-O'Connor, *Ties That Buy*, 70.
63. Thomas Corbyn to Elijah Collins, 13 Oct. 1746, MS 5442, Corbyn & Co. MSS.
64. Thomas Corbyn to Thomas Richardson, 25 Apr. 1748, MS 5442, Corbyn & Co. MSS.
65. Thomas Corbyn to Daniel & Joshua Lathrop, 1 July 1749, MS 5442, Corbyn & Co. MSS.
66. Thomas Corbyn to Elijah Collins, 20 Sept. 1752, MS 5442, Corbyn & Co. MSS (quotation); Thomas Corbyn to Joseph Jackman, 31 Jan. 1754, MS 5442, Corbyn & Co. MSS; Thomas Corbyn to Thomas Clark, 29 July 1754, MS 5442, Corbyn & Co. MSS.
67. Thomas Corbyn to Elijah Collins, 13 Oct. 1746, MS 5442, Corbyn & Co. MSS; Thomas Corbyn to Israel Pemberton Jr., 9 Sept. 1746, MS 5442, Corbyn & Co. MSS; Thomas Corbyn to Elijah Collins, 13 Oct. 1746, MS 5442, Corbyn & Co. MSS (quotation). On Pemberton, see Thayer, *King of the Quakers*.
68. Hancock, *Citizens of the World*, 81–82; Gauci, *Emporium of the World*, 128–33.
69. Quincy, *Complete English Dispensatory*, 108 ("wonders," "Holy-wood"), 180; Winterbottom, "China Root," 26.
70. Thomas Corbyn to Christopher Marshall, 18 Apr. 1750, MS 5442, Corbyn & Co. MSS.
71. Thomas Corbyn to Daniel & Joshua Lathrop, 12 Apr. 1753, MS 5442, Corbyn & Co. MSS (quotations). On coffee cultivation in Jamaica, see Anderson, *Mahogany*, 81–83; Shepherd and Monteith, "Pen-Keepers and Coffee Farmers," 85–89. Guaiacum prices per pound rose from 6/6 to 9/- in some cases. For price data, see Thomas Corbyn to John Wills, 4 Sept. 1751, MS 5442, Corbyn & Co. MSS; Thomas Corbyn to Reynolds Skeete, 18 Jan. 1753, MS 5442, Corbyn & Co. MSS.
72. Topik, "Historicizing Commodity Chains," 54–55.
73. Thomas Corbyn to Edward Penington, 1 June 1753, MS 5442, Corbyn & Co. MSS.
74. Thomas Corbyn to Edward Penington, 1 June 1753, MS 5442, Corbyn & Co. MSS.
75. Guaiacum prices fell to 4/4 per pound in 1754 by some accounts, a significant decline from the prices noted only several years earlier. Thomas Corbyn to Edward Penington, 6 Dec. 1753, MS 5442, Corbyn & Co. MSS; Thomas Corbyn to Thomas Richardson, 30 Apr. 1754, MS 5442, Corbyn & Co. MSS; Thomas Corbyn to Thomas Richardson, 10 Oct. 1754, MS 5442, Corbyn & Co. MSS.
76. Price, "Credit in the Slave Trade," 306, 309–10; Smith, *Gentry Capitalism*, 139–76; Martin, "Slavery's Invisible Engine," 826–28, 839–40; Hancock, "Financial Markets."
77. Davidoff and Hall, *Family Fortunes*, 198; Smith, *Gentry Capitalism*, 165–69; Price, "Credit in the Slave Trade," 323.
78. Thomas Corbyn to John Richards, 31 Jan. 1754, MS 5442, Corbyn & Co. MSS.

79. Thomas Corbyn to Josiah Beale, 8 Sept. 1748, MS 5442, Corbyn & Co. MSS; Thomas Corbyn to Robert Penny, 4 Oct. 1750, MS 5442, Corbyn & Co. MSS (quotation); Thomas Corbyn to Joseph Jackman, 24 Aug. 1752, MS 5442, Corbyn & Co. MSS; Thomas Corbyn to John Wills, 4 Sept. 1751, MS 5442, Corbyn & Co. MSS; Steele, *Atlantic Merchant-Apothecary*, xviii–xix.
80. Price, "Credit in the Slave Trade," 294, 317–18.
81. On sugar as a "food-drug," see Mintz, *Sweetness and Power*.
82. Corbyn's terms appear consistent with those of other merchants of medicines. Thomas Corbyn to Cadwalader Evans, 18 Apr. 1750, MS 5442, Corbyn & Co. MSS.
83. Thomas Corbyn to Robert James, 23 Dec. 1749, MS 5442, Corbyn & Co. MSS; Thomas Corbyn to Robert James, 17 Dec. 1750, MS 5442, Corbyn & Co. MSS.
84. Thomas Corbyn to Thomas Richardson, 7 Mar. 1747, MS 5442, Corbyn & Co. MSS; Letter Book (1710–1715), vault A #70, Richardson MSS; Letter Book (1716–1741), vault A #1595, Richardson MSS; Petty Account Book (1722–1754), vault A #487, Richardson MSS. On Ann Richardson, see Crane, *Poison Plot*, 129, 138–39.
85. Thomas Richardson to Thomas Corbyn, 18 June 1752, 10 Oct. 1753, Letter Book (1751–1761), Vault A #1596, Richardson MSS; Day Book, 4 Mar. 1747, vault A #542, Richardson MSS.
86. Thomas Corbyn to Thomas Richardson, 7 Sept. 1749, MS 5442, Corbyn & Co. MSS; Thomas Corbyn to Thomas Richardson, 18 Apr. 1750, MS 5442, Corbyn & Co. MSS.
87. Thomas Richardson to Thomas Corbyn, 26 Jan. 1759, Letter Book (1751–1761), Vault A #1596, Richardson MSS.
88. Steele, *Atlantic Merchant-Apothecary*, xix.
89. Thomas Corbyn to Thomas Richardson, 7 Mar. 1747 ("Tobacco dust"), 14 Mar. 1751, MS 5442, Corbyn & Co. MSS; Petty Account Book, p. 552, vault A #487, Richardson MSS. For background on the New England fur trade, see Cronon, *Changes in the Land*, 91–92, 97–99, 102, 105–7.
90. Thomas Corbyn to Thomas Richardson, 10 June 1752, MS 5442, Corbyn & Co. MSS; Thomas Richardson to Thomas Corbyn, 8 Apr. 1752, Letter Book (1751–1761), Vault A #1596, Richardson MSS; Thomas Corbyn to Robert James, 29 Aug. 1746, MS 5442, Corbyn & Co. MSS.
91. *Slavery and Justice*, 3, 10.
92. For example, Craine, *Poison Plot*.
93. Krumbhaar, "Doctor William Hunter," 506–13, 518–19; Inventory of the personal estate of the late William Hunter, 7 Feb. 1777, vault A, box 90, folder 9, Hunter MSS; Physician's Book (1765–1774), vault A #452, Hunter MSS.
94. Physician's Book (1765–1774), vault A #452, Hunter MSS; William Hunter Bill to Aaron Lopez, 1763–76, vault A, box 167, folder 1, Hunter MSS. On Lopez, see Platt, "Slave Trade of Aaron Lopez."
95. Inventory of the personal estate of the late William Hunter, 7 Feb. 1777, vault A, box 90, folder 9, Hunter MSS.
96. Hunter probably meant either old tenor paper bills or the lawful money bills issued to finance the war, both of which experienced massive depreciation. William Hunter to a Glasgow Merchant, 26 Jan. 1762, vault A, box 15, folder 1, Hunter MSS. On colonial currency in this period, see McCusker, *Money and Exchange*; and on the declining value of Rhode Island's old tenor currency, see Brock, *Currency of the American Colonies*, 325–34.
97. Platt, "Slave Trade of Aaron Lopez," 616–18.
98. Hancock, *Citizens of the World*, 52; Policy #49534, MS11936/30, Sun Insurance.

99. Joseph Gurney Bevan to Thomas Brown, 27 Oct. 1777, AH027, Allen & Hanburys MSS (quotation); for Bevan's acceptance of Caribbean sugars as remittance, see Joseph Gurney Bevan to Samuel Forte, 24 Jan. 1792, AH028, Allen & Hanburys MSS; Stander, "Transatlantic Trade in Pharmaceuticals," 329.
100. Price, "What Did Merchants Do," 278, 280, 282–84.
101. See, for example, the squabbling over debts after Corbyn's death. Estate of Thomas Corbyn Letters (1791–1822), CLC/B/136/MS18768, Janson, Cobb, Pearson and Company, London Metropolitan Archives (hereafter cited as Corbyn Estate MSS).
102. For an earlier example, see Shaw and Welch, *Making and Marketing Medicine*, 132, 134–35, 138–40.
103. For example: Wennerlind, *Casualties of Credit*; Desan, *Making Money*; Valenze, *Social Life of Money*; Finn, *Character of Credit*.
104. Thomas Corbyn to Thomas Richardson, 25 Aug. 1748, MS 5442, Corbyn & Co. MSS.
105. Thomas Corbyn to Thomas Richardson, 10 Oct. 1754, MS 5442, Corbyn & Co. MSS; "Amount of John Easton's Account of Sales for T. Corbyn," ca. 1755, MS 5442, Corbyn & Co. MSS.
106. Hoppit, *Risk and Failure*, 31–32.
107. Thomas Corbyn to Robert Penny, 4 Oct. 1750, MS 5442, Corbyn & Co. MSS.
108. Thomas Corbyn to Joseph Jackman, 24 Aug. 1752, MS 5442, Corbyn & Co. MSS; Thomas Corbyn to John Wills, 24 Aug. 1752, MS 5442, Corbyn & Co. MSS.
109. Procuration of Thomas Corbyn to William Logan, 8 Feb. 1773, box 1, folder 7, Logan Family Papers, coll. 2023, Historical Society of Pennsylvania (hereafter cited as Logan MSS).
110. Thomas Corbyn to Israel Pemberton Jr., 2 Apr. 1763, vol. 16, p. 67, Pemberton Family Papers, coll. 484A, Historical Society of Pennsylvania (hereafter cited as Pemberton MSS).
111. Thomas Corbyn to Israel Pemberton Jr., 2 Apr. 1763, vol. 16, p. 67, Pemberton MSS ("Mortgage or Sell"); Thomas Corbyn to Israel Pemberton Jr., 28 June 1763, vol. 16, p. 104, Pemberton MSS; Thomas Corbyn to Israel Pemberton Jr., 29 Apr. 1768, vol. 20, p. 24, Pemberton MSS ("Cause the greatest").
112. Price, "Credit in the Slave Trade," 324–25, 330; Pares, *Merchants and Planters*, 45–50; Smith, *Gentry Capitalism*, 140–42, 148, 150–52. On the increasing popularity of mortgages for merchants of medicines, see Cause: Bridge v. Bridge, Papers relating to the estate of Thomas Bridge, J 90/12–14, Supreme Court of Judicature and Former Superior Courts Records, National Archives of the UK.
113. Price, "Credit in the Slave Trade," 330, 339.
114. Joseph Gurney Bevan to George Pugh, 1 Apr. 1782, AH033, Allen & Hanburys MSS; Joseph Gurney Bevan to James Robertson, 1 Apr. 1782, AH033, Allen & Hanburys MSS (quotation). For demographic context, see Higman, *Slave Populations*, 92–94.
115. Sheridan, *Sugar and Slavery*, 274–78.
116. Joseph Gurney Bevan to George Pugh, 19 Aug. 1783, 22 Nov. 1783, AH033, Allen & Hanburys MSS.
117. Joseph Gurney Bevan to John Worrell, 6 Jan. 1789, AH035, Allen & Hanburys MSS; Joseph Gurney Bevan to James Waddell, 14 Feb. 1792, AH028, Allen & Hanburys MSS.

CHAPTER THREE

1. Kupperman, "Fear of Hot Climates"; Morgan, *Laboring Women*. For other perspectives on the influence of climate and nature on bodies in the Atlantic World, see Cañizares-Esguerra, *Nature, Empire, and Nation*.

2. For example: Blochwitz, *Anatomia Sambuci.*
3. For another version of this arrangement wherein colonial practitioners influenced medical change in Europe, see Harrison, *Commerce and Empire*, 23–24.
4. On the topics of touch and consent, see Owens, *Medical Bondage*, 81, 108; Spillers, *Black, White, and in Color*, chap. 8; Schiebinger, *Secret Cures of Slaves*, 8; Fuentes, *Dispossessed Lives*, 83–84.
5. Turner, *Contested Bodies*, 66; Fuentes, *Dispossessed Lives*, 16.
6. Charters, *Imperial State*, 15–16, 84, 142; Cook, "Markets and Cultures," 141; Cook, "Practical Medicine," 3–4, 12–13; Alsop, "British Imperial Medicine"; Maehle, *Drugs on Trial*, 270–71; on the siege of Havana, see McNeill, *Mosquito Empires*, 184–86.
7. For details of several of these facets, see Harrison, *Commerce and Empire*; Chakrabarti, *Medicine and Empire.*
8. In fact, aspects of the ontological view have often appeared when convenient throughout history, such as among the Methodic school in Greece and Rome (Webster, "Heuristic Medicine"). Some early proponents of the ontological view in England can be found among the iatrochemists and followers of Van Helmont (Cook, "Markets and Cultures," 129). For more background, see Siraisi, *Medieval & Early Renaissance Medicine.*
9. Journal and logbook of an anonymous Scotch sailor, 1726, HC 363/1299, Hispanic Society of America. For details of the South Sea Company's trade under the *Asiento*, see Wennerlind, *Casualties of Credit*, sec. 3; Sperling, *South Sea Company*; Palmer, *Human Cargoes*; Murphy, "Collecting Slave Traders," 660–61; Brown, "South Sea Company and Contraband Trade"; Nelson, "Contraband Trade under the Asiento." The transatlantic slave trade also produced perceptions of similarity between locations and people, for example, Seth, *Difference and Disease*, 283.
10. On the persistence of Hippocratic views, see Kupperman, "Fear of Hot Climates," 214; Seth, *Difference and Disease*, chaps. 1–2. For others: Charters, *Imperial State*, 4; Duden, *Woman beneath the Skin*; Berger, *Ayurveda Made Modern*; Mukharji, *Nationalizing the Body*; Osseo-Asare, *Bitter Roots.*
11. Wallis and Pirohakul, "Medical Revolutions"; Chakrabarti, *Medicine and Empire*; Wear, *Knowledge and Practice*; Cook, "Practical Medicine," 15.
12. Cook, "Medicine," 420–21; Cook, "Markets and Cultures," 142, 144; Wallis, "Exotic Drugs," 26–27, 29, 37.
13. Wallis, "Exotic Drugs," 20–21, 36–38; Schumpeter, *English Overseas Trade Statistics*, 10–11; Chakrabarti, *Medicine and Empire*, 20.
14. Cook, "Markets and Cultures," 129–30; Cook, "New Philosophy"; Maehle, *Drugs on Trial*, 27, 29. For examples of the social and geographical diffusion of medicines, see Dayton, "Taking the Trade"; Ulrich, *A Midwife's Tale*. On approaches to consumerism in the past, see the introduction, notes 21, 35.
15. See Sydenham, *Observationes medicae*; Boerhaave, *Aphorismi de cognoscendis*; Linnaeus, *Genera morborum*; Boissier de Sauvages, *Nosologia methodica*; Cook, "Practical Medicine," 15–16; Cook, "Markets and Cultures," 129; Clow and Clow, *Chemical Revolution*. Again, these views did not exclude others that emphasized climate in the understanding of disease. Harrison, "Treatment of Fevers," 91–93, 110.
16. Luyendijk-Elshout, "(Anatomy) as Underlying Principle," 34.
17. Maehle, *Drugs on Trial*, 1–6, 29, 106, 284.
18. Invoice of Books Sent to Silvester Gardiner [Boston] from London, 14 Mar. 1755, B MS c50.4, John Denison Hartshorn Papers, Center for the History of Medicine, Countway Library of Medicine (hereafter cited as Hartshorn MSS); James, *Pharma-*

copoeia Universalis, 332 ("Gonorrhea"), 663 ("Phthisic," "hectic fevers"); Boerhaave, *Boerhaave's Aphorisms*, 177 ("promise"), 273 ("infallible"), 367.

19. Steele, *Atlantic Merchant-Apothecary*, xiv–xv; Culpepper, *Pharmacopoeia Londinensis*.
20. Grainger, *Common West-India Diseases*; Clark, *Observations on the Diseases*.
21. Salmon, *Ars Chirurgica*, 14 ("greatest Specificks"), 86 ("a Specifick"), 180 ("against the Dropsy"), 217 ("given as a specifick").
22. Talbor, *Pyretologia*, 42 (quotation); Blochwitz, *Anatomia Sambuci*; Mullins, *Some Observations*.
23. For example, *Rational Discours touching the Universal Medicin*; Boulton, *Examination of Mr. John Colbatch*.
24. Maehle, *Drugs on Trial*, 258.
25. Charters, *Imperial State*, 8; John Rous to Navy Office, 7 Nov. 1746, ADM 106/1034/155, Navy Board Records, National Archives of the UK (hereafter cited at NBR); [Treatment certificate], B MS c50.5, Hartshorn MSS.
26. Szechi and Holmes, *Age of Oligarchy*, 63–67, 374; Chakrabarti, *Medicine and Empire*, 40–41; Charters, *Imperial State*, 3–6, 174.
27. Notebook containing invoices for medicines and materials for the use of regiments in America, 1774, Add. MS 73622, vol. 77, Barrington Papers, Archives and Manuscripts, British Library (hereafter cited as Barrington Papers).
28. Brunsman, *Evil Necessity*, 6–7, 106–7, 113–14, 147–48.
29. "Contents of 14 Chests of medicines put up for the use of His Excellency Wm Shirley & Gen Pepperrell's Regts in the Expedition to Niagara June 1755," B MS c50.4, Hartshorn MSS; Thomas Dove to Navy Office, 10 July 1750, ADM 106/1080/263, NBR; 28 Apr. 1756, Diary of John Denison Hartshorn, Center for the History of Medicine, Countway Library of Medicine (hereafter cited as Hartshorn Diary).
30. "Depot for Medicine Chests," 111888.O.3, Helfand Collection.
31. CUST 3.
32. Hunting, *Society of Apothecaries*, 164–68, 171, 174; Navy Stock Articles of Copartnership, 1703, box 64/10, Navy Stock Records, Worshipful Society of Apothecaries (hereafter cited as Navy Stock Records); Simmons, "Stocks and Science," 155; Court of Assistants Minutes, 27 Nov. 1766, ff. 206r–206v, Court Minute Book, MS8200/7, Worshipful Society of Apothecaries; "Copy Case laid before Counsel by the General Committee of the Navy Stock in the Month of March 1767 and of the Opinions of Counsel thereon," 1767, box 64/10, Navy Stock Records; Navy Stock Accounts, MS 8225, Navy Stock Records.
33. Office of the Sick and Hurt Board to the Board of Admiralty, 12 Mar. 1756, ADM/F/13, Admiralty Collection, Caird Library, National Maritime Museum, Greenwich (hereafter cited as ADM); Crimmin, "Sick and Hurt Board"; Hunting, *Society of Apothecaries*, 172.
34. Court Minute Book (1745–1767), ff. 96v, 103v, MS 8200/7, Worshipful Society of Apothecaries; Navy Stock Accounts, MS 8225, Navy Stock Records; Hunting, *Society of Apothecaries*, 175; Navy Stock Articles of Copartnership, 1778, box 64/10, Navy Stock Records.
35. McNeill, *Mosquito Empires*, 4.
36. Patients' Reports, 24, 28, 31 Jan. 1756, B MS c50.4, Hartshorn MSS.
37. 3 May–18 May 1756, 12 May ("write the cases," "same treatment"), 16 May ("contagion"), 18 May ("tired"), Hartshorn Diary. "Nurses" is Hartshorn's term to describe the women practitioners who interacted with the soldiers. Much about their role in this moment, such as training or compensation, remains unclear. For details of

women practitioners in nursing and midwifery roles, see Ulrich, "Midwifery and Mortality."

38. Shammas, "How Self-Sufficient"; Maehle, *Drugs on Trial*, 16. On the diversity of medical practices, see Gómez, *Experiential Caribbean*; Sheridan, *Doctors and Slaves*; Jenner and Wallis, *Medicine and the Market*.
39. Wilson, *Pious Traders*, 129–30; Renaudet Notebook.
40. William Logan to John Hunt, 28 Dec. 1752, Letter Book, William Logan Journals and Papers, coll. 3386, Historical Society of Pennsylvania (hereafter cited as Logan Letters); William Logan to Silvanus Bevan, 16 Apr. 1754, Logan Letters (quotations).
41. Renaudet Notebook; Shaw, *Pharmacopoeia Edinburgensis*, 94.
42. Chakrabarti, *Materials and Medicine*, 2.
43. On the persistence of Indian ontologies, see chapter 5.
44. Stern, *Company-State*, 3, 185, 191–92, 197, 213; Chakrabarti, *Medicine and Empire*, 102. On the transition to direct rule in India, see Jasanoff, *Edge of Empire*, 47–48.
45. Chaudhuri, *Trading World*, 220–27, 234–35.
46. Chakrabarti, "Medicine amidst War," 3; Harrison, *Anglo-Indian Preventive Medicine*, 7.
47. Charters, *Imperial State*, 151, 169–71; Marshall, *East Indian Fortunes*, 14–15. Exact figures of British regulars, Royal Navy sailors, and EIC troops in earlier years are not known. At first, European and Eurasian Christian mercenaries filled the Company's military ranks before the introduction of more Indian troops.
48. Charters, *Imperial State*, 153; Fort William to Court of Directors, 14 Mar. 1765, p. 329, in *Fort William-India House Correspondence*, vol. 4 (quotation).
49. Harrison, "Disease and Medicine," 95–98; Chakrabarti, *Medicine and Empire*, 11; see also Clark, *Observations on the Diseases*.
50. Fort William to Court of Directors, 28 Nov. 1766, p. 444, in *Fort William-India House Correspondence*, vol. 4 (quotations); William B. Sumner to Robert Clive, 25 June 1766 in Mss Eur G37/40/3, f. 76, India Office Records; Richard Smith to Robert Clive, 15 Feb. 1766 in Mss Eur G37/38/2, f. 51, India Office Records.
51. 24 Apr. 1766 in IOR/P/240/24, p. 184, Madras Proceedings, India Office Records.
52. Charters, *Imperial State*, 158–63; Harrison, "Disease and Medicine," 90–92.
53. Commerce Journals, 1735–90, IOR/L/AG/1/6/11–20, India Office Records.
54. The directors also limited the export of new medical instruments; instead ordering surgeons to send old ones back to England for repair. To Fort St. George Public Department, 12 Nov. 1783 in IOR/E/4/871, pp. 133–34, India Office Records; Bombay Circular, 20 Aug. 1784 in IOR/E/4/1002, p. 361, India Office Records.
55. CUST 3; Commerce Journals, 1735–90, IOR/L/AG/1/6/11–20, India Office Records; Jasanoff, *Edge of Empire*, 154.
56. To Bombay Public Department, 23 Mar. 1787 in IOR/E/4/1004, p. 109, India Office Records ("considerable expence," "strictest economy"); Sick and Wounded Commissioners to Admiralty, 24 Dec. 1762, ADM F/23 ("inconvenience").
57. Fort William to Court, 24 Mar. 1766, p. 414, in *Fort William-India House Correspondence*, vol. 4; 26 Feb. 1762 in P/240/20, p. 177, Madras Proceedings, India Office Records; 30 Oct. 1765 in IOR/P/240/23, p. 505, Madras Proceedings, India Office Records; "List of medicines required at Mocha factory," 5 Aug. 1722 in IOR/G/17/1 Part I, ff. 158r–159v, India Office Records; "List of medicines required by Mocha factory," 4 Aug. 1723, in IOR/G/17/1 Part II, ff. 258r–258v, India Office Records; Chakrabarti, "Medicine amidst War," 10; Extract Fort St. George Military Letter, 16 Oct. 1804 in IOR/F/4/184, pp. 1–201, esp. pp. 154–56 ("proper treatment" p. 154), 164–67, India Office Records. See chapter 5 for further discussion of Company surgeons using country medicines.

58. General Letter to Fort St. George, 15 Feb. 1760 in IOR/E/4/862, p. 17, India Office Records.
59. Charters, *Imperial State*, 14.
60. Court Minutes, 2 Aug. 1758 in IOR/B/75, p. 95, India Office Records; Bengal General Letter (sent to Fort William), 1 Apr. 1760 in IOR/E/4/617, pp. 40–42, India Office Records.
61. Bengal General Letter, 21 Nov. 1766 in IOR/E/4/618, pp. 467–68, India Office Records; Fort William to Court, 28 Nov. 1766, p. 444 in *Fort William-India House Correspondence*, vol. 4; Navy Stock Accounts, MS 8225, Navy Stock Records.
62. General Letter to Fort St. George, 5 July 1758 in IOR/E/4/861, p. 937, India Office Records; General Letter to Fort St. George, 10 Dec. 1773 in IOR/E/4/865, pp. 979–80, India Office Records; Bombay General Letter, 11 Feb. 1774 in IOR/E/4/999, pp. 507–9, India Office Records.
63. General Letter to Bengal, 21 Nov. 1766 in IOR/E/4/618, pp. 467–68, India Office Records ("report to us," "answer our"); General Letter to Bengal, 9 Dec. 1768 in IOR/E/4/619, p. 587, India Office Records ("whom to blame").
64. Court Minutes, 22 Oct. 1773 in IOR/B/89, pp. 452–53, India Office Records; Court of Directors to Fort William, 1 Apr. 1760, p. 20 in *Fort William-India House Correspondence*, vol. 3.
65. Directions to report on "goodness and quality" can often be found throughout the letters, for example: General Letter to Fort William, 1 Apr. 1760 in IOR/E/4/617, pp. 40–42, India Office Records; General Letter to Fort St. George, 21 Nov. 1766 in IOR/E/4/863, p. 437, India Office Records. Exporters' markings are mentioned in: General Letter to Fort St. George, 10 Dec. 1773 in IOR/E/4/865, pp. 980–81, India Office Records; Bombay General Letter, 11 Feb. 1774 in IOR/E/4/999, pp. 509–10, India Office Records.
66. Fort William to Court, 28 Nov. 1766, p. 444 in *Fort William-India House Correspondence*, vol. 4.
67. Bengal General Letter (sent to Fort William), 20 Nov. 1769 in IOR/E/4/619, p. 587, India Office Records; "To Our President and Council at Fort William in Bengal," 4 Jan. 1771 in IOR/E/4/620, p. 390, India Office Records; Navy Stock Accounts, MS 8225, Navy Stock Records.
68. Philip Hurlock to the Court, 21 Oct. 1773 in IOR/E/1/57, f. 383r, India Office Records.
69. Timothy Bevan to the Court, 22 Oct. 1773 in IOR/E/1/57, f. 385r, India Office Records.
70. Court Minutes, 22 Oct. 1773 in IOR/B/89, pp. 452–53, India Office Records.
71. Fort St. George General Letter, 1 Feb. 1771 in IOR/E/4/865, p. 93, India Office Records; Bombay General Letter Supplement, 25 Apr. 1771 in IOR/E/4/998, p. 1115, India Office Records.
72. Fort St. George General Letter, 22 Dec. 1786 in IOR/E/4/873, pp. 29–30, India Office Records.
73. Fort St. George General Letter, 23 Mar. 1770 in IOR/E/4/864, pp. 973–74, India Office Records.
74. Schiebinger, *Plants and Empire*, 164; Schiebinger, *Secret Cures of Slaves*, 92, 107; Parsons, "Discovery of Ginseng," 61. On experimentation in the Royal Navy, see Charters, *Imperial State*, 130–35; and among "buccaneer surgeons," see Sheridan, "Doctor and the Buccaneer," 76–87, esp. 80n10. On the history of drug trials, as well as the tensions between trials and clinical experience, see Leong and Rankin, *Testing Drugs*

and Trying Cures, esp. Rivest, "Testing Drugs." The French case seems to mirror the process of testing in the EIC: small-scale trials driven by military pressures to determine the granting of monopoly privileges or military contracts.

75. Medical Board, Fort St. George, 13 Oct. 1800, vol. 12 (n. 72), p. 106, Tamil Nadu State Archives, Chennai, p. 36 in Chakrabarti, "Medicine amidst War."
76. Turner, *Contested Bodies*, 73.
77. Weaver, *Medical Revolutionaries*; Gómez, "Circulation of Bodily Knowledge," 386, 392.
78. While the islands exhibited far from a self-sufficient polyculture, the extent of sugar monoculture across the early modern Caribbean has often been overemphasized at the expense of other crops, such as coffee, cotton, tobacco, and indigo; see Roberts, "Labor and Agriculture," 552.
79. On the history of the sugar complex, see McNeill, *Mosquito Empires*, chap. 2; Mintz, *Sweetness and Power*; Schwartz, *Tropical Babylons*; Curtin, *Plantation Complex*. On migration in the early modern Atlantic, see Eltis, *Rise of African Slavery*, 11.
80. McNeill, *Mosquito Empires*, 28, 47–49, 60–61.
81. Blackburn, *New World Slavery*, 460–61; Berlin, *Many Thousands Gone*, 369–71; Burnard, *Mastery, Tyranny, and Desire*, 181; Sheridan, *Sugar and Slaves*, 230.
82. Sheridan, *Doctors and Slaves*, 24–25, 188, 272; Hutson, *Treatment and Management*, xxii. On mortality in the Caribbean, see Eltis, *Rise of African Slavery*, 68, 108, 161; Brown, *Reaper's Garden*, 17, 24. For a discussion of infant and child mortality, see Patterson, *Sociology of Slavery*, 101–2; Higman, *Slave Population and Economy*, 47–49.
83. Burnard, *Mastery, Tyranny, and Desire*, 181–82.
84. Brown, *Reaper's Garden*, 51–52, 55–56; Dunn, "Slave Labor Pattern."
85. Gómez, "Incommensurable Epistemologies," 97, 104–6; Gómez, "Circulation of Bodily Knowledge," 392, 397; Gómez, *Experiential Caribbean*, 122–25.
86. Schiebinger, *Plants and Empire*, 80; Sheridan, *Doctors and Slaves*, 81; Fair Copy of Knight's *History of Jamaica*, ca. 1746, vol. 2, f. 90, Add MS 12419, Archives and Manuscripts, British Library.
87. Gómez, "Circulation of Bodily Knowledge," 386; Chakrabarti, *Medicine and Empire*, 7; Dunn, *Sugar and Slaves*, 250; Voeks, "African Medicine."
88. Turner, *Contested Bodies*, 115, 127; De Barros "Medicine in Magic in British Guiana."
89. Hillary, *Epidemical Diseases*, 341 (quotations); Handler and Lange, *Plantation Slavery*, 101. Enslaved men and women grew medicinal plants alongside food crops (Voeks, "African Medicine," 74).
90. Hening, *Statutes at Large*, 105 (quotation); Fett, *Working Cures*, 165–67; Savitt, *Medicine and Slavery*, 175; Morgan, *Slave Counterpoint*, 612–19; Schwarz, *Twice Condemned*, 102. Some of these laws remained part of slave statutes until the American Civil War. The fear of poisoning was pervasive among slave societies across the Americas, see also Wood and Clayton, "Golden Grove," 111.
91. *Notes on the Reports*, 56.
92. Harrison, *Commerce and Empire*, 13–14; Schiebinger, *Plants and Empire*, 131; Price and Price, *Stedman's Surinam*, 136; Owens, *Medical Bondage*, 52. For estimates of the number of doctors in Jamaica, see Sheridan, *Doctors and Slaves*, 52; Turner, *Contested Bodies*, 76.
93. [Invoice to William & Mary College from James Carter], 1767, box 5, folder 2, Office of the Bursar Records, Special Collections Research Center, Earl Gregg Swem Library, College of William & Mary.
94. Philip Pinnock with Alexander Johnston Account Currant, 1774, series 12a, box 54, folder 11, Powel Family Papers, coll. 1582, Historical Society of Pennsylvania (here-

after cited as Powel MSS); [Alexander Johnston Medical Record], 1773, series 12a, vol. 343, Powel MSS; Karras, "World of Alexander Johnston," 58–60. For background on the piecework system and other examples of it throughout the Caribbean, see Sheridan, *Doctors and Slaves*, chap. 11.

95. Hogarth, *Medicalizing Blackness*, 156–57; Savitt, *Medicine and Slavery*, 33–34.
96. Sheridan, *Doctors and Slaves*, 73, 319–20; Grainger, *Common West-India Diseases*, 41–42, 44–46, 71–73. For other examples of British medicines on plantations, see Robertson, *Detection of the State*, 12; Holder, *Short Essay on the Subject*, 20–21.
97. Caines, *Letters on the Cultivation*, 144–45, 148–49, 167–73.
98. Plough Court Order Book, 1776–77, AH140, Allen & Hanburys MSS. Many of these were also standard purgatives for different kinds of ailments, though in this eighteenth-century context they appear as wide-ranging medicines.
99. "Directions for the Use of Box of Medicines," 17 Apr. 1750, MS 5442, Corbyn & Co. MSS.
100. [Collins], *Practical Rules*, 249, 466–67.
101. Invoice of sundries shipped for Running Gut Estate, 16 Nov. 1801, series 13, box 63, folder 7, Powel MSS; Invoice of Higgin & Crawford bought of Hoiner & Hawkes, 22 Oct. 1801, series 13, box 63, folder 7, Powel MSS; Medicines bought for Running Gut Estate from John Walker & Co., 1801, series 13, box 63, folder 7, Powel MSS.
102. Plough Court Order Book, 1776–77, AH140, Allen & Hanburys MSS; Thomas Corbyn to John Wills, 4 Sept. 1751, MS 5442, Corbyn & Co. MSS.
103. Wood and Lynn, *Travel, Trade and Power*, 93; Claims 16908, 15984, 12003, 24498, 24499, 21393, 19284, Legacies of British Slave-Ownership Database; Plough Court Order Book, 1776–77, AH140, Allen & Hanburys MSS; Stander, "Transatlantic Trade in Pharmaceuticals," 336, 342; Vanneck-Arc/3C/1793/1, Vanneck-Arc/3C/1801/1, Vanneck-Arc/3C/1802/1, Taylor and Vanneck-Arcedeckne Papers, Institute of Commonwealth Studies, University of London (hereafter cited as Vanneck-Arc). For shipping times, see Elliott, *Empires of the Atlantic World*, 50; Steele, *English Atlantic*, 21–22, 30, figs. 2–3.
104. Sheridan, "Simon Taylor," 291–92; Wood and Clayton, "Golden Grove," 100–103.
105. Thomas Corbyn to John Pickering, 1 Sept. 1746, MS 5442, Corbyn & Co. MSS (quotation); Jenkins, *Quaker Experiment*, 7–10, 42, 51; Pettigrew, *John Coakley Lettsom*, 175–77.
106. Brandow, *Genealogies of Barbados Families*, 690; Ingram, *Practical Cases*, 111.
107. Hughes-Queree Collection, Barbados Department of Archives; Will of Joseph Jackman, RB6/20, pp. 135–38, Wills, Barbados Department of Archives (hereafter cited as BDA Wills).
108. "A list of Vessels which have entered inwards at the Naval Office in the Island of Barbados," 1785, RB9/1/4, Shipping Returns, Barbados Department of Archives (hereafter cited as BDA Returns); "A list of all Vessels which have cleared outwards at the Naval Office in the Island of Barbados," 1785, RB9/1/5, BDA Returns.
109. Hughes-Queree Collection, Barbados Department of Archives; "Tombs on Plantations," 26; Brandow, *Genealogies of Barbados Families*, 628–31; Claim 5920, Legacies of British Slave-Ownership Database. For other examples, see Will of William Astin, pp. 289–91, RB6/31, BDA Wills; Will of William Eversley, pp. 43–45, RB6/25, BDA Wills; Will of Reynold Skeete, pp. 463–66, RB6/28, BDA Wills; Will of Allen Sharrett, pp. 517–19, RB6/17, BDA Wills. For background on the island's leading families during the eighteenth century, see Beckles, *History of Barbados*, 43.
110. Vanneck-Arc/3C/1793/1; MS.523/176/1/13–14, reel 5, Newton Papers, University of London (hereafter cited as Newton Papers). For a later example, see the list of

medicines required for Worthy Park estate in 1824 (for approximately 484 enslaved individuals): Craton, *Invisible Man*, 398–99.

111. Morgan, "Business Networks," 37.
112. Brown, *Reaper's Garden*, 11–12, 15.
113. CUST 3/74–75. In 1775, Jamaica received £6,508 (364,448 pounds) of medicines, compared to £4,404.11.5 (246,656 pounds) of haberdashery and £11,329.3.7 of cotton. In 1774, Jamaica's enslaved population approached 193,000 compared to Barbados's of around 92,000 and more than 250,000 in Virginia and Maryland together. For these population figures: Berlin, *Many Thousands Gone*, 369–71; McCusker and Menard, *Economy of British America*, 153; Beckles, *History of Barbados*, 42; Newman, *Dark Inheritance*, 17. And, for a slightly later period, see Higman, *Slave Populations*.
114. CUST 3. By 1700, the British Caribbean had six times the enslaved population of mainland British North America. For population comparison, see data in Fogel and Engerman, *Time on the Cross*, 21–22; Blackburn, *New World Slavery*, 460; O'Malley, *Final Passages*. For estimates of the transatlantic slave trade, see Curtin, *Atlantic Slave Trade*, 52–64, 88–89, 119; Eltis, "Volume and Structure"; Slave Voyages; Dunn, *Sugar and Slaves*, 26–28, 87.
115. Beginning in the 1730s, medicine imports (pounds) outpaced slave trade disembarkations for Antigua, Barbados, and Jamaica, continuing to be significantly higher through the 1770s. Yearly medicine imports also grew more rapidly than either the enslaved or white populations on Barbados. In general, colonial population increased tenfold between 1700 and 1775. Such metrics suggest rising per capita consumption of imported medicines. CUST 3; Slave Voyages; Berlin, *Many Thousands Gone*, 369–71; McCusker and Menard, *Economy of British America*, 153; Beckles, *History of Barbados*, 42; Szechi and Holmes, *Age of Oligarchy*, 63–65.
116. Indiferente General 2726, Archivo General de Indias. Trade policy was a popular issue among Spanish officials in the early eighteenth century. In the later 1710s, a new tax policy sought to promote Spanish trade (which was taxed) versus the low-priced cargo of the English "Annual Ship" or other colonial traders. A new economic policy subsequently emerged for the Spanish Caribbean over the years 1716 to 1720, embodied by the *Real Proyecto* of 1720, which aimed to facilitate the sale of Spanish goods in America. Walker, *Spanish Politics*, 89, 110–11, 130.
117. *Daily Advertiser*, 2 July 1774.
118. Leslie, *New History of Jamaica*, 49 (quotation); Sheridan, *Doctors and Slaves*, 42–44.
119. Papers relating to the affairs of Thomas Downes, MS 5441/1/1, Corbyn & Co. MSS; Oliver, *History of the Island of Antigua*, cxvi; Thomas Downes to Thomas Corbyn, 19 Feb. 1757, MS 5441/1/2/1, Corbyn & Co. MSS.
120. Durham, *Caribbean Quakers*, 33–35; Besse, *Collection of the Sufferings*, vol. 2.
121. Thomas Downes to Thomas Corbyn, 14 Jan. 1758, MS 5441/1/2/6, Corbyn & Co. MSS.
122. Bonds, promissory notes, and receipts relating to debts of Thomas Downes, 1755–56, MS 5441/1/5/1–10, Corbyn & Co. MSS; Thomas Downes to Thomas Corbyn, 19 Feb. 1757, MS 5441/1/2/1, Corbyn & Co. MSS; Thomas Downes to Thomas Corbyn, 18 July 1757, MS 5441/1/2/4, Corbyn & Co. MSS.
123. Thomas Downes to Thomas Corbyn, 1 June 1758, MS 5441/1/2/7, Corbyn & Co. MSS; John Godfrey to Thomas Corbyn, 21 July 1758, MS 5441/1/3/1, Corbyn & Co. MSS.
124. For example: Account for goods exported by Thomas Corbyn to Robert James, druggist, in Antigua [ca. 1745], MS 5441/3/3, Corbyn & Co. MSS.

125. Will of Joseph Gamble Jackman, p. 44, RB6/48, p. 442, RB6/28, BDA Wills; "Joseph Gamble Jackman Inventory," 6 Oct. 1786, BDA Inventories. On Speightstown, see Fuentes, *Dispossessed Lives*, 31.
126. Turner, *Contested Bodies*, 80, 131.
127. Hutson, *Treatment and Management*, xviii–xix.
128. Schiebinger, *Plants and Empire*, 82; Long, *History of Jamaica*, 381 (quotations).
129. Sheridan, *Doctors and Slaves*, 70, 292, 300, 311, 335, 331, 28; Turner, *Contested Bodies*, 129, 136, 145.
130. Joseph Gurney Bevan to McIntyre & Moodie, 18 Oct. 1776, AH027, Allen & Hanburys MSS; Joseph Gurney Bevan to Jonas Langford Blizzard, 2 Sept. 1777, AH027, Allen & Hanburys MSS. On Blizzard's work, see Sheridan, *Doctors and Slaves*, 299–300. He provided care, medicines, and midwifery for more than 300 enslaved persons (for six shillings per head) and white people on Samuel Martin's plantations on Antigua.
131. Craton, *Invisible Man*, 129–30.
132. For example: Paugh, *Politics of Reproduction*, 87; Schiebinger, *Plants and Empire*.
133. Sheridan, *Doctors and Slaves*, 28–30.
134. Hogarth, *Medicalizing Blackness*, 111, 214n48. Other notable plantation manuals included Dr. Collins's *Practical Rules* (1803; 2nd ed. 1811), Thomas Dancer's *Medical Assistant* (1801; 2nd ed. 1809), James Thomson's *Treatise on the Diseases of Negroes* (1820).
135. Hutson, *Treatment and Management*, xiv–xv.
136. Grainger, *Common West-India Diseases*, 23–31, 55–60.
137. For example: *Medical Vade Mecum*.
138. For example: Belgrove, *Treatise upon Husbandry*, 42, 53–54; Grainger, *Common West-India Diseases*, 26–27 ("sufficient dose," p. 27), 41–42, 71–73. Grainger broadly settled on "a sufficient dose [of an emetic] for a grown up Negroe" in cases of particular fluxes.
139. Grainger, *Common West India Diseases*, 42. On the popularity of opium and mercury-based medicines, see Harrison, *Commerce and Empire*, chap. 8.
140. Grainger, *Common West-India Diseases*, 45–48.
141. [Martin], *Essay upon Plantership*, 2nd ed., vii.
142. McNeill, *Mosquito Empires*, 63, 74–77.
143. Kein, *Essay upon Pen-Keeping*, 14–15 ("wonderful cures," p. 14). Kein appears to have taken this phrasing from the second edition of Martin's text (p. vii) and in future editions (4th ed., p. xiv).
144. For example: Hughes, *Natural History*, 34–35.
145. Hogarth, *Medicalizing Blackness*, 98, 153–55; Sheridan, *Doctors and Slaves*, 279; [Collins], *Practical Rules*, 255, 265 ("place of security"); Weaver, *Medical Revolutionaries*, 45–46. On the forced ingestion of medicines, see Maxwell, "Pathological Inquiry," 418; Thomson, *Treatise on the Diseases of Negroes*, 46 ("frequent pukes"). Thomson talks of administering "frequent pukes," implying a lack of consent and use of force, in cases of dirt-eating among children, for example.
146. [Collins], *Practical Rules*, title page ("Professional Planter"), 248–50 ("expense of time," p. 249), 466–67.
147. Grainger, *More Common West-India Diseases*, 74–75.
148. Paugh, *Politics of Reproduction*, 92–94.
149. [Collins], *Practical Rules*, 231–36 ("anatomist," p. 232; "two varieties," pp. 235–36). For background on Collins, see Sheridan, *Doctors and Slaves*, 32; Hogarth, *Medicalizing Blackness*, 125. Collins's racial observations continued to be cited in other medical works well into the nineteenth century.
150. Schiebinger, *Secret Cures of Slaves*, 112–15; Seth, *Difference and Disease*.

151. Hogarth, *Medicalizing Blackness*, xi–xiv, 4.
152. [Collins], *Practical Rules*, 250.
153. Newman, *Dark Inheritance*, 13; Seth, *Difference and Disease*, 15–16, 21.
154. Harrison, *Commerce and Empire*, 17.
155. CUST 3; Wallis, "Exotic Drugs," 30; Plough Court Ledgers, AH181, AH207, AH228, AH226, Allen & Hanburys MSS.
156. Morris, Kendrick, et al., *Edinburgh Medical and Physical Dictionary*, s.v. "specifics." Morris and Kendrick appear to have borrowed the above quote about specifics from the tenth edition of John Quincy's *New Medicinal Dictionary* (1787). It also shows up in *The New Royal Encyclopaedia* (1789) and *Encyclopaedia Britannica* (1797).
157. For the trend in India toward seeing human characteristics as immutable, see Harrison, *Climates and Constitutions*.

CHAPTER FOUR

1. May 1753, Diary of William Jepson, R. Stanton Avery Special Collections Department, New England Historic Genealogical Society (hereafter cited as Jepson Diary) (quotation). For details of the *Kennebeck's* trips, see June 1753, Jepson Diary; 4 Sept. 1754, Hartshorn Diary; Jacob Wendell et al. to Silvester Gardiner, ca. 1752, box 1, folder 7, Kennebec Proprietors Papers, coll. 60, Maine Historical Society (hereafter cited as Kennebec Papers); Proprietors of the Plymouth Patent to Silvester Gardiner, Apr. 1754, box 1, folder 10, Kennebec Papers. Evidence of Gardiner's shipping investments is found in Naval Office Shipping Lists for Massachusetts, 1686–1765, pt. 2, box 2, pp. 880–880A, MS N-1635, Massachusetts Historical Society (hereafter cited as NOSL); NOSL, pt. 3, box 3, pp. 1051–51A.
2. Petty Ledger (1773–1777), vol. 10, Greenleaf Family Papers, MS L80, Boston Athenaeum (hereafter cited as Greenleaf MSS); Gardiner's estimates of his losses, 1783, II.15, Gardiner-Whipple-Allen Family Papers, MS N-1271, Massachusetts Historical Society (hereafter cited as GWA Papers).
3. Coclanis, "Introduction," xii; Shammas, *Pre-Industrial Consumer*, 270, 284–85.
4. The number of medical practitioners recorded in Massachusetts grew by an estimated 32.4 percent every ten years from 1700 to 1790, while the general population grew by only 24.3 percent over the same span. Christianson, "Medical Practitioners of Massachusetts," 54–55; Gevitz, "New England Apothecary," 23; Shryock, *Medicine and Society in America*, 12; Brown, "Emergence of Urban Society."
5. Dorner, "Deployment of Expertise."
6. Buck, "Tucker's Wife's Leg," 937 ("Physick"); Jepson Diary ("rules").
7. Milford, *Gardiners of Massachusetts*, 17–19. On the procedure of lithotomy, see Cook, *Trials of an Ordinary Doctor*. For record of John Gibbins, see *Boston News-Letter*, 26 June 1760, [3].
8. Packard, "William Cheselden," 537; [Boston (MA) Record Commissioners], *Boston Marriages*, 328; Christianson, "Medicine in New England," 58–59; Sonnedecker, *History of Pharmacy*, 155; Cash, "Professionalization of Boston Medicine," 70; Brock, "Influence of Europe," 107. Ann Gibbins Gardiner recorded a variety of recipes reflecting her family's connection to local and overseas commerce; see Gibbons, *Mrs. Gardiners Receipts*.
9. On lithotomy, particularly in colonial America, see Christianson, "Practice of Lithotomy," 104–5.
10. *Boston Evening-Post*, 9 Nov. 1741, [2] (quotation); *Boston Weekly News-Letter*, 13 Nov. 1741, [2].

11. *Boston Evening-Post,* 9 Nov. 1741, [2]; 15 Oct. 1754, Hartshorn Diary; Lithotomy Report, 15 Oct. 1754, B MS c50.4, Hartshorn MSS; Christianson, "Practice of Lithotomy," 108–9; Seybolt, "Lithotomies Performed," 109.
12. Christianson, "Practice of Lithotomy," 105–6; Dorner, "Deployment of Expertise," 302.
13. *Boston Gazette,* 19 June 1744, [2]; 29 May, 9 Dec. 1755, 23 Jan. 1756, Hartshorn Diary; Benjamin Stockbridge to Silvester Gardiner, 21 Aug. 1755, B MS c50.2, Hartshorn MSS; Gordon, *Æsculapius,* 89; Warden, "Medical Profession," 148–49; Milford, *Gardiners of Massachusetts,* 2, 20.
14. 19 June 1755, Hartshorn Diary; Invoice of books sent to Silvester Gardiner, 14 Mar. 1755, B MS c50.4, Hartshorn MSS; Recipes of Dr. Gardiner in the hand of John Dension Hartshorn, B MS c50.4, Hartshorn MSS.
15. Gevitz, "New England Apothecary," 23; Christianson, "Medical Practitioners of Massachusetts," 54–55; Brown, "Healing Arts," 40–42; Mutschler, " 'Social Credit.' "
16. For example: 10 Nov. 1754, 19 Nov. 1754, 1 Jan. 1755, Hartshorn Diary.
17. *Boston Gazette,* 19 June 1744, [2] ("all Sorts"; see also *Boston Weekly Post Boy,* 30 July 1744, [4]); *Boston Weekly News-Letter,* 28 June 1750, [2] ("the Sign"; see also *Boston Gazette,* 19 June 1744, [2]; *Boston Weekly Post Boy,* 30 July 1744, [4]); Recipes of Dr. Gardiner, B MS c50.4, Hartshorn MSS; Thomas Paine account with Silvester Gardiner, 12 Mar. 1746 [1747]–25 Apr. 1749, reel 9, Robert Treat Paine Papers, P–392, Massachusetts Historical Society (hereafter cited as Paine MSS); Silvester Gardiner to William Jepson, 21 May 1759, case 7, box 29, folder 5, Simon Gratz Autograph Collection, Historical Society of Pennsylvania (hereafter cited as Gratz Collection).
18. *Boston Weekly News-Letter,* 28 June 1750, [2] (quotation); Silvester Gardiner to Robert Southgate, 24 Jan. 1769, B MS Misc., Center for the History of Medicine, Countway Library of Medicine; Wallis, "Consumption, Retailing, and Medicine," 27, 30.
19. 14 Feb. 1754, 23 Mar. 1754, 21 Aug. 1755, Hartshorn Diary; Silvester Gardiner to Simon Tufts, 26 July 1751, Simon Tufts Correspondence, R. Stanton Avery Special Collections Department, New England Historic Genealogical Society.
20. 18 June 1754, Hartshorn Diary.
21. Silvester Gardiner to Robert Treat Paine, 1 June 1771, reel 3, Paine MSS; Recipes of Dr. Gardiner, B MS c50.4, Hartshorn MSS.
22. Supplement to the *Massachusetts Gazette & Boston News-Letter,* 1 Dec. 1763, [1].
23. *Boston Post-Boy,* 29 Oct. 1764, [1].
24. *Boston Evening-Post,* 23 June 1766, [3].
25. *Boston-Gazette, and Country Journal,* 23 Apr. 1764, [4]; *Boston Weekly News-Letter,* 10 June 1756, [2]; *Boston-Gazette, and Country Journal,* 10 Dec. 1759, [4]; *Boston-Gazette, and Country Journal,* 31 Oct. 1763, [1]; Cleary, *Elizabeth Murray,* 60.
26. Another example is Susanna Renken at her grocery shop in Fore Street near the drawbridge in Boston; see *Boston Evening-Post,* 29 Apr. 1765, [4]; *Boston-Gazette, and Country Journal,* 8 July 1765 [1]; for examples in Philadelphia, see Brandt, "Women's Medical Entrepreneurship."
27. CUST 3/22–74.
28. Wilson, *Pious Traders,* 129–30. For a picture of Gardiner's accounts with London suppliers, see Kilby, Barnard & Parker (hereafter KBP) to Silvester Gardiner, 31 July 1762, II.32, GWA Papers; Gardiner in account currant with Kilby, Barnard & Co., July 1762, II.95, GWA Papers.
29. Silvester Gardiner to William Jepson, 3 Mar. 1760, case 7, box 29, folder 6, Gratz Collection; Silvester Gardiner to William Jepson, 7 May 1760, case 7, box 29, folder 6,

Gratz Collection (quotation); Silvester Gardiner to William Jepson, 22 Aug. 1760, case 7, box 29, folder 6, Gratz Collection. On Captain Bradford's ordeal, see *Boston News-Letter*, 10 Apr. 1760, [3]; *Pennsylvania Gazette*, 24 Apr. 1760, [3]; *New-York Mercury*, 17 Mar. 1760, [3]; *New-York Mercury*, 12 May 1760, [3]; *New-Hampshire Gazette*, 11 Apr. 1760, [2].

30. "Invoice of merchandise ship'd by Harrison's & Ansley," 25 Feb. 1773, box 1, folder 3, Silvester Gardiner Papers, coll. 41, Maine Historical Society (hereafter cited as Gardiner MSS); Silvester Gardiner to Captain Kirkwood, 15 Nov. 1755, B MS c50.3, Hartshorn MSS; KBP to Silvester Gardiner, 31 July 1762, II.32, GWA Papers; 27 Aug. 1754, Hartshorn Diary.
31. Silvester Gardiner to William Jepson, 18 Aug. 1760, case 7, box 29, folder 6, Gratz Collection.
32. KBP in account currant with Silvester Gardiner, 1755–58, box 1, folder 1, Gardiner MSS. For example, Griffenhagen, "Bartholemew Browne."
33. KBP to Silvester Gardiner, 31 July 1762, II.32, GWA Papers; Silvester Gardiner in account currant with Kilby, Barnard & Co., July 1762, II.95, GWA Papers; KBP in account currant with Silvester Gardiner, 17 July 1763, box 1, folder 1, Gardiner MSS; KBP in account currant with Silvester Gardiner, 1755–58, box 1, folder 1, Gardiner MSS; Hughes & Whitlock to Silvester Gardiner, 1 Sept. 1775, box 1, folder 4, Gardiner MSS; Harrison & Ansley (hereafter HA) to Silvester Gardiner, 3 Jan. 1776, box 1, folder 4, Gardiner MSS; HA to Silvester Gardiner, 30 Apr. 1777, box 1, folder 4, Gardiner MSS; "Invoice of merchandise ship'd by HA," 25 Feb. 1773, box 1, folder 3, Gardiner MSS.
34. NOSL, pt. 2, box 2, pp. 880–880A; NOSL, pt. 3, box 3, pp. 1051–51A; 1754, Hartshorn Diary; Silvester Gardiner to Robert Southgate, 24 Jan. 1769, B MS Misc., Center for the History of Medicine, Countway Library of Medicine. The estimated per capita income in Massachusetts in 1771 was £12 per annum (Shammas, *Pre-Industrial Consumer*, 274).
35. Wilson, "Trading in Drugs," 355–56.
36. "Depot for Medicine Chests," 111888.O.3, Helfand Collection ("East & West"); Thomas Dove to Navy Office, 10 July 1750, ADM 106/1080/263, NBR; Stores shipped on the sloop Elizabeth, [no date], Massachusetts Archive Collection, vol. 291, p. 36, Massachusetts State Archives (hereafter cited as MA Archive Collection) ("cags and chests"); "Contents of 14 Chests of medicines put up for the use of his Excellency Wm Shirley & Gen Pepperrell's Regts in the Expedition to Niagara June 1755," B MS c50.4, Hartshorn MSS; [Invoice of Regimental Chests], B MS c50.4, Hartshorn MSS; 15 June 1755 ("Sunday"), 16 Sept. 1755 ("trooping"), 24 Apr. 1756, 28 Apr. 1756, Hartshorn Diary. For background on the military expeditions in 1755 and 1756, see Anderson, *Crucible of War*, 68–69, 110–23.
37. Thomas Corbyn to Jabez Bowen, 14 Aug. 1748, MS 5442, Corbyn & Co. MSS; Thomas Corbyn to John Greenleaf, 5 Sept. 1748, MS 5442, Corbyn & Co. MSS; Dorner, "Deployment of Expertise," 306.
38. Invoices (1748–1760), vol. 1, Greenleaf MSS; Account Book (1767–1768), vol. 5, Greenleaf MSS; Petty Accounts (1753–1773), vol. 2, Greenleaf MSS; Petty Ledger (1770–1771), vol. 7, Greenleaf MSS.
39. Petty Ledger (1764–1767), vol. 3, Greenleaf MSS; Petty Accounts (1753–1773), vol. 2, Greenleaf MSS; Account Book (1767–1768), vol. 5, Greenleaf MSS; Ledger (1770–1773), vol. 6, Greenleaf MSS; Alms House Book (1766–1772), vol. 4, Greenleaf MSS.
40. Martin Brimmer Account Ledger, Mss. Acc. 2012.359, Special Collections Research

Center, Earl Gregg Swem Library, College of William & Mary (hereafter cited as Brimmer Accounts).

41. *Boston Evening-Post,* 10 Dec. 1764, [4] ("general Assortment," "Country Practitioners"); *Boston Evening-Post,* 14 May 1764, [4] ("fresh Supply"); *Boston Evening-Post,* 4 July 1768, [4]; *Boston Evening-Post,* 5 Nov. 1770, [4]; *Boston Evening-Post,* 18 Jan. 1773, [4].
42. For example: Warren, *New England Bound;* Piersen, *Black Yankees.*
43. Interpretations of New England merchants have shifted over time; see Bailyn, *New England Merchants;* Wright and Viens, *Boston Business Community;* Valeri, *Heavenly Merchandize;* Kimball, "Slave Economies."
44. For a fuller picture of the Newport medical community, see Crane, *Poison Plot.*
45. Parsons, *Rhode Island Physicians,* 5–7.
46. Physician's Books (1765–1779), vault A, #452, #453, Hunter MSS; Accounts of Estate of William Hunter with Malbone Family, 1763–76, MS 549, folder 4, Malbone Family Collection, Rhode Island Historical Society; Inventory of the personal estate of the late William Hunter, 7 Feb. 1777, vault A, box 90, folder 9, Hunter MSS.
47. Bill: William Hunter to Aaron Lopez, 1763–76, vault A, box 167, folder 1, Hunter MSS.
48. 25 Nov. 1754, Hartshorn Diary; Crane, *Poison Plot,* 140–41.
49. Cornelia Dayton has analyzed this case at length in Dayton, "Taking the Trade," 35–39, while Elaine Crane elaborates the possible connections between Hallowell and Tweedy in Crane, *Poison Plot,* 118–22, 144.
50. Thomas Corbyn to John Tweedy, 10 June 1752, MS 5442, Corbyn & Co. MSS. For details of Tweedy's career, including the charges of bad medicines, see Crane, *Poison Plot,* 118–21, 135.
51. Silvester Gardiner to William Jepson, 26 May 1758, case 7, box 29, folder 5, Gratz Collection.
52. Silvester Gardiner to William Jepson, 8 Mar. 1764, case 7, box 29, folder 9, Gratz Collection.
53. Silvester Gardiner to William Jepson, 26 Apr. 1764, case 7, box 29, folder 9, Gratz Collection.
54. *Boston Post-Boy & Advertiser,* 4 May 1767, [3] ("very rough"); Silvester Gardiner to William Jepson, 28 Nov. 1765, case 7, box 29, folder 10, Gratz Collection ("declining").
55. Silvester Gardiner to William Jepson, 21 May 1759, case 7, box 29, folder 5, Gratz Collection.
56. Jepson Diary (quotation); Haggerty, *British-Atlantic Trading Community,* 54.
57. Silvester Gardiner to William Jepson, 12 Jan. 1758, case 7, box 29, folder 5, Gratz Collection; *True State of the Copartnership,* 1.
58. Silvester Gardiner to William Jepson, 26 May 1758, case 7, box 29, folder 5, Gratz Collection (quotation); Silvester Gardiner to William Jepson, 18 Apr. 1758, case 7, box 29, folder 5, Gratz Collection.
59. Silvester Gardiner to William Jepson, 7 June 1759, case 7, box 29, folder 5, Gratz Collection; Silvester Gardiner to William Jepson, 16 June 1759, case 7, box 29, folder 5, Gratz Collection (quotation).
60. Stanley, *Historic Sites of Norwich;* Flexner, *Traitor and the Spy,* 7, 9–11; Warren, *New England Bound,* 118, 125, 132, 161; Inventory of the personal estate of the late William Hunter, 7 Feb. 1777, vault A, box 90, folder 9, Hunter MSS; "Joseph Gamble Jackman Inventory," 6 Oct. 1786, BDA Inventories.
61. Silvester Gardiner to William Jepson, 27 Sept. 1758, case 7, box 29, folder 5, Gratz Collection; Silvester Gardiner to William Jepson, 7 June 1759, case 7, box 29, folder 5, Gratz Collection.

62. Silvester Gardiner to William Jepson, 30 May 1765, case 7, box 29, folder 10, Gratz Collection (quotations); Silvester Gardiner to William Jepson, 10 Dec. 1767, case 7, box 29, folder 12, Gratz Collection; Silvester Gardiner to William Jepson, 18 July 1765, case 7, box 29, folder 10, Gratz Collection; Silvester Gardiner to William Jepson, 26 Nov. 1767, case 7, box 29, folder 12, Gratz Collection; Potash Invoices and Accounts of Sales, 1764–72, box 26, Hancock Manuscripts, Baker Library, Harvard Business School; Roberts, "American Potash Manufacture"; "Mr. Stephen's Process for Making Plantation Pot-Ash." On potash's use as a substrate for making medicines, such as vitriolated tartar, see Lewis, *New Dispensatory*, 1st ed., 290–91.
63. Silvester Gardiner to William Jepson, 10 Dec. 1767, case 7, box 29, folder 12, Gratz Collection (quotation). On the importance of diversification, see Hancock, *Citizens of the World*, 84; Gordon, *Æsculapius*, 19.
64. Kershaw, *Kennebeck Proprietors*, 3–5.
65. On Dummer's War, its aftermath, and Abenaki sovereignty, see Saxine, "Performance of Peace"; Ghere, "Mistranslations and Misinformation."
66. Saxine, "Performance of Peace," 381–82, 410; Ghere, "Mistranslations and Misinformation," 16–20; see also Taylor, "Contest for Land."
67. Kershaw, *Kennebeck Proprietors*, xiv–xv, 26–27, 30, 35, 150; Taylor, "Contest for Land," 5, 13.
68. Kershaw, *Kennebeck Proprietors*, 44–45.
69. 4 Sept. 1754, Hartshorn Diary. For Apthorp's slave trading, see *Boston Gazette*, 1 Aug. 1737, [3]; Butterfield, *Diary and Autobiography of John Adams*, 54, 77, 151 (quotation).
70. David Jeffries to Silvester Gardiner, 1760, box 2, folder 7, Kennebec Papers; John Temple to the Kennebeck Proprietors, 6 July 1780, box 3, folder 10, Kennebec Papers (quotation); Kershaw, *Kennebeck Proprietors*, 45. Gardiner may be a source of the character "Dr. G" in Royall Tyler's quasi-memoir *The Bay Boy* written in 1824–25 but first published in 1978.
71. Leamon, *Revolution Downeast*, 7; Kershaw, *Kennebeck Proprietors*, 100; 24 July 1754, Hartshorn Diary; Silvester Gardiner to William Jepson, 18 May 1764, 13 Sept. 1764, 27 Sept. 1764, case 7, box 29, folder 9, Gratz Collection.
72. Hanson, *History of Gardiner*, 46–47, 57–63.
73. For example, Proprietors of the Plymouth Patent to Silvester Gardiner, 19 Jan. 1757, box 1, folder 17, Kennebec Papers; 1753, Jepson Diary; Benjamin Fitch deposition, 13 Dec. 1766, Misc. Bound Manuscripts, Massachusetts Historical Society; Records, vol. 2, p. 54, in Kennebec Papers, vol. 6. Company financial details can be found in Ledger (1754–1800), pp. 12, 38, 51, 58, 67, in Kennebec Papers, vol. 9; Records, vol. 2, pp. 58, 89, 92, 144, in Kennebec Papers, vol. 6; "A Summary of the Proportions of the several Proprietors brought forward," in Kennebec Papers, vol. 15, p. 16.
74. Gershom Flagg Bill, 28 Jan. 1755, box 1, folder 13, Kennebec Papers; "The Committee of the Plimouth Compy to Gershom Flagg," May 1754, box 1, folder 10, Kennebec Papers; Records, vol. 1, p. 48 (ca. 1750s), in Kennebec Papers, vol. 3; Kershaw, *Kennebeck Proprietors*, 113; Leamon, *Revolution Downeast*, 6–7, 9.
75. Kershaw, *Kennebeck Proprietors*, 100; Records, vol. 1, p. 115, in Kennebec Papers, vol. 3; Records, vol. 2, pp. 58–59 (ca. 1750s), in Kennebec Papers, vol. 6.
76. For details of the sloop's trips, see June 1753, Jepson Diary; 4 Sept. 1754, 31 May 1755, Hartshorn Diary; Jacob Wendell et al. to Silvester Gardiner, ca. 1752, box 1, folder 7, Kennebec Papers; Proprietors of the Plymouth Patent to Silvester Gardiner, Apr. 1754, box 1, folder 10, Kennebec Papers; NOSL, pt. 2, box 2, pp. 880–880A.
77. Proprietors of the Plymouth Patent to Silvester Gardiner, 29 June 1757, box 1,

folder 17, Kennebec Papers; "The Kennebeck Company in Acct with Silvester Gardiner," 9 Mar. 1770, box 3, folder 1, Kennebec Papers.

78. Silvester Gardiner to William Jepson, 13 Apr. 1761, case 7, box 29, folder 7, Gratz Collection; Silvester Gardiner to William Jepson, 29 Apr. 1762, case 7, box 29, folder 7, Gratz Collection (quotation).

79. *True State of the Copartnership,* 3 ("Produce of that Colony"); Silvester Gardiner to William Jepson, 12 Feb. 1767, case 7, box 29, folder 12, Gratz Collection; Silvester Gardiner to William Jepson, 13 Apr. 1761, case 7, box 29, folder 7, Gratz Collection; Silvester Gardiner to William Jepson, 10 June 1762, case 7, box 29, folder 7, Gratz Collection; Silvester Gardiner to William Jepson, 29 Dec. 1760, case 7, box 29, folder 6, Gratz Collection; Silvester Gardiner to William Jepson, 28 Mar. 1764, case 7, box 29, folder 9, Gratz Collection; Silvester Gardiner to William Jepson, 29 Apr. 1762, case 7, box 29, folder 7, Gratz Collection; Silvester Gardiner to William Jepson, 27 May 1762, case 7, box 29, folder 7, Gratz Collection. On the currency situation, see McCusker, *Money and Exchange,* 135; Brock, *Currency of the American Colonies,* 289–90, 318–20.

80. Silvester Gardiner to William Jepson, 2 May 1760, case 7, box 29, folder 6, Gratz Collection; Silvester Gardiner to William Jepson, 24 Nov. 1760, case 7, box 29, folder 6, Gratz Collection.

81. Silvester Gardiner to William Jepson, 24 Oct. 1765, case 7, box 29, folder 10, Gratz Collection; Silvester Gardiner to William Jepson, 27 May 1762, case 7, box 29, folder 7, Gratz Collection (quotations); NOSL, pt. 3, box 3, pp. 1051–51A, 1074–74A.

82. Silvester Gardiner to William Jepson, 29 Oct. 1766, case 7, box 29, folder 11, Gratz Collection.

83. Silvester Gardiner to William Jepson, 23 June 1763, case 7, box 29, folder 8, Gratz Collection.

84. Donald Cummings to Silvester Gardiner, 2 Jan. 1770, box 1, folder 3, Gardiner MSS; Donald Cummings to Silvester Gardiner, 9 Dec. 1767, box 1, folder 2, Gardiner MSS.

85. Donald Cummings to Silvester Gardiner, 10 June 1766, box 1, folder 2, Gardiner MSS.

86. Donald Cummings to Silvester Gardiner, 8 May 1766, box 1, folder 2, Gardiner MSS.

87. Samuel Watts in Account with Silvester Gardiner 1758–1772, Jan. 1772, vol. 2, p. 101 oversized, Chelsea (MA) Papers, MS N-2008, Massachusetts Historical Society (quotations); Donald Cummings to Silvester Gardiner, 2 Jan. 1770, box 1, folder 3, Gardiner MSS.

88. Silvester Gardiner to William Jepson, 31 Oct. 1765, case 7, box 29, folder 10, Gratz Collection.

89. Silvester Gardiner to William Jepson, 23 June 1763, case 7, box 29, folder 8, Gratz Collection; Silvester Gardiner to William Jepson, 27 Sept. 1764, case 7, box 29, folder 9, Gratz Collection; Silvester Gardiner to William Jepson, 1 Aug. 1765, case 7, box 29, folder 10, Gratz Collection.

90. Albion, *Forests and Sea Power,* 31, 230–80; Barry and Peabody, *Tate House,* 11–12. These policies were enumerated in the Massachusetts Charter of 1691 and strengthened by legislation in 1711, 1722, and 1729.

91. Scammell to Wentworth, 20 July 1772, T 1/496/62–63, Treasury Board Papers; Albion, *Forests and Sea Power,* 270–72.

92. Kershaw, *Kennebeck Proprietors,* 201–2, 206; Albion, *Forests and Sea Power,* 258–59, 267.

93. Silvester Gardiner to William Jepson, 31 Oct. 1765, case 7, box 29, folder 10, Gratz Collection.

94. Notes from the author's visits to the Pownalborough Court House in Dresden, Maine, managed by the Lincoln County (ME) Historical Association. For other strategies of circumventing the policy, see Roberts, *Pines, Profits, and Politics*, 99.
95. Kershaw, *Kennebeck Proprietors*, 220–23.
96. Lumber remained a crucial part of the New England economy even though it was often overshadowed by southern staples among colonial exports. Ledgers of Imports and Exports, America, CUST 16/1, National Archives of the UK; Albion, *Forests and Sea Power*, 259–60.
97. Albion, *Forests and Sea Power*, 32–33, 38; Stearns, *American Herbal*, 258–59, 333; Lacey, *Micmac Medicines*, 36, 55; Chamberlain, "Plants Used." Stearns also discusses the resin of the tacamahaca tree, a balsam poplar, "employed by the Indians externally for . . . tumours, abating pains of the limbs" (Stearns, *American Herbal*, 322).
98. Recipes of Dr. Gardiner, B MS c50.4, Hartshorn MSS; Estes, "Therapeutic Practice," 343–44; Belknap, *History of New-Hampshire*, 73–94.
99. Albion, *Forests and Sea Power*, 273–75; "Timothy Parsons to Samuel P. Savage" (quotations).
100. Silvester Gardiner to William Jepson, 30 May 1765, case 7, box 29, folder 10, Gratz Collection; Silvester Gardiner to William Jepson, 18 July 1765, case 7, box 29, folder 10, Gratz Collection; Silvester Gardiner to William Jepson, 13 Feb. 1766, case 7, box 29, folder 11, Gratz Collection; Silvester Gardiner to William Jepson, 2 Oct. 1766, case 7, box 29, folder 11, Gratz Collection; Silvester Gardiner to William Jepson, 26 Nov. 1767, case 7, box 29, folder 12, Gratz Collection; Roberts, "American Potash Manufacture," 392–93.
101. Silvester Gardiner to William Jepson, 1 June 1768, case 7, box 29, folder 13, Gratz Collection.
102. Silvester Gardiner to William Jepson, 2 Oct. 1766, case 7, box 29, folder 11, Gratz Collection; Kershaw, *Kennebeck Proprietors*, 58; Silvester Gardiner to William Jepson, 18 July 1765, case 7, box 29, folder 10, Gratz Collection (quotation).
103. Silvester Gardiner to William Jepson, 16 Aug. 1764, case 7, box 29, folder 9, Gratz Collection; Silvester Gardiner to William Jepson, 5 Dec. 1764, case 7, box 29, folder 9, Gratz Collection ("no more right"); Silvester Gardiner to William Jepson, 25 Apr. 1765, case 7, box 29, folder 10, Gratz Collection; Silvester Gardiner to William Jepson, 12 Jan. 1758, case 7, box 29, folder 5, Gratz Collection ("one moment more"). For other examples of suits as part of medical disputes, see Crane, *Poison Plot*.
104. Silvester Gardiner to William Jepson, 19 Feb. 1762, case 7, box 29, folder 7, Gratz Collection.
105. Silvester Gardiner to William Jepson, 26 Feb. 1759, case 7, box 29, folder 5, Gratz Collection; Silvester Gardiner to William Jepson, 23 June 1763, case 7, box 29, folder 8, Gratz Collection; Silvester Gardiner to William Jepson, 7 June 1764, case 7, box 29, folder 9, Gratz Collection.
106. *True State of the Copartnership*, 1, 4.
107. Silvester Gardiner to William Jepson, 9 Oct. 1763, case 7, box 29, folder 8, Gratz Collection; Silvester Gardiner to William Jepson, 25 Apr. 1765, case 7, box 29, folder 10, Gratz Collection; Silvester Gardiner to William Jepson, 10 July 1765, case 7, box 29, folder 10, Gratz Collection.
108. Silvester Gardiner to William Jepson, 7 Dec. 1769, case 7, box 29, folder 13, Gratz Collection; Silvester Gardiner to William Jepson, 12 Feb. 1767, case 7, box 29, folder 12, Gratz Collection; *True State of the Copartnership*, 4.

109. Silvester Gardiner to William Jepson, 5 Dec. 1764, case 7, box 29, folder 9, Gratz Collection.
110. On the anger building in Maine, see Taylor, *Liberty Men*.
111. Bartlet, *Frontier Missionary*, 112, 351 (quotations); for more details on Bailey, see Leamon, *Reverend Jacob Bailey*; Allen, "Loyalists of the Kennebec," 625. Religious and political tensions continued in the Kennebec; see Vassall Family Papers (1768–1790), MS S-111, Massachusetts Historical Society.
112. Following Ann Gibbins Gardiner's death in 1771, Gardiner married the widow Abigail Pickman Eppes. Milford, *Gardiners of Massachusetts*, 37.
113. Carbone et al., *American Paintings in the Brooklyn Museum*; additional provenance information courtesy Seattle Art Museum.

CHAPTER FIVE

1. For Philadelphia's medical and trading communities, see Brandt, "Women's Medical Entrepreneurship."
2. Fort St. George Military Department to the Honorable Court of Directors, 9 Oct. 1800 in IOR/E/4/327, India Office Records; Fort St. George Military Department to the Honorable Court of Directors, 18 Mar. 1801 in IOR/E/4/328, India Office Records.
3. CUST 3; Commerce Journals, 1735–90, IOR/L/AG/1/6/11–20, India Office Records; To Bombay Public Department, 23 Mar. 1787 in IOR/E/4/1004, p. 109, India Office Records. The EIC recorded its spending on medicines in the Commerce Journals, while official values of medicine exports were compiled in the government's customs ledgers (CUST 3). Despite nominally tracking the same thing—medicines sent to Company settlements—the values in the EIC's Commerce Journals are larger than the official values in the customs ledgers. Since the values in the two sources were calculated using different constants, comparisons of the Company and official export values cannot be quantified. For a lengthier discussion of these datasets, see Dorner, "Manufacturing Pharmaceuticals," app. C.
4. For medical self-sufficiency in the Russian context, see Monahan, "Locating Rhubarb," 234. For concurrent debates about the uses of nature in political economy and local manufacture, see Jonsson, *Enlightenment's Frontier*; Koerner, *Nature and Nation*.
5. Schiebinger, *Plants and Empire*, 73.
6. Chakrabarti, *Materials and Medicine*, 105.
7. 18 Oct. 1765 in IOR/P/240/23, pp. 491–92, Madras Proceedings, India Office Records. For Company spending on medicine exports at this time, see Commerce Journals, 1735–90, IOR/L/AG/1/6/11–20, India Office Records.
8. 18 Oct. 1765 in IOR/P/240/23, pp. 494–95, Madras Proceedings, India Office Records (quotations); "General Letter to the Honble the Court of Directors," 18 Oct. 1765 in IOR/E/4/301, f. 180r, India Office Records.
9. For a richer discussion of this topic, see Drayton, *Nature's Government*; Irving, *Natural Science*.
10. Baber, *Science of Empire*, 166–68.
11. James Anderson to Andrew Berry, 25 Oct. 1789 in IOR/P/241/15, pp. 3188–95 ("plants used here," p. 3195), India Office Records; James Anderson to Governor & Council of Madras, 17 Nov. 1789 in IOR/P/241/15, pp. 3185–87 ("such useful plants," p. 3187), India Office Records.
12. For other gardens with medicinal plants for political-economic purposes, see Dorner, "Medicines and Mercantilism."

13. General Letter to Fort St. George, 4 Mar. 1767 in IOR/E/4/863, pp. 509–10, India Office Records.
14. General Letter to Fort St. George, 21 Nov. 1766 in IOR/E/4/863, p. 437, India Office Records.
15. General Letter to Fort St. George, 4 Mar. 1767 in IOR/E/4/863, p. 495, India Office Records; Commerce Journals, 1735–90, IOR/L/AG/1/6/11–20, India Office Records.
16. 5 Feb. 1774 in IOR/P/240/37, p. 78, Madras Proceedings, India Office Records.
17. Fort St. George Military Department to the Honorable Court of Directors, 9 Oct. 1800 in IOR/E/4/327, India Office Records.
18. Chakrabarti, "Medicine amidst War," 10–16 (esp. 13), 19; Chakrabarti, *Materials and Medicine*, 37–39, 41. The spice trade in which Europeans had participated since the sixteenth century also laid the groundwork for the networks of bazaar medicine that developed later. For details about the country trade, see Chaudhuri, *Trading World of Asia*, 191–213. See also Holden Furber's work on the expansion of Company trade along the Coromandel Coast and in Bengal during the eighteenth century (Furber, *Rival Empires of Trade*). Timothy Walker and Hugh Cagle point to the long-standing roles of female folk healers in the Lusophone Atlantic and Indian Ocean worlds; see Owens and Mangan, *Women of the Iberian Atlantic*.
19. Chakrabarti, *Medicine and Empire*, 14.
20. Chakrabarti, "Medicine amidst War," 17–18, 23, 25–26.
21. Censure of Madras Medical Board, Extract of Fort St. George Military Consultations, 29 May 1804 in IOR/F/4/184/3700, pp. 5 (quotation), 11, India Office Records; Censure of Madras Medical Board, Extract of a letter from the Deputy Medical Storekeeper to the superintending Surgeon Northern Division, 25 Dec. 1803 in IOR/F/4/184/3700, pp. 101–2, India Office Records.
22. Censure of Madras Medical Board, James Gilmour to Alexander Boswell, 27 Nov. 1803 in IOR/F/4/184/3700, pp. 62–63, India Office Records.
23. Charters, *Imperial State*, 148.
24. To Bombay Public Department, 8 Apr. 1789 in IOR/E/4/1005, p. 216, India Office Records.
25. Censure of Madras Medical Board, Extract from Fort St. George Military Consultations, 19 June 1804 in IOR/F/4/184/3700, pp. 154–55, India Office Records.
26. Censure of Madras Medical Board, "List and Rates of Country Medicines &c supplied by the Medical Storekeeper" in IOR/F/4/184/3700, pp. 163–67 ("Market Prices," p. 163) India Office Records; Censure of Madras Medical Board, "Mr. Chamier (of the Medical Board) records his objections to each of the Resolutions from the Censure," 5 June 1804 in IOR/F/4/184/3700, pp. 134–36, India Office Records.
27. "A comparative Statement of the Prices of Europe Medicines charged by the Surgeons with their Price in the Companys Invoice," ca. 1781 in IOR/H/159, pp. 199–201, India Office Records.
28. Chakrabarti, "Medicine amidst War," 10–12, 19.
29. "A comparative Statement of the Prices of Europe Medicines charged by the Surgeons with their Price in the Companys Invoice," ca. 1781 in IOR/H/159, pp. 199–201, India Office Records.
30. To Madras Military Department, 8 July 1795 in IOR/E/4/881, p. 1035, India Office Records.
31. To Bombay Military Department, 30 June 1797 in IOR/E/4/1012, ff. 611–12, India Office Records.
32. New regulations were disseminated in two papers: "Regulations respecting Military

Hospitals in India" and "Forms & Regulations for the Apothecary in Charge of the Medicine Stores at the different Presidencies in India." Bombay General Letter, 21 Sept. 1785 in IOR/E/4/1002, pp. 1033–37, India Office Records; To Fort St. George Military Department, 21 May 1794 in IOR/E/4/880, pp. 521–22, India Office Records; To Madras Military Department, 8 July 1795 in IOR/E/4/881, p. 1035, India Office Records. For these broader trends, see Chakrabarti, *Medicine and Empire*, xii, 107; Chakrabarti, *Materials and Medicine*, 100–102; Charters, *Imperial State*, 12–16, 84, 142, 149–50; Harrison, "Treatment of Fevers," 109–10; Chakrabarti, "Medicine amidst War," 34–35; Mukharji, *Nationalizing the Body*; India Orders (1827–1828), MS 8261, Worshipful Society of Apothecaries.

33. For exports to Philadelphia, see McCusker and Menard, *Economy of British America*, 189–90, 208; CUST 3. On the city's medical community, see Tolles, *Meeting House and Counting House*, 226–28; Brandt, "Women's Medical Entrepreneurship," 798–800.
34. Kremers, "Historical Fragments," 682; Wilson, "Trading in Drugs," 361.
35. Financial Records Collection, Section I, Series 2, Records of the Pennsylvania Hospital, Pennsylvania Hospital Historic Collections (hereafter cited as PA Hospital Collection); Morton, *Pennsylvania Hospital*, 25. On the hospital's founding, see Tolles, *Meeting House and Counting House*, 228–29; Franklin, *Account of the Pennsylvania Hospital*, 3–7 ("for the Reception," p. 5), 39–40; Gordon, *Æsculapius*, 450–56. Though some large almshouses existed in the American colonies during the eighteenth century, there were few formal institutions devoted to poor relief. For example: Katz, *Shadow of the Poorhouse*, chap. 1; Lyons, *Sex among the Rabble*; Rockman, *Scraping By*, esp. chap. 7. For Corbyn's correspondence with Evans while he was in Jamaica, see chap. 2, notes 21, 82.
36. Morton, *Pennsylvania Hospital*, 16–17, 24–25; Financial Records Collection, Section I, Series 2, Subseries K: Loan Office of 1773 (boxes 11, 56), PA Hospital Collection; Financial Records Collection, Section I, Series 2, Subseries M: PA Land Company (box 57), PA Hospital Collection.
37. Board of Managers, Minutes, vol. 1, pp. 65–66, 7 & 12 Dec. 1752, reel 1, Pennsylvania Hospital Archives, American Philosophical Society (hereafter cited as PA Hospital Papers).
38. Minutes, vol. 1, p. 233, 28 Mar. 1757, reel 1, PA Hospital Papers; Minutes, vol. 2, p. 17, 31 Oct. 1757, reel 1, PA Hospital Papers.
39. Silvanus & Timothy Bevan to Managers of the Pennsylvania Hospital, 20 June 1759, reel 28, PA Hospital Papers; Minutes, vol. 2, pp. 258–61, 26 Oct. 1761, reel 1, PA Hospital Papers; Silvanus & Timothy Bevan to Managers of the Pennsylvania Hospital, 11 July 1761, Minutes, vol. 2, pp. 258–61, reel 1, PA Hospital Papers; "List of Drugs & Medicines Sent to Timothy Bevan & Sons," 30 Jan. 1766, reel 28, PA Hospital Papers; Estes, *Dictionary of Protopharmacology*, 34, 42, 137; Brookes, *Introduction to Physic and Surgery*, 263, 324.
40. Managers to Silvanus & Timothy Bevan, 7 July 1764, reel 28, PA Hospital Papers; Morton, *Pennsylvania Hospital*, 358–59; Hospital's account with Silvanus & Timothy Bevan, 3 Oct. 1764, reel 28, PA Hospital Papers.
41. Timothy Bevan to Israel Pemberton Jr., 21 May 1768, vol. 20, p. 36, Pemberton MSS (quotations); Managers to Silvanus & Timothy Bevan, 20 May 1765, Minutes, vol. 3, pp. 118–19, reel 1, PA Hospital Papers.
42. Managers to Silvanus & Timothy Bevan, 20 May 1765, Minutes, vol. 3, pp. 118–19, reel 1, PA Hospital Papers.
43. Timothy Bevan to Israel Pemberton Jr., 26 July 1765, Minutes, vol. 3, pp. 151–52, reel 1, PA Hospital Papers.

44. Minutes, vol. 3, pp. 501, 509, 29 Aug. 1768, reel 1, PA Hospital Papers; Minutes, vol. 4, p. 55, April 1770, reel 2, PA Hospital Papers.
45. Timothy Bevan to Israel Pemberton Jr., 13 Feb. 1771, vol. 22, p. 101, Pemberton MSS (quotation); Minutes, vol. 4, pp. 55, 115, 167, 238, 280, 286, 350, 407, reel 2, PA Hospital Papers; "List of Medicines wanting for the Hospital," 1774, reel 28, PA Hospital Papers; Estes, *Dictionary of Protopharmacology*, 55, 82–83; Lewis, *New Dispensatory*, 6th ed., 442.
46. "List of Medicines sent out," Oct. 1773, reel 28, PA Hospital Papers.
47. "Inventory of drugs, medicines, chymical apparatus, etc.," 30 Mar. 1773, reel 28, PA Hospital Papers.
48. Christopher Marshall's Bills of Lading (1762–1768), Am.916, Historical Society of Pennsylvania.
49. Lawall and Lawall, "Revolutionary Account Book," 302; Kremers, "Historical Fragments," 683, 685–90.
50. On the significance of botanical gardens in various facets of the broader imperial project, see Drayton, *Nature's Government*; Grove, *Green Imperialism*.
51. Wilson, "Trading in Drugs," 360; Kremers, "Historical Fragments," 691; Isaac Bartram Account Book (1790–1803), MS 3371, William L. Clements Library (hereafter cited as Bartram Account Book); Brandt, "Women's Medical Entrepreneurship," 801–2; *Pennsylvania Journal; and the Weekly Advertiser*, 20 Feb. 1772, [4] (quotation).
52. Brandt, "Women's Medical Entrepreneurship," 775–76, 779, 783–84, 799–802. Women faced other structural disadvantages in the medicine trade, such as the legal challenges to inheriting property.
53. Price, "Credit in the Slave Trade," 319, 323, 326.
54. Smith, *Wealth of Nations*, 567–68.
55. Nash, "Organization of Trade," 104, 123–27; CUST 3.
56. Sheridan, "Credit Crisis," 162–64.
57. Atton and Holland, *King's Customs*, 204, 272, 276; Holmes and Szechi, *Age of Oligarchy*, 285–87; Brewer, "Wilkites and the Law," 137–38; Eacott, *Selling Empire*, 136, 156–58. For a fuller picture of the period, see Griffin, *Townshend Moment*.
58. Davis, "Rise of Protection," 308, 317.
59. Schumpeter, *English Overseas Trade*, 71; Wallis, "Exotic Drugs," 23; Davis, "Rise of Protection," 307–8.
60. Report on Two Memorials, 21 Mar. 1769, T 1/469/157–158, Treasury Board Papers. On smuggling after the Seven Years' War, see Klooster, "Inter-Imperial Smuggling," 172–75. Compare the prices of jalap, 3/- in September 1768 and 2/6 in June 1770, and ipecac, 7/3 in September 1768 and 6/6 in October 1769, to the rates in February 1769 (see table 5.1). Jalap: 6/- (Sept. 1751), 4/6 (May 1764), 3/- (Sept. 1768), 2/6 (June 1770); Ipecac: 12/- (May 1764), 7/3 (Sept. 1768), 6/6 (Oct. 1769). Price data compiled from MSS 3181, 5439, 5441–42, 5448, Wellcome Library.
61. Report on Two Memorials, 21 Mar. 1769, pp. 195–96, T 1/470/194–199, Treasury Board Papers.
62. Report on Two Memorials, 21 Mar. 1769, T 1/469/157–158, Treasury Board Papers.
63. Report on Two Memorials, 21 Mar. 1769, pp. 195–96, T 1/470/194–199, Treasury Board Papers.
64. O'Brien, "Central Government"; Davis, "Rise of Protection," 311–12. On 16 April 1744, the House of Commons passed a bill allowing the same drawbacks on refined borax and camphor as were given on unrefined (CJ XXIV, 655: 16 Apr. 1744).

65. Report on Two Memorials, 21 Mar. 1769, p. 194, T 1/470/194–199, Treasury Board Papers.
66. Sheridan, "Credit Crisis," 170, 173; Adams, *Works of John Adams*, 265 (quotation).
67. Sheridan, "Credit Crisis," 167, 169–70, 174–76, 184. For medicine exports during this period, see CUST 3/74–76.
68. Resolution of Non-Importation Made by the Citizens of Philadelphia, 25 Oct. 1765, Am.340, Historical Society of Pennsylvania; Brandt, "Women's Medical Entrepreneurship," 786; Merritt, "Tea Trade"; Oaks, "Philadelphia Merchants." On the Stamp Act, see Anderson, *Crucible of War*, chap. 66.
69. "Whereas this province labours under a heavy debt," Houghton Library, Harvard University; "Address of the Merchants and Others, of Boston to Gov. Hutchinson," 30 May 1774, pp. 123–25, in Stark, *Loyalists of Massachusetts*.
70. *Boston Chronicle*, 17 Aug. 1769, [262]; *Boston Chronicle*, 21 Aug. 1769, [265].
71. Cleary, *Elizabeth Murray*, 101, 133, 140–42, 148–49; *Boston-Gazette, and Country Journal*, 22 Aug. 1768, [3]; *Boston Evening-Post*, 11 Dec. 1769, [4]; *Boston Evening-Post*, 24 Dec. 1770, [1]. Eustis announced her departure on the front page of the *Boston Gazette* (*Boston-Gazette, and Country Journal*, 6 Nov. 1769, [1]).
72. "A List of the Sons of Liberty who dined at the Liberty Tree Tavern in Dorchester, August 14, 1769," Misc. Bound Manuscripts, Massachusetts Historical Society. For accounts of the 1769 celebration, see Hutchinson, *History of the Colony*, 88–89, 146; Butterfield, *Diary and Autobiography of John Adams*, 341–42; *Boston-Gazette and Country Journal*, 21 Aug. 1769, [1] (quotations). The list of sixty-one can be found in "February Meeting, 1898," 139–42. On the 1765 riots, see Anderson, *Crucible of War*, 664–71.
73. Brimmer Accounts; Stark, *Loyalists of Massachusetts*, 123–40.
74. Milford, *Gardiners of Massachusetts*, 37; Kershaw, *Kennebeck Proprietors*, 223, 275–76, 284. After some loyalist proprietors fled in 1776, the company continued to meet, however, now with little control over the Kennebec Valley.
75. Stark, *Loyalists of Massachusetts*, 313–18; "List of the addresses to the late Gov. Hutchinson" (quotations).
76. Gardiner's estimates of his losses, 1783, II.15, GWA Papers; "Committee's Return of Leasing Out an Estate Belonging to a Tory," 30 Oct. 1776, vol. 154, p. 72, MA Archive Collection; Milford, *Gardiners of Massachusetts*, 37–41. The *Newport Mercury* (14 Aug. 1786, [2]) carried his obituary, noting "in the Line of his chirurgical and medical Profession he long stood foremost" and his "uncommon Vigour and Activity of Mind, and by unremitting Diligence and Attention acquired a large Property." The cenotaph was written in Latin and erected in black marble by Gardiner's grandson, Robert Hallowell Gardiner (Hanson, *History of Gardiner*, 91–92).
77. Parsons, *Rhode Island Physicians*, 5–7.
78. CUST 3/76–79.
79. Petty Accounts (1775–1776), vol. 11, Greenleaf MSS.
80. *Boston Evening-Post*, 10 Dec. 1764, [4]; *Boston Evening-Post*, 14 May 1764, [4]; Brimmer Accounts; *Boston Evening-Post*, 4 July 1768, [4]; *Boston Evening-Post*, 5 Nov. 1770, [4]; *Boston Evening-Post*, 18 Jan. 1773, [4]; *Continental Journal & Weekly Advertiser*, 22 Aug. 1776, [2].
81. Sonnedecker, *History of Pharmacy*, 166.
82. Brandt, "Women's Medical Entrepreneurship," 791; Klooster, "Inter-Imperial Smuggling," 177.

83. McNeill, *Mosquito Empires*, 4, 200–203, 211–12. To compare death rates, see Duncan, *Medical Men*, 371.
84. Notebook containing invoices for medicines and materials for the use of regiments in America, 1774, Add. MS 73622, vol. 77, Barrington Papers; Notebook containing a copy of medicines and materials delivered by the Apothecary General of the Army for the Service of H.M.'s Hospitals in America, 1777, Add. MS 73623, vol. 78, Barrington Papers.
85. Griffenhagen, "Drug Supplies," 129–30; Brown, "Healing Arts," 44–45; Warden, "Medical Profession," 153; Frey, *British Soldier*, 47–52; McNeill, *Mosquito Empires*, 212.
86. Petty Accounts (1775–1776), vol. 11, Greenleaf MSS.
87. Griffenhagen, "Drug Supplies," 129–30; Petty Ledger (1773–1777), vol. 10, Greenleaf MSS. On the prize ships, see *Independent Chronicle and the Universal Advertiser*, 10 Apr. 1777, [3]; *Continental Journal, and Weekly Advertiser*, 24 Apr. 1777, [2].
88. Petty Ledger (1773–1777), vol. 10, Greenleaf MSS; Gardiner's estimates of his losses, 1783, II.15, GWA Papers.
89. Petty Ledger (1773–1777), vol. 10, Greenleaf MSS; Sonnedecker, *History of Pharmacy*, 163–64. French medical practitioners also shared knowledge, see Cash, "Professionalization of Boston Medicine," 82.
90. "Philadelphia Business of the Olden Time," 179; Lawall and Lawall, "Revolutionary Account Book," 302.
91. Christopher Junr. and Charles Marshall Waste Book (1776), vol. 15, Christopher Marshall Papers, coll. 395, Historical Society of Pennsylvania (hereafter cited as Marshall MSS); Add MSS 73622–23, vols. 77–78, Barrington Papers; Griffenhagen, "Drug Supplies," 130–33.
92. Benjamin Marshall & Brothers Waste Book (1775–1796), Daniel Parker Papers, coll. 1587, Historical Society of Pennsylvania (hereafter cited as Parker MSS); Griffenhagen, "Drug Supplies," 129–30.
93. Sonnedecker, *History of Pharmacy*, 163–65, 167; Griffenhagen, "Drug Supplies," 128–29. Inflation in 1778 lessens the scale of this spending increase, but it nonetheless represented a change from previous years.
94. Petty Ledger (1773–1777), vol. 10, Greenleaf MSS.
95. Brandt, "Women's Medical Entrepreneurship," 791–93.
96. Minutes, vol. 5, pp. 14, 59, 103, 157, 205, 251, 305, 351, reel 2, PA Hospital Papers; "Inventory of the Drugs, Medicines, Instruments, & Shop Furniture belonging to the Pennsylvania Hospital," 3 Dec. 1782, reel 28, PA Hospital Papers. An inventory taken by James Hartley in 1784 confirms that more medicines were being made at the hospital as the 1780s wore on; see "Inventory of the Drugs, Medicines, Instruments, and Shop Furniture belonging to the Pennsylvania Hospital," 5 May 1784, reel 28, PA Hospital Papers.
97. Minutes, vol. 5, pp. 286, 350, reel 2, PA Hospital Papers.
98. Minutes, vol. 6, pp. 39, 93, 149, 159, 215, 255, 295, 356, 416, 478, 533, reel 2, PA Hospital Papers; Christopher Jr. & Charles Marshall to Managers, 22 Oct. 1787, reel 28, PA Hospital Papers.
99. Brown, *Pharmacopoeia simpliciorum et efficaciorum*; Kremers, "Lititz Pharmacopoeia," 5, 28–29; Sonnedecker, *History of Pharmacy*, 169–70; Griffenhagen, "Drug Supplies," 130.
100. Griffenhagen and Bogard, *Drug Containers*, 107–14; McCusker and Menard, *Economy of British America*, 363.

101. *OED Online*, s.v. "revolution," June 2017. Ralph Davis used the phrase "a revolution in trade" referring to the growth of English trade in the second half of the seventeenth century; see Davis, "English Foreign Trade, 1660–1700," 162.
102. Stander, "Transatlantic Trade in Pharmaceuticals," 333; Joseph Gurney Bevan to John Freebastin, 29 Apr. 1778, AH246, Allen & Hanburys MSS; Joseph Gurney Bevan to Barwick Bruce, 2 Dec. 1785, AH031, Allen & Hanburys MSS.
103. For an overview of this new political and economic arrangement, see the essays in Gould and Onuf, *Empire and Nation*; also, Crouzet, "Crisis of the British Imperial Economy," 290–96, 305–15.
104. Joseph Gurney Bevan to Drummond & Kincastle, 2 Feb. 1778, AH027, Allen & Hanburys MSS; Joseph Gurney Bevan to John Rocke, 4 Nov. 1778, AH246, Allen & Hanburys MSS.
105. Joseph Gurney Bevan to William Hutchinson, 4 Oct. 1781, AH033, Allen & Hanburys MSS.
106. Commutation for Duties on Sundry Drugs, 10 Feb. 1786, ff. 52r ("it may be said"), 57r, Add MS 38409, vol. 220, Liverpool Papers, Archives and Manuscripts, British Library (hereafter cited as Liverpool Papers). Parliament removed some of the discounts allowed on customs duties in 1781, which necessitated a rewriting of portions of the *Book of Rates*, but these drugs seem to have been unaffected (Atton and Holland, *King's Customs*, 311, 314). On smuggling rates, see Schumpeter, *English Overseas Trade*, 6–7; Cole, "Trends in Eighteenth-Century Smuggling"; Mui and Mui, " 'Trends in Eighteenth-Century Smuggling' Reconsidered."
107. Commutation for Duties on Sundry Drugs, 10 Feb. 1786, ff. 56r, 53v, Add MS 38409, vol. 220, Liverpool Papers. William Pitt the Younger's Commutation Act in 1784 had improved some rates (notably the duties on tea) and offered the possibility of fixing what many saw as an outdated customs system. Evans, *William Pitt the Younger*, xiii; Mui and Mui, "Enforcement of the Commutation Act."
108. Griffenhagen, *Medicine Tax Stamps*, 3. The United States applied its own stamp tax on proprietary medicine and toiletries in 1862 to raise revenue to meet the mounting expenses of the American Civil War.
109. Joseph Gurney Bevan to Gera Alton, 9 Oct. 1788, AH035, Allen & Hanburys MSS; *Medical Register for the Year 1783*, 170.
110. Petition of George Webster, 24 July 1772, T 1/495/21–22, Treasury Board Papers.
111. See chapter 5, note 60.
112. Invoice Book (1753–1806), pp. 25, 29, 35–37, 49, 108–13, MS 5878, Primatt & Maud MSS; Invoices, accounts, drafts of out-letters, and rough notes and calculations, 1768–1800, MS 5880, Primatt & Maud MSS; Johann Georg Schneider to Thomas Corbyn & John Brown, 5 May 1778, MS 5439/4/9, Corbyn & Co. MSS; Manufacturing Recipe Book (ca. 1782), MS 5447, Corbyn & Co. MSS; Add MS 38409, vol. 220, Liverpool Papers.
113. Joseph Gurney Bevan to Warner & Blizzard, 26 May 1784, AH031, Allen & Hanburys MSS.
114. Joseph Gurney Bevan to Thomas Brown, 2 July 1782, AH033, Allen & Hanburys MSS.
115. Joseph Gurney Bevan to James Malcolm, 2 July 1782, AH033, Allen & Hanburys MSS; Stander, "Transatlantic Trade in Pharmaceuticals," 329. On mortgages, see Price, "Credit in the Slave Trade," 324; Martin, "Slavery's Invisible Engine," 818–19, 839–40. Opinions differ as to where the money came from that underwrote the expansion of the sugar economy in the eighteenth century. Richard Pares argued that reinvested

plantation profits and short-term mercantile credit fed expansion, whereas Adam Smith and S. D. Smith emphasized long-term capital imports from Britain. They all agreed that short-term credit extensions by merchants to planters sustained plantation agriculture, but disagreed on the use of mortgages over the longer term. For a summary of this debate, see Smith, *Gentry Capitalism*, 140–42.

116. See Turner, *Contested Bodies*, esp. chaps. 2–3; Wood and Clayton, "Golden Grove," 113; Craton, *Invisible Man*, 119, 398–99; Paugh, *Politics of Reproduction*, esp. chaps. 1, 5.

117. Joseph Gurney Bevan to Samuel Forte, 4 Oct. 1780, AH033, Allen & Hanburys MSS; Plough Court Ledger (1781–1784), ff. 283–84, 334, 342, AH207, Allen & Hanburys MSS.

118. Plough Court Ledger (1781–1784), ff. 13, 39–40, AH207, Allen & Hanburys MSS. On the Lascelles family, see Smith, *Gentry Capitalism*, 58–59, 74–78.

119. Benjamin Marshall & Brothers Waste Book (1775–1796), Parker MSS; Joseph Gurney Bevan to Christopher Jr. & Charles Marshall, 31 May 1784, AH031, Allen & Hanburys MSS; Joseph Gurney Bevan to Christopher Jr. & Charles Marshall, 18 Aug. 1786, AH035, Allen & Hanburys MSS; Joseph Gurney Bevan to Christopher Jr. & Charles Marshall, 13 July 1787, AH035, Allen & Hanburys MSS; Joseph Gurney Bevan to Christopher Jr. & Charles Marshall, 1 Dec. 1789, AH035, Allen & Hanburys MSS; Joseph Gurney Bevan to Christopher Jr. & Charles Marshall, 7 Aug. 1794, AH028, Allen & Hanburys MSS.

120. Christopher Junr. & Charles Marshall Waste Book (1797–1804), Parker MSS; Joseph Gurney Bevan to Christopher Jr. & Charles Marshall, 2 Nov. 1791, AH028, Allen & Hanburys MSS; Joseph Gurney Bevan to Christopher Jr. & Charles Marshall, 16 July 1792, AH028, Allen & Hanburys MSS.

121. Bartram Account Book; *Biographical Annals of Lancaster County Pennsylvania*, 669.

122. *New Complete Guide*, 214; *Compleat Compting-House Companion*, 192; Bartram Account Book; Pountney, *Old Bristol Potteries*, 253; Joseph Gurney Bevan to Isaac Bartram, 3 Dec. 1789, AH035, Allen & Hanburys MSS (quotation).

123. *Catalogus Medicamentorum Chemicorum*, 1764, Worshipful Society of Apothecaries; Turlington's Balsam Tracts, ca. 1800, 111829.O.1–2, Helfand Collection; Estes, "Therapeutic Practice," 344–45. On oversight, see Sonnedecker, *History of Pharmacy*, 181.

124. See Abrams, *Revolutionary Medicine*, chap. 5, though the account emphasizes notable men and women in shaping the medical discourse of the new nation.

125. Peter Bartlett to William Stearns, 27 May 1783, MS 5855/1, William Stearns Papers, Wellcome Library (hereafter cited as Stearns MSS); David Taylor & Sons to William Stearns, 26 Apr. 1803, MS 5855/4, Stearns MSS. On uses of these plants, see Stearns, *American Herbal*, 259–60; Elliot, *Medical Pocket-Book*, 56; Quincy, *Complete English Dispensatory*, 183. Other claims about snakeroot's medicinal potential appear in John Tennent's *Epistle to Dr. Richard Mead* (1738). For a few examples of snakeroot's and pink root's prevalence, see *Pennsylvania Journal; and the Weekly Advertiser*, 19 Apr. 1770, [3]; *Georgia Gazette*, 25 Apr. 1770, [3]; *Virginia Journal and Alexandria Advertiser*, 4 Aug. 1785, [1]; *Salem Chronicle, and Essex Advertiser*, 13 Apr. 1786, [2]; *Augusta Chronicle*, 12 May 1792, [3].

126. Heard Family Business Records, MS 766, Baker Library, Harvard Business School.

127. Wilson, "Trading in Drugs," 358, 361–63; Gifford, "Botanic Remedies," 285; Stearns, *American Herbal*.

128. Crowley, "Neo-Mercantilism," 347; Burnett, "Observations of London Merchants," 778–79; Fichter, *So Great a Proffit*, 35; McCusker and Menard, *Economy of British America*, 361, 371, 375. By some measurements, US exports shrank per capita from

1775 to 1790 but then experienced a period of expansion from 1790 to 1805. This trend continued into the first decade of the nineteenth century even though serious economic difficulties arose across the British empire. For other views of the postwar economic landscape, see Crouzet, "Crisis of the British Imperial Economy"; Emmer et al., *Deux Ex Machina Revisited*.

129. Account of pharmaceuticals received of Ebenezer Seccomb, ca. 1800, MS 5859/36, Stearns MSS; David Taylor & Sons to William Stearns, 26 Apr. 1803, MS 5855/4, Stearns MSS; David Taylor to William Stearns, 20 Aug. 1805, MS 5855/10, Stearns MSS.
130. For comparison, see Kupperman, "Fear of Hot Climates," 226–27n44.
131. Joseph Gurney Bevan to Armstrong & King, 14 Mar. 1780, AH033, Allen & Hanburys MSS; Plough Court Ledger (1781–1784), ff. 232, 354–68, AH207, Allen & Hanburys MSS.
132. See the Stearns and Sprague Family Papers, Phillips Library, Peabody Essex Museum.
133. Joseph Gurney Bevan to John Worrell, 6 Jan. 1789, AH035, Allen & Hanburys MSS; Joseph Gurney Bevan to James Waddell, 14 Feb. 1792, AH028, Allen & Hanburys MSS; Joseph Gurney Bevan to Samuel Bell, 8 Nov. 1794, AH028, Allen & Hanburys MSS (quotation).
134. On these nineteenth-century developments, see Starr, *Social Transformation*; McCoy, *Elusive Republic*; Baptist, *Making of American Capitalism*.

CONCLUSION

1. Commutation for Duties on Sundry Drugs, 10 Feb. 1786, f. 67r, Add MS 38409, vol. 220, Liverpool Papers; *Medical Register for the Year 1783*, 151–62, 169–72. By many accounts, British overseas trade and the annual net produce of the customs increased after passage of the Customs Consolidation Act of 1787, but the desires of many merchants of medicines largely went unattended. 27 Geo. III, c. 13, pp. 23–175, in Pickering, *Statutes at Large*, vol. 36; Atton and Holland, *King's Customs*, 372–73, 411–12.
2. Sales of rhubarb accounted for only 4.77 percent of the yearly Company sales from 1774 to 1783, whereas they accounted for 13.61 percent of the duties paid. Commerce Journal, 1 July 1735–30 June 1742, IOR/L/AG/1/6/11, p. 198, India Office Records; Commerce Journal, 1 July 1763–30 June 1769, IOR/L/AG/1/6/15, pp. 89, 130, 134, India Office Records; General Ledger, 1 July 1793–30 June 1796, IOR/L/AG/1/1/27, ff. 45–46, India Office Records; Atton and Holland, *King's Customs*, 284, 408; Commutation for Duties on Sundry Drugs, 10 Feb. 1786, f. 56r, Add MS 38409, vol. 220, Liverpool Papers. A parliamentary committee corroborated these fears: "Reports from the Committee on Illicit Practices," 291.
3. Robert Wigram to William Richardson, ca. 1800, MS 7226/1, Wellcome Library. On Wigram's career, see Lubbock, *Blackwall Frigates*.
4. Stander, "Transatlantic Trade in Pharmaceuticals," 327–28, 335–36; "Dr. Hooper's Female Pills," 111888.O.12, Helfand Collection; Turlington's Balsam Tracts, ca. 1800, 111829.O.1–2, Helfand Collection; Young, *Toadstool Millionaires*. Before 1800, American pharmacists imported most of their specialized glassware from Britain. The war had prevented the importation of bottles, so American products had to be put in reused British bottles. This practice continued after the war, especially in the case of patent medicines; see Griffenhagen and Bogard, *Drug Containers*, 75.
5. Competition from American shipping increased in the 1790s; see Crouzet, "Crisis of the British Imperial Economy," 281–83, 301–6. On the history of the medicine

trade in the nineteenth century, see Poynter, *Evolution of Pharmacy*; Zebroski, *Brief History of Pharmacy*; Chakrabarti, *Medicine and Empire*; Harrison, *Commerce and Empire*. For example, Papers Related to Society's Negotiations to Supply Army, 1819–21, Box 65/5, Worshipful Society of Apothecaries; India Orders (1827–1828), MS 8261, Worshipful Society of Apothecaries.

6. Scientific or medical institutions were not separated from the extractive drive of early modern empire as is often the case in the narratives championing innovation that emerged from the Enlightenment. Rather, they could not have developed as they did without the others. On such paradigms, see Cook, "Moving About," 102–5.
7. Frost, *Recollections of James Jenkins*, 225 (quotations); Harrison, *Memoir of William Cookworthy*, 132–33. On Corbyn's religious thoughts and gardening habits, see Thomas Corbyn to William Logan, 10 Mar. 1771, vol. 3, Logan MSS. On Corbyn's estate, see "Probate of the Will of Thomas Corbyn, deceased, 25 Feb. 1791," CLC/B/136/MS18768, Corbyn Estate MSS.
8. Letters (1791–1822), CLC/B/136/MS18768, Corbyn Estate MSS; Porter and Porter, "English Drugs Industry," 293; Foreign Letter Book (1809–1851), MS 5443, Corbyn & Co. MSS.
9. Deed of partnership of William Allen and Luke Howard, 8 Mar. 1798, ACC/1037/1, Howards and Sons Limited Records, London Metropolitan Archives. On Allen's and Howard's other activities, see Fayle, *Spitalfields Genius*, 36–37; Howard, *Modifications of Clouds*; Allen, *West India Produce*.
10. Letters from Suppliers and Export Letters (1795–1823), AH128, Allen & Hanburys MSS; Order Book No. 12 (Jan.–June 1801), AH140, Allen & Hanburys MSS; Plough Court Ledgers, AH207, AH226, AH228, Allen & Hanburys MSS. The pharmacy would soon become connected to the wealthy, transatlantic tobacco merchant Osgood Hanbury under the name Allen & Hanburys (Price, *Capital and Credit*, 21).
11. Chapman-Huston and Cripps, *Through a City Archway*; Cripps, *Plough Court*; Richmond, Stevenson, and Turton, *Pharmaceutical Industry*.
12. Chakrabarti, *Medicine and Empire*, 48–51.
13. Mark Harrison has been down this path before, though he emphasizes the influence of British colonial expansion on European "rational" medicine; see Harrison, *Commerce and Empire*, 2–4. Suman Seth has discussed the contributions of medicine to several categories of difference central to colonialism in the later eighteenth century, see Seth, *Difference and Disease*, 11–12.
14. Baucom, *Specters of the Atlantic*, 16–18. See also Pocock, *Virtue, Commerce, and History*, 37–51.

BIBLIOGRAPHY

PRIMARY SOURCES

Archival Collections

American Philosophical Society. Philadelphia, PA.

Pennsylvania Hospital Archives

Archivo General de Indias. Seville, Spain.

Indiferente General

Baker Library, Harvard Business School. Boston, MA.

Hancock Manuscripts
Heard Family Business Records

Barbados Department of Archives. Black Rock, St. James.

Hughes-Queree Collection
Shipping Returns and Miscellaneous Accounts
Wills, Deeds, Inventories of Property and Personal Effects

Boston Athenaeum. Boston, MA.

Greenleaf Family Papers (MS L80)

British Library. London, United Kingdom.

Barrington Papers, vols. 77–78 (Add MS 73622–73623)
Fair Copy of Knight's *History of Jamaica*, vol. 2 (Add MS 12419)
India Office Records and Private Papers
Liverpool Papers, vol. 220 (Add MS 38409)
Tracts Relating to Trade (816.m.12, 1–163)
Trade Cases (1887.b.60)
Petitions (CUP.645.b.11, 1–45)

Caird Library, National Maritime Museum. Greenwich, London, United Kingdom.

Admiralty Collection
Green Blackwall Collection
Sir Robert Wigram Papers

Center for the History of Medicine, Countway Library of Medicine. Boston, MA.

Diary of John Denison Hartshorn (B MS b118.2)
John Denison Hartshorn Papers (B MS c50)
Misc. Manuscripts

GlaxoSmithKline. Brentford, London, United Kingdom.

Allen & Hanburys Manuscripts

Hispanic Society of America. New York, NY.

Journal and logbook of an anonymous Scotch sailor (HC 363/1299)

Historical Society of Pennsylvania. Philadelphia, PA.

Christopher Marshall's Bills of Lading (Am.916)
Christopher Marshall Papers (coll. 395)
Daniel Parker Papers (coll. 1587)
Logan Family Papers (coll. 2023)
Pemberton Family Papers (coll. 484A)
Penington Family Papers (coll. 1435)
Powel Family Papers (coll. 1582)
"Resolution of Non-Importation Made by the Citizens of Philadelphia" (Am.340)
Simon Gratz Autograph Collection
William Logan Journals and Papers (coll. 3386)

Houghton Library, Harvard University. Cambridge, MA.

"Whereas this province labours under a heavy debt, incurred in the course of the late war: and the inhabitants by this means must be for some time subject to very burthensome taxes . . ." Boston: [s.n.], 1767.

Library Company of Philadelphia. Philadelphia, PA.

Peter Renaudet medical apprentice notebook
William H. Helfand Popular Medicine Collection

Lionel Pincus and Princess Firyal Map Division, New York Public Library. New York, NY.

G. Jones, "London" (1815)

London Metropolitan Archives. London, United Kingdom.

Howards and Sons Limited (ACC 1037)
Janson, Cobb, Pearson and Company (CLC/B/136)
Royal and Sun Alliance Insurance Group (CLC/B/192)

Maine Historical Society. Portland, ME.

Kennebec Proprietors Papers (coll. 60)
Misc. Gardiner Family Papers (coll. S-6563)
Silvester Gardiner Papers (coll. 41)

Massachusetts Historical Society. Boston, MA.

Chelsea (MA) Papers (MS N-2008)
Gardiner-Whipple-Allen Family Papers (MS N-1271)
Misc. Bound Manuscripts
Naval Office Shipping Lists for Massachusetts (MS N-1635)
Quincy, Wendell, Holmes, and Upham Family Papers (P-347)
Robert Treat Paine Papers (P-392)
Vassall Family Papers (MS S-111)

Massachusetts State Archives. Boston, MA.

Massachusetts Archive Collection, vols. 154, 291

National Archives of the UK. Kew, London, United Kingdom.

Ledgers of Imports and Exports, America (CUST 16/1)
Ledgers of Imports and Exports, England and Wales (CUST 3/1–82)
Navy Board Records
Supreme Court of Judicature and Former Superior Courts Records (J 90)
Treasury Board Papers and In-Letters (T 1)

Newport Historical Society. Newport, RI.

Thomas Richardson Papers
William Hunter Papers

Pennsylvania Hospital Historic Collections. Philadelphia, PA.

Records of the Pennsylvania Hospital

Phillips Library, Peabody Essex Museum. Salem, MA.

Stearns and Sprague Family Papers

Rhode Island Historical Society. Providence, RI.

Bowen Family Papers (MS 301)
Malbone Family Collection (MS 549)

Royal College of Physicians. London, United Kingdom.

Censors' Visitations of Apothecaries' Shops

R. Stanton Avery Special Collections Department, New England Historic Genealogical Society. Boston, MA.

Diary of William Jepson (MS C 3095)
Simon Tufts Correspondence (MS C 1355)

Special Collections Research Center, Earl Gregg Swem Library, College of William & Mary. Williamsburg, VA.

Martin Brimmer Account Ledger
Office of the Bursar Records

University of London. London, United Kingdom.

Newton Papers (MS 523)
Taylor and Vanneck-Arcedeckne Papers

Wellcome Library. London, United Kingdom.

Corbyn & Company Papers (MSS 5436–5460)
Gilbert Laurie & Company Papers (MS 3181)
Primatt & Maud Papers (MSS 5878–5886)
Robert Wigram Account Book (MS 832)
Robert Wigram to India House (MS 7226)
William Stearns Papers (MSS 5855–5859)

William L. Clements Library. Ann Arbor, MI.

Humphry and Moses Marshall Papers
Isaac Bartram Account Book

Worshipful Society of Apothecaries. London, United Kingdom.

Catalogus Medicamentorum Chemicorum
Court Minutes
India Orders
Laboratory Stock Records
Navy Stock Records
Papers Related to Society's Negotiations to Supply Army
Thomas Mayleigh Account Book

Databases

Legacies of British Slave-Ownership Database (ucl.ac.uk/lbs)
OED Online (oed.com)
Trans-Atlantic Slave Trade Database (slavevoyages.org)

Newspapers

Augusta Chronicle
Boston Chronicle
Boston Evening-Post
Boston News-Letter (*Boston Weekly News-Letter*)
Boston Weekly Post-Boy (*Boston Post-Boy & Advertiser*)
Boston-Gazette, and Country Journal
Continental Journal and Weekly Advertiser (Boston)
Daily Advertiser (London)
Georgia Gazette (Savannah, GA)
Independent Chronicle and the Universal Advertiser (Boston)
New-Hampshire Gazette (Portsmouth, NH)
New-York Mercury
Newport Mercury
Pennsylvania Gazette (Philadelphia)
Pennsylvania Journal; and the Weekly Advertiser (Philadelphia)
Salem Chronicle, and Essex Advertiser (Salem, MA)
Virginia Journal and Alexandria Advertiser (Alexandria, VA)

Printed

Adams, Charles Francis. *The Works of John Adams, Second President of the United States . . .* Vol. 2. Boston, 1850.

"Address of the Merchants and Others, of Boston to Gov. Hutchinson" [May 30, 1774]. In Stark, *Loyalists of Massachusetts*, 123–25.

Allen, William. *The Duty of Abstaining from the Use of West India Produce, a Speech, Delivered at Coach-Maker's-Hall, Jan. 12, 1792*. 2nd ed. London, 1792.

The Apothecary Display'd: Or, An Answer to the Apothecary's Pamphlet, Called Frauds detected in Drugs. London, 1748.

Belgrove, William. *A Treatise upon Husbandry or Planting*. Boston, 1755.

Belknap, Jeremy. *The History of New-Hampshire*. 2nd ed. Vol. 3. Boston, 1813.

Besse, Joseph. *A Collection of the Sufferings of the People Called Quakers*. Vol. 2. London, 1753.

Blochwitz, Martin. *Anatomia Sambuci, or, The Anatomy of the Elder: Cutting out of it Plain, Approved, and Specifick Remedies for most and chiefest Maladies . . .* London, 1677.

Boerhaave, Herman. *Aphorismi de cognoscendis et curandis morbis . . .* Leiden, 1709.

———. *Boerhaave's Aphorisms: Concerning the Knowledge and Cure of Diseases*. London, 1735. Translated from Latin editions printed at Leiden. In many editions.

Boissier de Sauvages [de Lacroix], [François]. *Nosologia methodica: Sistens morborum classes, genera et species; Juxtà Sydenhami mentem & Botanicorum ordinem.* [Amsterdam], 1763.

[Boston Records Commissioners.] *A Report of the Record Commissioners of the City of Boston: Containing Boston Marriages from 1700 to 1751.* Vol 29 of the Commissioners' Reports. Boston, 1898.

Boulton, Richard. *An Examination of Mr. John Colbatch His Books* . . . London, 1699.

Brookes, Richard. *An Introduction to Physic and Surgery.* 2nd ed. London, 1763.

Brown, William. *Pharmacopoeia simpliciorum et efficaciorum, in usum nosocomii militaris, ad exercitum foederatarum Americae civitatum pertinentis* . . . Philadelphia, 1778.

Butterfield, L. H., ed. *Diary and Autobiography of John Adams.* Vol. 1, *Diary 1755–1770.* Cambridge, MA: Harvard University Press, 1961.

Caines, Clement. *Letters on the Cultivation of the Otaheite Cane.* London, 1801.

Campbell, R. *The London Tradesman. Being a Compendious View of all the Trades, Professions, Arts, both Liberal and Mechanic, now practised in the Cities of London and Westminster.* London, 1747.

Catesby, Mark. *The Natural History of Carolina, Florida and the Bahama Islands* . . . Vol. 1. London, 1731.

Clark, John. *Observations on the Diseases in Long Voyages to Hot Countries* . . . London, 1773.

[Collins, Dr.]. *Practical Rules for the Management and Medical Treatment of Negro Slaves in the Sugar Colonies.* London, 1803.

The Compleat Compting-House Companion: Or, Young Merchant and Tradesman's Sure Guide. London, 1763.

Craskell, Thomas. *This map of the County of Surry in the Island of Jamaica laid down from the papers and under the direction of Henry Moore, Esqr. His Majesty's Lieutenant Governor and Commander in Chief of that island, in the years 1756, 57, 58, 59, 60 & 61 from a great number of actual surveys performed by the publishers* . . . London, 1763.

Crouch, Henry. *A Complete View of the British Customs.* London, 1745.

Culpepper, Nicholas. *Pharmacopoeia Londinensis; or, the London Dispensatory; Further adorned by the Studies and Collections of the Fellows* . . . London, 1653; Boston, 1720.

Dancer, Thomas. *The Medical Assistant; or Jamaica Practice of Physic: Designed chiefly for the Use of Families and Plantations.* 2nd ed. Jamaica, 1809.

Elliot, John. *The Medical Pocket-Book; For Those Who Are, and For All Who Wish, To Be, Physicians.* Philadelphia, 1784.

Encyclopaedia Britannica: or, A Dictionary of Arts, Sciences, and Miscellaneous Literature . . . 3rd ed. Edinburgh, 1797.

Fox, George. *The Line of Righteousness and Justice Stretched Forth Over All Merchants, etc.* London, 1661.

———. *A Warning to All the Merchants in London and Such as Buy and Sell with an Advertisement to Them to Lay Aside Their Superfluity and with it Nourish the Poor.* London, 1658.

Franklin, Benjamin. *Some Account of the Pennsylvania Hospital* . . . Philadelphia, 1754.

Frost, William J., ed. *The Records and Recollections of James Jenkins Respecting Himself and Others from 1761 to 1821.* New York: Edwin Mellen Press, 1984.

Grainger, James. *An Essay on the More Common West-India Diseases* . . . London, 1764.

Harrison, George. *Memoir of William Cookworthy, Formerly of Plymouth Devonshire.* London, 1854.

Hening, William Waller, ed. *The Statutes at Large; Being a Collection of the Laws of Virginia from the First Session of the Legislature, in the Year 1619.* Vol. 6. Richmond, 1819.

Hillary, William. *Observations on the Changes of the Air and the Concomitant Epidemical Diseases in the Island of Barbadoes.* 2nd ed. London, 1766.

Holder, Henry Evans. *A Short Essay on the Subject of Negro Slavery, with a Particular Reference to the Island of Barbadoes*. London, 1788.

Howard, Luke. *On the Modifications of Clouds: and on the Principles of their Production, Suspension, and Destruction . . .* London, 1803.

Hughes, Griffith. *The Natural History of the Island of Barbados*. London, 1750.

Hutchinson, Thomas. *The History of the Colony and Province of Massachusetts-Bay*. Vol. 3, edited by Lawrence Shaw Mayo. New York: Kraus Reprint, 1970.

Ingram, Dale. *Practical Cases and Observations in Surgery . . .* London, 1751.

James, Robert. *Pharmacopoeia Universalis: Or, A New Universal English Dispensatory . . .* London, 1747.

Journal of the House of Commons (CJ).

Journal of the House of Lords (LJ).

Kein, Patrick. *An Essay upon Pen-Keeping and Plantership*. Kingston, Jamaica, 1796.

Leslie, Charles. *A New History of Jamaica, from the Earliest Accounts, to the Taking of Porto Bello . . .* London, 1740.

Lewis, William. *Commercium Philosophico-Technicum; or, The Philosophical Commerce of Arts: Designed as an Attempt to Improve Arts, Trades, and Manufactures*. London, 1763.

———. *The Edinburgh New Dispensatory*. Edinburgh, 1786.

———. *The New Dispensatory*. London, 1753; 6th ed. London, 1799.

Linné [Linnaeus], Car. V. *Genera morborum; in auditorum usum*. [Uppsula], 1763.

"A list of the addresses to the late Gov. Hutchinson, Taken from the London Gazetteer, and New Daily advertiser, on Saturday, September 24th, 1774." Boston, 1774.

Long, Edward. *The History of Jamaica*. Vol. 2. London, 1774.

[Martin, Samuel]. *An Essay upon Plantership, Humbly inscrib'd to all the Planters of the British Sugar-Colonies in America*. 2nd ed. Antigua, 1750; 4th ed. London, 1765.

Maxwell, James. "Pathological Inquiry into the Nature of Cachexia Africana, as it is generally connected with Dirt Eating." *Jamaica Physical Journal* 2 (Nov.–Dec. 1835): 409–35.

The Medical Register for the Year 1783. London, 1783.

The Medical Vade Mecum. Charleston, 1800.

Morris, Robert, and James Kendrick, et al. *The Edinburgh Medical and Physical Dictionary, Containing an Explanation of the Terms of Art in Anatomy, Pharmacy, Physiology, Materia Medica, Pathology, Botany, Therapeutics, Chemistry, Surgery, Natural History, Midwifery . . .* Vol. 2. Edinburgh, 1807.

"Mr. Stephen's Process for Making Plantation Pot-Ash, For Which He Was Rewarded by the British Parliament." *Gentleman's Magazine* 25 (1755): 361–63.

Mullins, James. *Some Observations Made upon the Cylonian Plant. Shewing Its Admirable Virtues against Deafness*. London, 1695.

[National Archives of India.] *Fort William-India House Correspondence and Other Contemporary Papers Relating Thereto*. Vol. 3, *1760–1763*. Delhi: Manager of Publications, Government of India, 1968.

[———.] *Fort William-India House Correspondence and Other Contemporary Papers Relating Thereto*. Vol. 4, *1764–1766*. Delhi: Manager of Publications, Government of India, 1962.

The New Complete Guide to all Persons who have any Trade or Concern with the City of London, and Parts Adjacent. 16th ed. London, 1783.

The New Royal Encyclopaedia: Or, Complete Modern Dictionary of Arts and Sciences, on an Improved Plan . . . 3 vols. London, [1789].

Notes on the Reports from the Committee of the Honourable House of Assembly of Jamaica . . . London, 1789.

The Origin, Progress and Present State of the Various Establishments for Conducting Chemical Processes, and Other Medicinal Preparations, at Apothecaries Hall. London, 1823.

Pettigrew, Thomas Joseph, ed. *Memoirs of the Life and Writings of the Late John Coakley Lettsom*. Vol. 1. London, 1817.

[Philadelphia Yearly Meeting of the Religious Society of Friends.] *A Collection of Memorials Concerning Divers deceased Ministers and others of the People called Quakers, in Pennsylvania, New-Jersey, and Parts adjacent, from nearly the first Settlement thereof to the Year 1787*. Philadelphia, 1787.

Pickering, Danby. *The Statutes at Large* . . . Vol. 14, *1718–1721*. London, 1765.

———. *The Statutes at Large* . . . Vol. 18, *1742–1746*. London, 1765.

———. *The Statutes at Large* . . . Vol. 36, *1787–1789*. London, 1787.

Quincy, John. *Lexicon Physico-Medicum: Or, a New Medicinal Dictionary*. 10th ed. London, 1787.

———. *Pharmacopoeia Officinalis & Extemporanea. Or, a Complete English Dispensatory, in Four Parts*. 5th ed. London, 1724.

A Rational Discours touching the Universal Medicin. [Unknown], 1664.

"Reports from the Committee on Illicit Practices Used in Defrauding the Revenue [1783, 1784]." In *Reports from the Committees of the House of Commons*, vol. 11, *1782–1799*, pp. 228–99. London, 1803.

Richards, Ruth, R. H. Gardiner, and W. T. Gardiner, eds. *Mrs. Gardiner's Receipts from 1763*. Hallowell, ME: White and Horne, 1938.

Roberts, Lewes. *Merchants Map of Commerce*. London, 1671.

Robertson, Robert. *A Detection of the State and Situation Of the Present Sugar Planters, of Barbadoes and the Leward Islands* . . . London, 1732.

Salmon, William. *Ars Chirurgica. A Compendium of the Theory and Practice of Chirurgery* . . . London, 1698.

Seally, J., et al. *A Complete Geographical Dictionary* . . . Vol. 1. London, 1781.

Shaw, Peter. *Pharmacopoeia Edinburgensis: Or, the Dispensatory of the Royal College of Physicians in Edinburgh*. 3rd ed. London, 1737.

Smith, Adam. *An Inquiry into the Nature and Causes of the Wealth of Nations*. Cannan Ed. New York: Modern Library, 1937.

Spilsbury, Francis. *The Friendly Physician. A New Treatise* . . . London, 1773.

Stearns, Samuel. *The American Herbal, or Materia Medica*. Walpole, NH, 1801.

Steel, Henry Draper. *Portable Instructions for Purchasing the Drugs and Spices of Asia and the East-Indies* . . . London, 1779.

Sydenham, Thomas. *Observationes medicae circa morborum acutorum historiam et curationem*. London, 1676. Many editions and translations.

Talbor, Robert. *Pyretologia, A Rational Account of the Cause & Cure of Agues, With their Signes Diagnostick & Prognostick* . . . London, 1672.

Tennent, John. *An Epistle to Dr. Richard Mead, Concerning the Epidemical Diseases of Virginia* . . . Edinburgh, 1738.

Thomson, James. *A Treatise on the Diseases of Negroes, as They Occur in the Island of Jamaica: with Observations on the Country Remedies*. Jamaica, 1820.

"Timothy Parsons to Samuel P. Savage." *Proceedings of the Massachusetts Historical Society* 44 (Oct. 1910–June 1911): 84–85.

A True State of the Copartnership of Gardiner and Jepson. Boston, 1771.

Tyler, Royall. *The Bay Boy or the Autobiography of a Youth of Massachusetts Bay*. Edited by Martha R. Wright. Hicksville, NY: Exposition Press, 1978.

SECONDARY SOURCES

Abrams, Jeanne E. *Revolutionary Medicine: The Founding Fathers and Mothers in Sickness and in Health*. New York: New York University Press, 2013.

Ackrill, Margaret, and Leslie Hannah. *Barclays: The Business of Banking 1690–1996*. New York: University of Cambridge Press, 2001.

Albion, Robert Greenhalgh. *Forests and Sea Power: The Timber Problem of the Royal Navy 1652–1862*. Cambridge, MA: Harvard University Press, 1926.

Allen, Charles E. "Loyalists of the Kennebec." *New England Magazine* 37 (Sept. 1907–Feb. 1908): 623–29.

Allen, Katherine. "Hobby and Craft: Distilling Household Medicine in Eighteenth-Century England." *Early Modern Women* 11, no. 1 (2016): 90–114.

Allen, Robert C. "The Great Divergence in European Wages and Prices from the Middle Ages to the First World War." *Explorations in Economic History* 38, no. 4 (2001): 411–47.

Alpe, E. N. *Handy Book of Medicine Stamp Duty with the Statutes and Appendices*. London, 1880.

Alsop, J. D. "Warfare and the Creation of British Imperial Medicine." In Hudson, *British Military and Naval Medicine*, 23–50.

Amussen, Susan Dwyer. *Caribbean Exchanges: Slavery and the Transformation of English Society, 1640–1700*. Chapel Hill: University of North Carolina Press, 2007.

Anderson, Fred. *Crucible of War: The Seven Years' War and the Fate of Empire in British North America, 1754–1766*. New York: Vintage Books, 2000.

Anderson, Jennifer L. *Mahogany: The Costs of Luxury in Early America*. Cambridge, MA: Harvard University Press, 2012.

Anderson, Stuart, ed. *Making Medicines: A Brief History of Pharmacy and Pharmaceuticals*. Chicago: Pharmaceutical Press, 2005.

Anishanslin, Zara. *Portrait of a Woman in Silk: Hidden Histories of the British Atlantic World*. New Haven, CT: Yale University Press, 2016.

Arnold, David, ed. *Warm Climates and Western Medicine: The Emergence of Tropical Medicine 1500–1900*. Atlanta: Rodopi, 1996.

Ashworth, William J. *The Industrial Revolution: The State, Knowledge and Global Trade*. London: Bloomsbury, 2017.

"Attempted Legislation in 1748." *Chemist and Druggist* 105 (July 31, 1926): 198–200.

Atton, Henry, and Henry Hurst Holland. *The King's Customs*. London: John Murray, 1908.

Baber, Zaheer. *The Science of Empire: Scientific Knowledge, Civilization, and Colonial Rule in India*. Albany: State University of New York Press, 1996.

Bailyn, Bernard. *The New England Merchants in the Seventeenth Century*. Cambridge, MA: Harvard University Press, 1955.

Bailyn, Bernard, and Patricia L. Denault, eds. *Soundings in Atlantic History: Latent Structures and Intellectual Currents, 1500–1830*. Cambridge, MA: Harvard University Press, 2009.

Bair, Jennifer, ed. *Frontiers of Commodity Chain Research*. Stanford, CA: Stanford University Press, 2009.

Bamji, Alexandra. "Medical Care in Early Modern Venice." *Journal of Social History* 49, no. 3 (Spring 2016): 483–509.

Baptist, Edward. *The Half Has Never Been Told: Slavery and the Making of American Capitalism*. New York: Basic Books, 2014.

Barrett, Charles Raymond Booth. *The History of the Society of Apothecaries of London*. London: E. Stock, 1905.

Barry, William David, and Frances W. Peabody. *Tate House: Crown of the Maine Mast Trade*. Portland, ME: National Society of Colonial Dames of America, 1982.

Bartlet, William S. *The Frontier Missionary: A Memoir of the Life of the Rev. Jacob Bailey . . .* Boston, 1853.

Bashford, Alison. "Medicine, Gender, and Empire." In Levine, *Gender and Empire*, 112–33.

Baucom, Ian. *Specters of the Atlantic: Finance Capital, Slavery, and the Philosophy of History*. Durham, NC: Duke University Press, 2005.

Beckert, Sven. *Empire of Cotton: A Global History*. New York: Alfred A. Knopf, 2014.

Beckert, Sven, and Christine Desan, eds. *American Capitalism: New Histories*. New York: Columbia University Press, 2018.

Beckert, Sven, and Seth Rockman, eds. *Slavery's Capitalism: A New History of American Economic Development*. Philadelphia: University of Pennsylvania Press, 2016.

Beckles, Hilary. *A History of Barbados: From Amerindian Settlement to Nation-State*. New York: Cambridge University Press, 1990.

Bell, Jacobus, and Theophilus Redwood. *Historical Sketch of the Progress of Pharmacy in Great Britain*. London, 1880.

Ben-Amos, Ilana Krausman. *The Culture of Giving: Informal Support and Gift-Exchange in Early Modern England*. New York: Cambridge University Press, 2008.

Berg, Maxine. *The Age of Manufactures, 1700–1820: Industry, Innovation, and Work in Britain*. 2nd ed. New York: Routledge, 1994.

Berger, Rachel. *Ayurveda Made Modern: Political Histories of Indigenous Medicine in North India, 1900–1955*. New York: Palgrave Macmillan, 2013.

Berlin, Ira. *Many Thousands Gone: The First Two Centuries of Slavery in North America*. Cambridge, MA: Harvard University Press, 1998.

Berlin, Michael. "Guilds in Decline? London Livery Companies and the Rise of a Liberal Economy, 1600–1800." In Epstein and Prak, *Guilds, Innovation, and the European Economy*, 316–41.

Beukers, H., A. M. Luyendijk-Elshout, M. E. van Opstall, and F. Vos, eds. *Red-Hair Medicine: Dutch-Japanese Medical Relations*. Atlanta: Rodopi, 1991.

Bhattarcharya, Nandini. "From *Materia Medica* to the Pharmacopoeia: Challenges of Writing the History of Drugs in India." *History Compass* 14, no. 4 (2016): 131–39.

Biographical Annals of Lancaster County Pennsylvania. Lancaster, PA: J. H. Beers, 1903.

Blackburn, Robin. *The Making of New World Slavery: From the Baroque to the Modern, 1492–1800*. New York: Verso, 1997.

Bleichmar, Daniela. "Atlantic Competitions: Botany in the Eighteenth-Century Spanish Empire." In Delbourgo and Dew, *Science and Empire*, 225–52.

Brandow, James C. *Genealogies of Barbados Families*. Baltimore: Genealogical, 1983.

Brandt, Susan. "'Getting into a Little Business': Margaret Hill Morris and Women's Medical Entrepreneurship During the American Revolution." *Early American Studies* 13, no. 4 (2015): 774–807.

Breen, Benjamin. "Drugs and Early Modernity." *History Compass* 15, no. 4 (2017): 1–9.

Breen, T. H. "'Baubles of Britain': The American and Consumer Revolutions of the Eighteenth Century." *Past & Present* 119, no. 1 (1988): 73–104.

———. *The Marketplace of Revolution: How Consumer Politics Shaped American Independence*. New York: Oxford University Press, 2004.

Brewer, John. *The Sinews of Power: War, Money, and the English State, 1688–1783*. London: Unwin Hyman, 1989.

———. "The Wilkites and the Law, 1763–74: A Study of Radical Notions of Governance." In Brewer and Styles, *Ungovernable People*, 128–71.

Brewer, John, and Roy Porter, eds. *Consumption and the World of Goods*. New York: Routledge, 1993.

Brewer, John, and John Styles, eds. *An Ungovernable People: The English and Their Law in the Seventeenth and Eighteenth Centuries*. New Brunswick, NJ: Rutgers University Press, 1980.

Brock, C. Helen. "The Influence of Europe on Colonial Massachusetts Medicine." In [Colonial Society], *Medicine in Colonial Massachusetts*, 101–16.

Brock, Leslie V. *The Currency of the American Colonies, 1700–1764: A Study in Colonial Finance and Imperial Relations*. New York: Arno Press, 1975.

Brockliss, Laurence, and Colin Jones. *The Medical World of Early Modern France*. New York: Oxford University Press, 1997.

Brown, Richard D. "The Emergence of Urban Society in Rural Massachusetts, 1760–1820." *Journal of American History* 61 (1974): 29–51.

———. "The Healing Arts in Colonial and Revolutionary Massachusetts: The Context for Scientific Medicine." In [Colonial Society], *Medicine in Colonial Massachusetts*, 35–47.

———. *Knowledge is Power: The Diffusion of Information in Early America, 1700–1865*. New York: Oxford University Press, 1989.

Brown, Vera Lee. "The South Sea Company and Contraband Trade." *American Historical Review* 31, no. 4 (July 1926): 662–78.

Brown, Vincent. *The Reaper's Garden: Death and Power in the World of Atlantic Slavery*. Cambridge, MA: Harvard University Press, 2008.

Brunsman, Denver. *The Evil Necessity: British Naval Impressment in the Eighteenth-Century Atlantic World*. Charlottesville: University of Virginia Press, 2013.

Buck, Robert W. "Tucker's Wife's Leg." *New England Journal of Medicine* 246, no. 24 (June 1952): 937–38.

Burke, Victoria E., and Jonathan Gibson, eds. *Early Modern Women's Manuscript Writing: Selected Papers from the Trinity/Trent Conference*. Burlington, VT: Ashgate, 2004.

Burnard, Trevor. *Mastery, Tyranny, and Desire: Thomas Thistlewood and His Slaves in the Anglo-Jamaican World*. Chapel Hill: University of North Carolina Press, 2004.

———. *Planters, Merchants, and Slaves: Plantation Societies in British America, 1650–1820*. Chicago: University of Chicago Press, 2015.

Burnby, Juanita G. L. "Printer's Ink and Patent Medicines: The Story of the Diceys." *Pharmaceutical Journal* 229 (Aug. 1982): 162–63, 169.

———. *A Study of the English Apothecary from 1660 to 1760*. London: Wellcome Institute for the History of Medicine, 1983.

———. "A Study of the English Apothecary from 1660–1760 with Special Reference to the Provinces." PhD diss., University of London, 1979.

Burnett, Edmund C. "Observations of London Merchants on American Trade, 1783." *American Historical Review* 18, no. 4 (1913): 769–80.

Cagle, Hugh Glenn. "The Botany of Colonial Medicine: Gender, Authority, and Natural History in the Empires of Spain and Portugal." In Owens and Mangan, *Women of the Iberian Atlantic*, 174–95.

Cañizares-Esguerra, Jorge. *Nature, Empire, and Nation: Explorations of the History of Science in the Iberian World*. Stanford, CA: Stanford University Press, 2006.

Cantor, Geoffrey. "Quakers in the Royal Society, 1660–1750." *Notes and Records of the Royal Society of London* 51, no. 2 (1997): 175–93.

Carbone, Teresa A., et al. *American Paintings in the Brooklyn Museum: Artists Born by 1876*. Vol. 1. New York: Brooklyn Museum, 2006.

Cash, Philip. "The Professionalization of Boston Medicine, 1760–1803." In [Colonial Society], *Medicine in Colonial Massachusetts*, 69–100.

Cavallo, Sandra. *Artisans of the Body in Early Modern Italy: Identities, Families and Masculinities*. Manchester: Manchester University Press, 2007.

Chakrabarti, Pratik. *Materials and Medicine: Trade, Conquest and Therapeutics in the Eighteenth Century*. Manchester: Manchester University Press, 2010.

———. *Medicine and Empire, 1600–1960*. New York: Palgrave Macmillan, 2014.

———. "'Neither of Meate Nor Drinke, but What the Doctor Alloweth': Medicine amidst War and Commerce in Eighteenth-Century Madras." *Bulletin of the History of Medicine* 80, no. 1 (2006): 1–38.

Chamberlain, Lucia Sarah. "Plants Used by the Indians of Eastern North America." *American Naturalist* 35, no. 409 (Jan. 1901): 1–10.

Chapman-Huston, Desmond, and Ernest C. Cripps. *Through a City Archway: The Story of Allen and Hanburys 1715–1954*. London: John Murray, 1954.

Charters, Erica. *Disease, War, and the Imperial State: The Welfare of the British Armed Forces During the Seven Years' War*. Chicago: University of Chicago Press, 2014.

Chaudhuri, K. N. *The Trading World of Asia and the English East India Company 1660–1760*. New York: Cambridge University Press, 1978.

Christianson, Eric H. "The Colonial Surgeon's Rise to Prominence: Dr. Silvester Gardiner (1707–1786) and the Practice of Lithotomy in New England." *New England Historical and Genealogical Register* 136 (1982): 104–14.

———. "The Medical Practitioners of Massachusetts, 1630–1800: Patterns of Change and Continuity." In [Colonial Society], *Medicine in Colonial Massachusetts*, 49–67.

———. "Medicine in New England." In Leavitt and Numbers, *Sickness and Health in America*, 47–71.

Clark, Gregory. "What Were the British Earnings and Prices Then? (New Series)." Measuring Worth, 2016.

Cleary, Patricia. *Elizabeth Murray: A Woman's Pursuit of Independence in Eighteenth-Century America*. Amherst: University of Massachusetts Press, 2000.

Clow, Archibald, and Nan L. Clow. *The Chemical Revolution: A Contribution to Social Technology*. London: Batchworth Press, 1952.

Cockerell, H. A. L., and Edwin Green. *The British Insurance Business, 1547–1970: An Introduction and Guide to Historical Records in the United Kingdom*. London: Heinemann Educational, 1976.

Coclanis, Peter, ed. *The Atlantic Economy During the Seventeenth and Eighteenth Centuries: Organization, Practice, and Personnel*. Columbia: University of South Carolina Press, 2005.

———. "Introduction." In Coclanis, *Atlantic Economy*, xi–xix.

Cole, W. A. "Trends in Eighteenth-Century Smuggling." *Economic History Review* 10, no. 3 (1958): 395–410.

[Colonial Society of Massachusetts], ed. *Medicine in Colonial Massachusetts, 1620–1820: A Conference Held 25 and 26 May 1978 by the Colonial Society of Massachusetts*. Boston: Colonial Society of Massachusetts, 1980.

Cook, Harold J. *The Decline of the Old Medical Regime in Stuart London*. Ithaca, NY: Cornell University Press, 1986.

———. "Early Modern Science and Monetized Economies: The Co-Production of Commensurable Materials." In Füssel, Knäble, and Elsemann, *Wirtschaft Und Wissen*, 97–114.

———. "Good Advice and Little Medicine: The Professional Authority of Early Modern English Physicians." *Journal of British Studies* 33, no. 1 (1994): 1–31.

———. "Markets and Cultures: Medical Specifics and the Reconfiguration of the Body in Early Modern Europe." *Transactions of the Royal Historical Society* 21 (2011): 123–45.

———. *Matters of Exchange: Commerce, Medicine, and Science in the Dutch Golden Age*. New Haven, CT: Yale University Press, 2007.

———. "Medicine." In Daston and Park, *Cambridge History of Science*, 3:407–34.

———. "Moving About and Finding Things Out: Economies and Sciences in the Period of the Scientific Revolution." *Osiris* 27, no. 1 (2012): 101–32.

———. "The New Philosophy and Medicine in Seventeenth-Century England." In Lindberg and Westman, *Reappraisals of the Scientific Revolution*, 397–436.

———. "Practical Medicine and the British Armed Forces After the 'Glorious Revolution.'" *Medical History* 34, no. 1 (1990): 1–26.

———. "The Rose Case Reconsidered: Physic and the Law in Augustan England." *Journal of the History of Medicine* 45, no. 4 (1990): 527–55.

———. "The Society of Chemical Physicians, the New Philosophy, and the Restoration Court." *Bulletin of the History of Medicine* 61 (1987): 61–77.

———. *Trials of an Ordinary Doctor: Joannes Groenevelt in Seventeenth-Century London*. Baltimore: Johns Hopkins University Press, 1994.

Cook, Harold J., and Timothy D. Walker. "Circulation of Medicine in the Early Modern Atlantic World." *Social History of Medicine* 26, no. 3 (2013): 337–51.

Court, William E. "Pharmacy from the Ancient World to 1100 AD." In Anderson, *Making Medicines*, 21–36.

Crane, Elaine Forman. *The Poison Plot: A Tale of Adultery and Murder in Colonial Newport*. Ithaca, NY: Cornell University Press, 2018.

Craton, Michael. *Searching for the Invisible Man: Slaves and Plantation Life in Jamaica*. Cambridge, MA: Harvard University Press, 1978.

Crawford, Catherine. "Patients' Rights and the Law of Contract in Eighteenth-Century London." *Social History of Medicine* 13, no. 3 (2000): 381–410.

Crawford, Matthew James. *The Andean Wonder Drug: Cinchona Bark and Imperial Science in the Spanish Atlantic, 1630–1800*. Pittsburgh: University of Pittsburgh Press, 2016.

Crellin, J. K. "Dr. James's Fever Powder." *Transactions of the British Society for the History of Pharmacy* 1 (1974): 136–43.

Crellin, J. K., and J. R. Scott. "Lionel Lockyer and his Pills." In *Proceedings of the XXIII International Congress of the History of Medicine*, 1182–86.

———. *Glass and British Pharmacy 1600–1900*. London: Wellcome Institute of the History of Medicine, 1972.

Cressy, David. "Saltpetre, State Security and Vexation in Early Modern England." *Past & Present* 212 (Aug. 2011): 73–111.

Crimmin, P. K. "The Sick and Hurt Board and the Health of Seamen c. 1700–1806." *Journal for Maritime Research* 1, no. 1 (1999): 48–65.

Cripps, Ernest C. *Plough Court: The Story of a Notable Pharmacy 1715–1927*. London: Allen & Hanburys, 1927.

Cronon, William. *Changes in the Land: Indians, Colonists, and the Ecology of New England*. New York: Hill & Wang, 1983/2003.

Crosland, Maurice. "Early Laboratories, c. 1600–c. 1800 and the Location of Experimental Science." *Annals of Science* 62, no. 2 (2005): 233–53.

Crouzet, François. "America and the Crisis of the British Imperial Economy, 1803–1807." In McCusker and Morgan, *Early Modern Atlantic Economy*, 278–315.

Crowley, John E. "Neo-Mercantilism and the Wealth of Nations: British Commercial Policy After the American Revolution." *Historical Journal* 33, no. 2 (1990): 339–60.

Curran, Andrew. *The Anatomy of Blackness: Science and Slavery in an Age of Enlightenment*. Baltimore: Johns Hopkins University Press, 2011.

Curth, Louise Hill, ed. *From Physick to Pharmacology: Five Hundred Years of British Drug Retailing*. Burlington, VT: Ashgate, 2006.

Curtin, Philip D. *The Atlantic Slave Trade; A Census*. Madison: University of Wisconsin Press, 1969.

———. *The Rise and Fall of the Plantation Complex: Essays in Atlantic History*. 2nd ed. New York: Cambridge University Press, 1998.

Daston, Lorraine, and Katherine Park, eds. *The Cambridge History of Science. Vol. 3, Early Modern Europe*. New York: Cambridge University Press, 2006.

Davidoff, Leonore, and Catherine Hall. *Family Fortunes: Men and Women of the English Middle Class 1780–1850*. New York: Routledge, 2002.

Davids, Karel. *The Rise and Decline of Dutch Technological Leadership: Technology, Economy and Culture in the Netherlands, 1350–1800*. Leiden: Brill, 2008.

Davis, David Brion. *Inhuman Bondage: The Rise and Fall of Slavery in the New World*. New York: Oxford University Press, 2006.

Davis, Ralph. "English Foreign Trade, 1660–1700." *Economic History Review* 7, no. 2 (1954): 150–66.

———. "English Foreign Trade, 1700–1774." *Economic History Review* 15, no. 2 (1962): 285–303.

———. "The Rise of Protection in England, 1689–1786." *Economic History Review* 19, no. 2 (1966): 306–17.

Dayton, Cornelia Hughes. "Taking the Trade: Abortion and Gender Relations in an Eighteenth-Century New England Village." *William & Mary Quarterly* 48, no. 1 (1991): 19–49.

De Barros, Juanita. "'Setting Things Right': Medicine and Magic in British Guiana, 1803–38." *Slavery & Abolition* 25, no. 1 (2004): 28–50.

De Vos, Paula. "Natural History and the Pursuit of Empire in Eighteenth-Century Spain." *Eighteenth-Century Studies* 40, no. 2 (Winter 2007): 209–39.

De Vries, Jan. *The Industrious Revolution: Consumer Behavior and the Household Economy, 1650 to the Present*. New York: Cambridge University Press, 2008.

Delbourgo, James. *Collecting the World: Hans Sloane and the Origins of the British Museum*. Cambridge, MA: Harvard University Press, 2017.

Delbourgo, James, and Nicholas Dew, eds. *Science and Empire in the Atlantic World*. New York: Routledge, 2008.

Deneweth, Heidi, and Patrick Wallis. "Households, Consumption and the Development of Medical Care in the Netherlands, 1650–1900." *Journal of Social History* 49, no. 3 (2016): 532–57.

Desan, Christine. *Making Money: Coin, Currency, and the Coming of Capitalism*. New York: Oxford University Press, 2014.

Digby, Anne. *Making a Medical Living: Doctors and Patients in the English Market for Medicine, 1720–1911*. New York: Cambridge University Press, 1994.

Dopson, Laurence. "State of London Chemists' Shops in the 18th and Early 19th Centuries." *Chemist and Druggist* 163 (June 25, 1955): 718–21.

Dorner, Zachary. "From Chelsea to Savannah: Medicines and Mercantilism in the Atlantic World." *Journal of British Studies* 58, no. 1 (January 2019): 28–57.

———. "Manufacturing Pharmaceuticals, Credit, and Empire in the Eighteenth-Century British Atlantic." PhD diss., Brown University, 2016.

———. "'No One Here Knows Half So Much of This Matter as Yourself': The Deployment of Expertise in Silvester Gardiner's Surgical, Druggist, and Land Speculation Networks, 1734–83." *William & Mary Quarterly* 72, no. 2 (2015): 287–322.

Drayton, Richard. *Nature's Government: Science, Imperial Britain and the "Improvement" of the World*. New Haven, CT: Yale University Press, 2000.

Duden, Barbara. *The Woman Beneath the Skin: A Doctor's Patients in Eighteenth-Century Germany*. Translated by Thomas Dunlap. Cambridge, MA: Harvard University Press, 1991.

Dugan, Holly. *The Ephemeral History of Perfume: Scent and Sense in Early Modern England*. Baltimore: Johns Hopkins University Press, 2011.

Duncan, Louis C. *Medical Men in the American Revolution, 1775–1783*. Carlisle Barracks, PA: Medical Field Service School, 1931.

Dunn, Richard S. "'Dreadful Idlers' in the Cane Fields: The Slave Labor Pattern on a Jamaican Sugar Estate, 1762–1831." In Solow and Engerman, *British Capitalism and Caribbean Slavery*, 163–90.

———. *Sugar and Slaves: The Rise of the Planter Class in the English West Indies, 1624–1713*. Chapel Hill: Published for the Institute of Early American History and Culture at Williamsburg, VA, by the University of North Carolina Press, 1972.

DuPlessis, Robert S. *The Material Atlantic: Clothing, Commerce, and Colonization in the Atlantic World, 1650–1800*. Cambridge: Cambridge University Press, 2016.

Durham, Harriet Frorer. *Caribbean Quakers*. Hollywood, FL: Dukane Press, 1972.

Eacott, Jonathan. *Selling Empire: India in the Making of Britain and America, 1600–1830*. Chapel Hill: Published for the Omohundro Institute of Early American History & Culture, Williamsburg, VA, by the University of North Carolina Press, 2016.

"Eighteenth-Century London Chemists." *Chemist and Druggist* 127 (August 14, 1937): 178–79.

Elliott, J. H. *Empires of the Atlantic World: Britain and Spain in America 1492–1830*. New Haven, CT: Yale University Press, 2006.

Eltis, David. *The Rise of African Slavery in the Americas*. New York: Cambridge University Press, 2000.

———. "The Volume and Structure of the Transatlantic Slave Trade: A Reassessment." *William & Mary Quarterly* 58, no. 1 (2001): 17–46.

Emmer, P. C., O. Pétré-Grenouilleau, and J. V. Roitman, eds. *A Deus Ex Machina Revisited: Atlantic Colonial Trade and European Economic Development*. Boston: Brill, 2006.

Epstein, S. R., and Maarten Prak, eds. *Guilds, Innovation, and the European Economy, 1400–1800*. New York: Cambridge University Press, 2008.

Estes, J. Worth. *Dictionary of Protopharmacology: Therapeutic Practices, 1700–1850*. Canton, MA: Science History Publications, 1990.

———. "Therapeutic Practice in Colonial New England." In [Colonial Society], *Medicine in Colonial Massachusetts*, 289–383.

Evans, Chris. "The Plantation Hoe: The Rise and Fall of an Atlantic Commodity, 1650–1850." *William & Mary Quarterly* 69, no. 1 (2012): 71–100.

Evans, Eric J. *William Pitt the Younger*. New York: Routledge, 2002.

Fayle, Joshua. *The Spitalfields Genius: The Story of William Allen* . . . London, 1884.

"February Meeting, 1898." *Proceedings of the Massachusetts Historical Society*, 2nd ser., 12 (1897–1899): 138–54.

Fett, Sharla M. *Working Cures: Healing, Health, and Power on Southern Slave Plantations*. Chapel Hill: University of North Carolina Press, 2002.

Fichter, James R. *So Great a Proffit: How the East Indies Trade Transformed Anglo-American Capitalism*. Cambridge, MA: Harvard University Press, 2010.

Findlen, Paula, ed. *Early Modern Things: Objects and Their Histories, 1500–1800*. New York: Routledge, 2013.

Finn, Margot C. *The Character of Credit: Personal Debt in English Culture, 1740–1914*. New York: Cambridge University Press, 2003.

Flexner, James Thomas. *The Traitor and the Spy: Benedict Arnold and John Andre*. New York: Harcourt, Brace, 1953.

Floud, Roderick, and Donald McCloskey, eds. *The Economic History of Britain Since 1700. Vol. 1, 1700–1860*. New York: Cambridge University Press, 1994.

Fogel, Robert William, and Stanley L. Engerman. *Time on the Cross: The Economics of American Negro Slavery*. Boston: Little, Brown, 1974.

"Forum: The Paper Technologies of Capitalism." *Technology and Culture* 58, no. 2 (April 2017): 487–569.

French, Roger, and Andrew Wear, eds. *The Medical Revolution of the Seventeenth Century*. New York: Cambridge University Press, 1989.

Frey, Sylvia R. *The British Soldier in America: A Social History of Military Life in the Revolutionary Period*. Austin: University of Texas Press, 1981.

Froide, Amy M. *Silent Partners: Women as Public Investors during Britain's Financial Revolution, 1690–1750*. New York: Oxford University Press, 2017.

Fuentes, Marisa J. *Dispossessed Lives: Enslaved Women, Violence, and the Archive*. Philadelphia: University of Pennsylvania Press, 2016.

Fumerton, Patricia. *Unsettled: The Culture of Mobility and the Working Poor in Early Modern England*. Chicago: University of Chicago Press, 2006.

Furber, Holden. *Rival Empires of Trade in the Orient, 1600–1800*. Minneapolis: University of Minnesota Press, 1976.

Füssel, Marian, Philip Knäble, and Nina Elsemann, eds. *Wirtschaft Und Wissen: Expertenkulturen Und Märkte vom 13. bis 18. Jahrhundert*. Gottingen: Vandenhoeck & Ruprecht, 2017.

Gauci, Perry. *Emporium of the World: The Merchants of London 1660–1800*. London: Hambledon Continuum, 2007.

———. *The Politics of Trade: The Overseas Merchant in State and Society, 1660–1720*. New York: Oxford University Press, 2001.

———, ed. *Regulating the British Economy, 1660–1850*. Burlington, VT: Ashgate, 2011.

Gentilcore, David. *Healers and Healing in Early Modern Italy*. Manchester: Manchester University Press, 1998.

Gevitz, Norman. "'Pray Let the Medicines Be Good': The New England Apothecary in the Seventeenth and Early Eighteenth Centuries." In Higby and Stroud, *Apothecaries and the Drug Trade*, 5–28.

Ghere, David L. "Mistranslations and Misinformation: Diplomacy on the Maine Frontier, 1725 to 1755." *American Indian Culture and Research Journal* 8, no. 4 (1984): 3–26.

Gibbs, F. W. "William Lewis, M.B., F.R.S. (1708–1781)." *Annals of Science* 8, no. 2 (1952): 122–51.

Gifford Jr., George E. "Botanic Remedies in Colonial Massachusetts, 1620–1820." In [Colonial Society], *Medicine in Colonial Massachusetts*, 263–88.

Golinski, Jan. "Chemistry." In Porter, *Cambridge History of Science*, 4:375–96.

Gómez, Pablo F. "The Circulation of Bodily Knowledge in the Seventeenth-Century Black Spanish Caribbean." *Social History of Medicine* 26, no. 3 (2013): 383–402.

———. *The Experiential Caribbean: Creating Knowledge and Healing in the Early Modern Atlantic*. Chapel Hill: University of North Carolina Press, 2017.

———. "Incommensurable Epistemologies?: The Atlantic Geography of Healing in the Early Modern Caribbean." *Small Axe* 18, no. 2 (2014): 95–107.

Gordon, Maurice Bear. *Æsculapius Comes to the Colonies: The Story of the Early Days of Medicine in the Thirteen Original Colonies*. Ventnor, NJ: Ventnor, 1949.

Gould, Eliga H., and Peter S. Onuf, eds. *Empire and Nation: The American Revolution in the Atlantic World*. Baltimore: Johns Hopkins University Press, 2005.

Granovetter, Mark S. "Economic Action and Social Structure: The Problem of Embeddedness." *American Journal of Sociology* 91, no. 3 (1985): 481–510.

———. "The Impact of Social Structure on Economic Outcomes." *Journal of Economic Perspectives* 19, no. 1 (2005): 33–50.

Grassby, Richard. *The Business Community of Seventeenth-Century England*. New York: Cambridge University Press, 1995.

Greif, Avner. "Institutions and Impersonal Exchange: From Communal to Individual Responsibility." *Journal of Institutional and Theoretical Economics* 158, no. 1 (March 2002): 168–204.

Griffenhagen, George. "Bartholemew Browne, Pharmaceutical-Chemist of Salem, Massachusetts, 1698–1704." *Essex Institute Historical Collections* 97 (1961): 19–30.

———. "Drug Supplies in the American Revolution." *Contributions from the Museum of History and Technology* [Smithsonian Institution] 225 (1963): 110–33.

———. *Medicine Tax Stamps Worldwide*. Milwaukee: American Topical Association, 1971.

Griffenhagen, George, and Mary Bogard. *History of Drug Containers and Their Labels*. Madison: American Institute of the History of Pharmacy, 1999.

Griffin, Patrick. *The Townshend Moment: The Making of Empire in Revolution in the Eighteenth Century*. New Haven, CT: Yale University Press, 2017.

Grove, Richard. *Green Imperialism: Colonial Expansion, Tropical Island Edens and the Origins of Environmentalism, 1600–1800*. New York: Cambridge University Press, 1995.

Haggerty, Sheryllynne. *The British-Atlantic Trading Community, 1760–1810: Men, Women, and the Distribution of Goods*. Boston: Brill, 2006.

———. "A Link in the Chain: Trade and the Transhipment of Knowledge in the Late Eighteenth Century." *International Journal of Maritime History* 14, no. 1 (2002): 157–72.

Hancock, David. "'Capital and Credit with Approved Security': Financial Markets in Montserrat and South Carolina, 1748–1775." *Business and Economic History* 23 (Winter 1994): 61–84.

———. *Citizens of the World: London Merchants and the Integration of the British Atlantic Community, 1735–1785*. New York: Cambridge University Press, 1995.

———. "'Domestic Bubbling': Eighteenth-Century London Merchants and Individual Investment in the Funds." *Economic History Review* 47, no. 4 (November 1994): 679–702.

———. *Oceans of Wine: Madeira and the Emergence of American Trade and Taste*. New Haven, CT: Yale University Press, 2009.

———. "'A World of Business to Do': William Freeman and the Foundations of England's Commercial Empire, 1645–1707." *William & Mary Quarterly* 57, no. 1 (2000): 3–34.

Handler, Jerome S., and Frederick W. Lange. *Plantation Slavery in Barbados: An Archaeological and Historical Investigation*. Cambridge, MA: Harvard University Press, 1978.

Hanson, John Wesley. *History of Gardiner, Pittston and West Gardiner, with a Sketch of the Kennebec Indians, & New Plymouth Purchase, Comprising Historical Matter from 1602 to 1852; with Genealogical Sketches of Many Families*. Gardiner, ME, 1852.

Harris, J. R. *Industrial Espionage and Technology Transfer: Britain and France in the Eighteenth Century*. Brookfield, VT: Ashgate, 1998.

Harris, Stephen J. "Long-Distance Corporations, Big Sciences, and the Geography of Knowledge." *Configurations* 6, no. 2 (1998): 269–304.

Harrison, Mark. *Climates and Constitutions: Health, Race, Environment and British Imperialism in India, 1600–1850*. New York: Oxford University Press, 1999.

———. "Disease and Medicine in the Armies of British India, 1750–1830: The Treatment of Fevers and the Emergence of Tropical Therapeutics." In Hudson, *British Military and Naval Medicine*, 87–120.

———. *Medicine in an Age of Commerce and Empire: Britain and Its Tropical Colonies, 1660–1830*. New York: Oxford University Press, 2010.

———. *Public Health in British India: Anglo-Indian Preventive Medicine, 1859–1914*. New York: Cambridge University Press, 1994.

Hartigan-O'Connor, Ellen. *The Ties That Buy: Women and Commerce in Revolutionary America*. Philadelphia: University of Pennsylvania Press, 2009.

Haulman, Kate. *The Politics of Fashion in Eighteenth-Century America*. Chapel Hill: University of North Carolina Press, 2011.

Haycock, David Boyd, and Patrick Wallis, eds. *Quackery and Commerce in Seventeenth-Century London: The Proprietary Medicine Business of Anthony Daffy*. London: Wellcome Trust Centre for the History of Medicine at UCL, 2005.

Higby, Gregory J., and Elaine C. Stroud, eds. *Apothecaries and the Drug Trade: Essays in Celebration of the Work of David L. Cowen*. Madison, WI: American Institute of the History of Pharmacy, 2001.

Higman, Barry W. *Slave Population and Economy in Jamaica, 1807–1834*. Kingston: University of the West Indies Press, 1995.

———. *Slave Populations of the British Caribbean, 1807–1834*. Baltimore: Johns Hopkins University Press, 1984.

Hogarth, Rana. *Medicalizing Blackness: Making Racial Difference in the Atlantic World, 1780–1840*. Chapel Hill: University of North Carolina Press, 2017.

Hoppit, Julian. *Britain's Political Economies: Parliament and Economic Life, 1660–1800*. New York: Cambridge University Press, 2017.

———. *Risk and Failure in English Business 1700–1800*. New York: Cambridge University Press, 1987.

Howard-Jones, N. "John Quincy, M.D. [d. 1722], Apothecary and Iatrophysical Writer." *Journal of the History of Medicine and Allied Sciences* 6, no. 2 (Spring 1951): 149–75.

Hudson, Briony, ed. *English Delftware Drug Jars: The Collection of the Museum of the Royal Pharmaceutical Society of Great Britain*. Chicago: Pharmaceutical Press, 2006.

Hudson, Geoffrey L., ed. *British Military and Naval Medicine, 1600–1830*. New York: Rodopi, 2007.

Hunt, Margaret R. *The Middling Sort: Commerce, Gender, and the Family in England, 1680–1780*. Berkeley: University of California Press, 1996.

Hunting, Penelope. *A History of the Society of Apothecaries*. London: Society of Apothecaries, 1998.

Hutson, J. Edward, ed. *On the Treatment and Management of the More Common West-India Diseases (1750–1802)*. Kingston: University of the West Indies Press, 2005.

Irving, Sarah. *Natural Science and the Origins of the British Empire*. London: Pickering & Chatto, 2008.

Jacob, Margaret C., and Matthew Kadane. "Missing, Now Found in the Eighteenth Century: Weber's Protestant Capitalist." *American Historical Review* 108, no. 1 (2003): 20–49.

Jarvis, Michael J. *In the Eye of All Trade: Bermuda, Bermudians, and the Maritime Atlantic World, 1680–1783*. Chapel Hill: Published for the Omohundro Institute of Early American History & Culture, Williamsburg, VA, by the University of North Carolina Press, 2010.

Jasanoff, Maya. *Edge of Empire: Lives, Culture, and Conquest in the East, 1750–1850*. New York: Alfred A. Knopf, 2005.

———. *Liberty's Exiles: American Loyalists in the Revolutionary World*. New York: Alfred A. Knopf, 2011.
Jenkins, Charles F. *Tortola: A Quaker Experiment of Long Ago in the Tropics*. London: Friends' Bookshop, 1923.
Jenner, Mark S. R., and Patrick Wallis, eds. *Medicine and the Market in England and Its Colonies, c. 1450–c. 1850*. New York: Palgrave Macmillan, 2007.
Jonsson, Fredrik Albritton. *Enlightenment's Frontier: The Scottish Highlands and the Origins of Environmentalism*. New Haven, CT: Yale University Press, 2013.
Karras, Alan. "The World of Alexander Johnston: The Creolization of Ambition, 1762–1787." *Historical Journal* 30, no. 1 (1987): 53–76.
Katz, Michael B. *In the Shadow of the Poorhouse: A Social History of Welfare in America*. 10th ed. New York: Basic Books, 1996.
Kennedy, William. *English Taxation 1640–1799*. New York: A. M. Kelley, 1964.
Kent, Stephen A. "The Quaker Ethic and the Fixed Price Policy: Max Weber and Beyond." *Sociological Inquiry* 53, no. 1 (1983): 16–32.
Kershaw, Gordon E. *The Kennebeck Proprietors, 1749–1775: "Gentlemen of Large Property & Judicious Men."* Portland: Maine Historical Society, 1975.
Kett, Joseph F. "Provincial Medical Practice in England 1730–1815." *Journal of the History of Medicine and Allied Sciences* 19, no. 1 (1964): 17–29.
Kimball, Eric. "'What have we to do with slavery?' New Englanders and the Slave Economies of the West Indies." In Beckert and Rockman, *Slavery's Capitalism*, 181–94.
King, Steven. "Accessing Drugs in the Eighteenth-Century Regions." In Curth, *From Physick to Pharmacology*, 49–78.
Klooster, Wim. "Inter-Imperial Smuggling in the Americas." In Bailyn and Denault, *Soundings in Atlantic History*, 141–80.
Koerner, Lisbet. *Linnaeus: Nature and Nation*. Cambridge, MA: Harvard University Press, 1999.
Kohler, Robert E. "Lab History: Reflections." *Isis* 99, no. 4 (2008): 761–68.
Kremers, Edward. "Historical Fragments No. 22 Two Invoices of 1785." *Journal of the American Pharmaceutical Association* 20, no, 7 (1931): 682–95.
———. "The Lititz Pharmacopoeia." *Badger Pharmacist* nos. 22–25 (June–December 1938): 1–70.
Krumbhaar, E. B. "Doctor William Hunter of Newport." *Annals of Surgery* 101, no. 1 (1935): 506–28.
Kupperman, Karen Ordahl. "Fear of Hot Climates in the Anglo-American Colonial Experience." *William & Mary Quarterly* 41, no. 2 (April 1984): 213–40.
Kuriyama, Shigehisa. *The Expressiveness of the Body and the Divergence of Greek and Chinese Medicine*. New York: Zone Books, 1999.
Lacey, Laurie. *Micmac Medicines: Remedies and Recollections*. Halifax, Nova Scotia: Nimbus, 1993.
Landes, Jordan. *London Quakers in the Trans-Atlantic World: The Creation of an Early Modern Community*. New York: Palgrave Macmillan, 2015.
Latour, Bruno. *We Have Never Been Modern*. Translated by Catherine Porter. Cambridge, MA: Harvard University Press, 1993.
Lawall, Charles H., and Millicent R. Lawall. "Revolutionary Account Book of Christopher, Jr., and Charles Marshall." *Journal of the American Pharmaceutical Association* 24, no. 4 (1935): 302–6.
Leamon, James S. *The Reverend Jacob Bailey Maine Loyalist: For God, King, Country, and for Self*. Amherst: University of Massachusetts Press, 2012.

———. *Revolution Downeast: The War for American Independence in Maine*. Amherst: University of Massachusetts Press, 1993.

Leavitt, Judith Walzer, and Ronald L. Numbers, eds. *Sickness and Health in America: Readings in the History of Medicine and Public Health*. 3rd ed. Madison: University of Wisconsin Press, 1997.

Leong, Elaine. "Making Medicines in the Early Modern Household." *Bulletin of the History of Medicine* 82, no. 1 (2008): 145–68.

Leong, Elaine, and Alisha Rankin, eds. *Secrets and Knowledge in Medicine and Science, 1500–1800*. Burlington, VT: Ashgate, 2011.

———, eds. *"Testing Drugs and Trying Cures." Special issue, Bulletin of the History of Medicine* 91, no. 2 (Summer 2017).

Leong, Elaine, and Sara Pennell. "Recipe Collections and the Currency of Medical Knowledge in the Early Modern 'Medical Marketplace.'" In Jenner and Wallis, *Medicine and the Market*, 133–52.

Levine, Philippa, ed. *Gender and Empire*. New York: Oxford University Press, 2004.

———. "Introduction: Why Gender and Empire?" In Levine, *Gender and Empire*, 1–13.

Lindberg, David C., and Robert S. Westman, eds. *Reappraisals of the Scientific Revolution*. New York: Cambridge University Press, 1990.

Lindemann, Mary. *Health and Healing in Eighteenth-Century Germany*. Baltimore: Johns Hopkins University Press, 1996.

Littleton, A. C. "Evolution of the Journal Entry." In Littleton and Yamey, *History of Accounting*, 223–35.

Littleton, A. C., and Basil S. Yamey, eds. *Studies in the History of Accounting*. New York: Arno Press, 1978.

Long, Pamela. *Openness, Secrecy, Authorship: Technical Arts and the Culture of Knowledge from Antiquity to the Renaissance*. Baltimore: Johns Hopkins University Press, 2001.

Loudon, Irvine. *Medical Care and the General Practitioner, 1750–1850*. New York: Oxford University Press, 1986.

Lubbock, Basil. *The Blackwall Frigates*. Glasgow: James Brown & Son, 1922.

Luyendijk-Elshout, A. M. "'Ontleedinge' (Anatomy) as Underlying Principle of Western Medicine in Japan." In Beukers et al., *Red-Hair Medicine*, 27–36.

Lyons, Clare A. *Sex among the Rabble: An Intimate History of Gender and Power in the Age of Revolution, Philadelphia, 1730–1830*. Chapel Hill: Published for the Omohundro Institute of Early American History & Culture, Williamsburg, VA, by the University of North Carolina Press, 2006.

MacLeod, Christine. "Accident or Design? George Ravenscroft's Patent and the Invention of Lead-Crystal Glass." *Technology & Culture* 28, no. 4 (1987): 776–803.

Madge, A. G. M. "Cookworthy-Chemist and Potter." *Pharmaceutical Historian* 23, no. 2 (1993): 3–4.

Maehle, Andreas-Holger. *Drugs on Trial: Experimental Pharmacology and Therapeutic Innovation in the Eighteenth Century*. Atlanta: Rodopi, 1999.

Margócy, Dániel. *Commercial Visions: Science, Trade, and Visual Culture in the Dutch Golden Age*. Chicago: University of Chicago Press, 2014.

Marshall, P. J. *East Indian Fortunes: The British in Bengal in the Eighteenth Century*. New York: Oxford University Press, 1976.

Martin, Ann Smart. *Buying into the World of Goods: Early Consumers in Backcountry Virginia*. Baltimore: Johns Hopkins University Press, 2005.

Martin, Bonnie. "Slavery's Invisible Engine: Mortgaging Human Property." *Journal of Southern History* 76, no. 4 (November 2010): 817–66.

Mathias, Peter. "Risk, Credit and Kinship in Early Modern Enterprise." In McCusker and Morgan, *Early Modern Atlantic Economy*, 15–35.

Matson, Cathy, ed. *Special issue, Journal of the Early Republic* 36, no. 4 (Winter 2016).

McCoy, Drew. *The Elusive Republic: Political Economy in Jeffersonian America*. Chapel Hill: Published for the Omohundro Institute of Early American History & Culture, Williamsburg, VA, by the University of North Carolina Press, 1980.

McCusker, John J. "The Current Value of English Exports, 1697 to 1800." *William & Mary Quarterly* 28, no. 4 (1971): 607–28.

———. *Money and Exchange in Europe and America, 1600–1775: A Handbook*. Chapel Hill: Published for the Omohundro Institute of Early American History & Culture, Williamsburg, VA, by the University of North Carolina Press, 1978.

———. "The Tonnage of Ships Engaged in British Colonial Trade during the Eighteenth Century." *Research in Economic History* 6 (1981): 73–105.

———. "Weights and Measures in the Colonial Sugar Trade: The Gallon and the Pound and Their International Equivalents." *William & Mary Quarterly* 30, no. 4 (1973): 599–624.

McCusker, John J., and Russell R. Menard. *The Economy of British America, 1607–1789*. Chapel Hill: Published for the Omohundro Institute of Early American History & Culture, Williamsburg, VA, by the University of North Carolina Press, 1985.

McCusker, John J., and Kenneth Morgan, eds. *The Early Modern Atlantic Economy*. New York: Cambridge University Press, 2000.

McKendrick, Neil, John Brewer, and J. H. Plumb, eds. *The Birth of a Consumer Society: The Commercialization of Eighteenth-Century England*. Bloomington: Indiana University Press, 1982.

McNeill, J. R. *Mosquito Empires: Ecology and War in the Greater Caribbean, 1620–1914*. New York: Cambridge University Press, 2010.

Merritt, Jane T. "Tea Trade, Consumption, and the Republican Paradox in Prerevolutionary Philadelphia." *Pennsylvania Magazine of History and Biography* 128 (2004): 117–48.

Milford, T. A. *The Gardiners of Massachusetts: Provincial Ambition and the British-American Career*. Durham: University of New Hampshire Press, 2005.

Mintz, Sidney. *Sweetness and Power: The Place of Sugar in Modern History*. New York: Viking, 1985.

Mokyr, Joel. *A Culture of Growth: The Origins of the Modern Economy*. Princeton, NJ: Princeton University Press, 2016.

———. *The Enlightened Economy: An Economic History of Britain 1700–1850*. New Haven, CT: Yale University Press, 2009.

Monahan, Erika. "Locating Rhubarb: Early Modernity's Relevant Obscurity." In Findlen, *Early Modern Things*, 227–51.

Morgan, Jennifer L. *Laboring Women: Reproduction and Gender in New World Slavery*. Philadelphia: University of Pennsylvania Press, 2004.

Morgan, Kenneth. "Business Networks in the British Export Trade to North America, 1750–1800." In McCusker and Morgan, *Early Modern Atlantic Economy*, 36–62.

Morgan, Philip. *Slave Counterpoint: Black Culture in the Eighteenth-Century Chesapeake and Low-Country*. Chapel Hill: Published for the Omohundro Institute of Early American History & Culture, Williamsburg, VA, by the University of North Carolina Press, 1998.

Morris, Gerald E., ed. *The Maine Bicentennial Atlas: An Historical Survey*. Portland: Maine Historical Society, 1976.

Morris, John S. "Silvanus Bevan the 'Quaker FRS' (1691–1765) Apothecary with a Note on His Contribution to the Founding of the Pharmaceutical Company Allen and Hanbury." *Journal of Medical Biography* 19, no. 1 (2011): 2–4.

Mortimer, Ian. *The Dying and the Doctors: The Medical Revolution in Seventeenth-Century England*. Rochester, NY: Boydell Press, 2009.

———. "The Rural Medical Marketplace in Southern England c. 1570–1720." In Jenner and Wallis, *Medicine and the Market*, 69–87.

Morton, Thomas G. *The History of the Pennsylvania Hospital 1751–1895*. Rev. ed. Philadelphia, 1897.

Mui, Hoh-Cheung, and Lorna H. Mui. "'Trends in Eighteenth-Century Smuggling' Reconsidered." *Economic History Review* 28, no. 1 (1975): 28–43.

———. "William Pitt and the Enforcement of the Commutation Act, 1784–1788." *English Historical Review* 76, no. 300 (1961): 447–65.

Mukharji, Projit Bihari. *Nationalizing the Body: The Medical Market, Print and Daktari Medicine*. London: Anthem Press, 2009.

———. "Vishalyakarani as 'Eupatonum Ayapana': Retrobotanizing, Embedded Traditions, and Multiple Historicities of Plants in Colonial Bengal, 1890–1940." *Journal of Asian Studies* 71, no. 1 (2014): 65–87.

Muldrew, Craig. *The Economy of Obligation: The Culture of Credit and Social Relations in Early Modern England*. New York: Palgrave, 1998.

Murphy, Anne L. *The Origins of English Financial Markets: Investment and Speculation before the South Sea Bubble*. New York: Cambridge University Press, 2009.

Murphy, Kathleen S. "Collecting Slave Traders: James Petiver, Natural History, and the British Slave Trade." *William & Mary Quarterly* 70, no. 4 (2013): 637–70.

Mutschler, Ben. "Illness in the 'Social Credit' and 'Money' Economies of Eighteenth-Century New England." In Jenner and Wallis, *Medicine and the Market*, 175–95.

Nagy, Doreen Evenden. *Popular Medicine in Seventeenth-Century England*. Bowling Green, OH: Bowling Green State University Popular Press, 1988.

Nash, R. C. "The Organization of Trade and Finance in the British Atlantic Economy, 1600–1830." In Coclanis, *Atlantic Economy*, 95–151.

Nelson, George H. "Contraband Trade under the Asiento, 1730–1739." *American Historical Review* 51, no. 1 (October 1945): 55–67.

Newman, Brooke N. *A Dark Inheritance: Blood, Race, and Sex in Colonial Jamaica*. New Haven, CT: Yale University Press, 2018.

Newman, William R. "From Alchemy to 'Chymistry.'" In Daston and Park, *Cambridge History of Science*, 3:497–517.

Newson, Linda A. *Making Medicines in Early Colonial Lima, Peru: Apothecaries, Science and Society*. Boston: Brill, 2017.

Norton, Marcy. *Sacred Gifts, Profane Pleasures: A History of Tobacco and Chocolate in the Atlantic World*. Ithaca, NY: Cornell University Press, 2008.

Numbers, Ronald L., ed. *Medicine in the New World: New Spain, New France, and New England*. Knoxville: University of Tennessee Press, 1987.

Oaks, Robert F. "Philadelphia Merchants and the Origins of American Independence." *Proceedings of the American Philosophical Society* 121 (1977): 407–36.

O'Brien, Patrick. "Central Government and the Economy, 1688–1815." In Floud and McCloskey, *Economic History of Britain*, 205–41.

Oliver, Vere Langford. *The History of the Island of Antigua*. London, 1894.

Olson, Alison Gilbert. *Making the Empire Work: London and American Interest Groups, 1690–1790*. Cambridge, MA: Harvard University Press, 1992.

O'Malley, Gregory E. *Final Passages: The Intercolonial Slave Trade of British America, 1619–1807*. Chapel Hill: Published for the Omohundro Institute of Early American History & Culture, Williamsburg, VA, by the University of North Carolina Press, 2014.

O'Neill, Lindsay. *The Opened Letter: Networking in the Early Modern British World*. Philadelphia: University of Pennsylvania Press, 2015.

Osseo-Asare, Abena Dove. *Bitter Roots: The Search for Healing Plants in Africa*. Chicago: University of Chicago Press, 2014.

Owens, Deirdre Cooper. *Medical Bondage: Race, Gender, and the Origins of American Gynecology*. Athens: University of Georgia Press, 2017.

Owens, Sarah E., and Jane E. Mangan, eds. *Women of the Iberian Atlantic*. Baton Rouge: Louisiana State University Press, 2012.

Packard, Francis R. "William Cheselden, Some of His Contemporaries, and Their American Pupils." *Annals of Medical History* 9 (1937): 533–48.

Palmer, Colin. *Human Cargoes: The British Slave Trade to Spanish America, 1700–1739*. Urbana: University of Illinois Press, 1981.

Pares, Richard. *Merchants and Planters*. Cambridge: Published for the Economic History Review at the University Press, 1960.

Parsons, Christopher M. "The Natural History of Colonial Science: Joseph-François Lafitau's Discovery of Ginseng and Its Afterlives." *William & Mary Quarterly* 73, no. 1 (2016): 37–72.

Parsons, Usher. *Sketches of Rhode Island Physicians, Deceased Prior to 1850*. Providence, 1859.

Patterson, Orlando. *The Sociology of Slavery: An Analysis of the Origins, Development and Structure of Negro Slave Society in Jamaica*. London: MacGibbon & Kee, 1967.

Paugh, Katherine. *The Politics of Reproduction: Race, Medicine, and Fertility in the Age of Abolition*. Oxford: Oxford University Press, 2017.

Pelling, Margaret. *The Common Lot: Sickness, Medical Occupations and the Urban Poor in Early Modern England*. New York: Longman, 1998.

———. *Medical Conflicts in Early Modern London: Patronage, Physicians and Irregular Practitioners, 1550–1640*. New York: Oxford University Press, 2003.

Pennell, Sara. "Perfecting Practice? Women, Manuscript Recipes and Knowledge in Early Modern England." In Burke and Gibson, *Early Modern Women's Manuscript Writing*, 237–58.

Peterson, Mark A. "The World in a Shilling: Silver Coins and the Challenge of Political Economy in the Early Modern Atlantic World." In Findlen, *Early Modern Things*, 252–71.

Pettegree, Andrew. *The Invention of News: How the World Came to Know About Itself*. New Haven, CT: Yale University Press, 2014.

"A Philadelphia Business of the Olden Time." *Druggists Circular* 51 (January 1907): 179.

Piersen, William D. *Black Yankees: The Development of an Afro-American Subculture in Eighteenth-Century New England*. Amherst: University of Massachusetts Press, 1988.

Platt, Virginia Bever. "'And Don't Forget the Guinea Voyage': The Slave Trade of Aaron Lopez of Newport." *William & Mary Quarterly* 32, no. 4 (1975): 601–18.

Pocock, J. G. A. *Virtue, Commerce, and History: Essays on Political Thought and History, Chiefly in the Eighteenth Century*. New York: Cambridge University Press, 1985.

Pollock, Linda. *With Faith and Physic: The Life of a Tudor Gentlewoman Lady Grace Mildmay 1552–1620*. London: Collins & Brown, 1993.

Pomeranz, Kenneth, and Steven Topik. *The World that Trade Created: Society, Culture, and the World Economy 1400 to the Present*. 2nd ed. Armonk, NY: M. E. Sharpe, 2006.

Porter, Roy. *Bodies Politic: Disease, Death, and Doctors in Britain, 1650–1900*. Ithaca, NY: Cornell University Press, 2001.

———, ed. *The Cambridge History of Science. Vol. 4, Eighteenth-Century Science*. New York: Cambridge University Press, 2003.

———. *Disease, Medicine, and Society in England, 1550–1860*. 2nd ed. Basingstoke: Macmillan, 1993.

———. *Health for Sale: Quackery in England 1660–1850*. Manchester: Manchester University Press, 1989.

Porter, Roy, and Dorothy Porter. "The Rise of the English Drugs Industry: The Role of Thomas Corbyn." *Medical History* 33, no. 3 (1989): 277–95.

Pountney, W. J. *Old Bristol Potteries, Being an Account of the Old Potters and Potteries of Bristol and Brislington, between 1650 and 1850*. New York: E. P. Dutton, 1920.

Poynter, F. N. L., ed. *The Evolution of Pharmacy in Britain*. Springfield, IL: Charles C. Thomas, 1965.

Price, Jacob M. *Capital and Credit in British Overseas Trade: The View from the Chesapeake, 1700–1776*. Cambridge, MA: Harvard University Press, 1980.

———. "Credit in the Slave Trade and Plantation Economies." In Solow, *Rise of the Atlantic System*, 293–340.

———. "What Did Merchants Do? Reflections on British Overseas Trade, 1660–1790." *Journal of Economic History* 49, no. 2 (1989): 267–84.

Price, Richard, and Sally Price, eds. *Stedman's Surinam: Life in an Eighteenth-Century Slave Society*. Baltimore: Johns Hopkins University Press, 1992.

Proceedings of the XXIII International Congress of the History of Medicine. London: Wellcome Institute for the History of Medicine, 1972.

Raj, Kapil. *Relocating Modern Science: Circulation and the Construction of Knowledge in South Asia and Europe, 1650–1900*. New York: Palgrave Macmillan, 2007.

Rankin, Alisha. *Panaceia's Daughters: Noblewomen as Healers in Early Modern Germany*. Chicago: University of Chicago Press, 2013.

Rappaport, Erika. *A Thirst for Empire: How Tea Shaped the Modern World*. Princeton, NJ: Princeton University Press, 2017.

Reddy, William. *The Rise of Market Culture: The Textile Trade and French Society, 1750–1900*. New York: Cambridge University Press, 1984.

Reiser, Stanley Joel. *Medicine and the Reign of Technology*. New York: Cambridge University Press, 1978.

"Retail Pharmacy over One Hundred Years." *Pharmaceutical Journal* 141 (April 12, 1941): 130–63.

Richmond, Lesley, Julie Stevenson, and Alison Turton, eds. *The Pharmaceutical Industry: A Guide to Historical Records*. Burlington, VT: Ashgate, 2003.

Rivest, Justin. "Testing Drugs and Attesting Cures: Pharmaceutical Monopolies and Military Contracts in Eighteenth-Century France." *Bulletin of the History of Medicine* 91, no. 2 (Summer 2017): 362–90.

Roberts, Justin. "Working between the Lines: Labor and Agriculture on Two Barbadian Sugar Plantations, 1796–97." *William & Mary Quarterly* 63, no. 3 (2006): 551–86.

Roberts, R. S. "The Early History of the Import of Drugs into Britain." In Poynter, *Evolution of Pharmacy*, 165–85.

Roberts, Strother E. "Pines, Profits, and Popular Politics: Responses to the White Pine Acts in the Colonial Connecticut River Valley." *New England Quarterly* 83, no. 1 (March 2010): 73–101.

Roberts, William I., III. "American Potash Manufacture Before the American Revolution." *Proceedings of the American Philosophical Society* 116, no. 5 (October 1972): 383–95.

Rockman, Seth. *Scraping By: Wage Labor, Slavery, and Survival in Early Baltimore*. Baltimore: Johns Hopkins University Press, 2009.

Ross, Lester A. *Archaeological Metrology: English, American, and Canadian Systems of Weights and Measures for North American Historical Archaeology*. Ottawa, Ontario: Parks Canada, 1983.

Rutten, A. M. G. *Blue Ships: Dutch Ocean Crossing with Multifunctional Drugs and Spices in the Eighteenth Century*. Translated by J. Wormer. Rotterdam: Erasmus, 2008.

Satia, Priya. *Empire of Guns: The Violent Making of the Industrial Revolution*. New York: Penguin Press, 2018.

Savitt, Todd L. *Medicine and Slavery: The Diseases and Health Care of Blacks in Antebellum Virginia*. Urbana: University of Illinois Press, 1978.

Saxine, Ian. "The Performance of Peace: Indians, Speculators, and the Politics of Property on the Maine Frontier, 1735–1737." *New England Quarterly* 87, no. 3 (2014): 379–411.

Schiebinger, Londa. *The Mind Has No Sex? Women in the Origins of Modern Science*. Cambridge, MA: Harvard University Press, 1991.

———. *Plants and Empire: Colonial Bioprospecting in the Atlantic World*. Cambridge, MA: Harvard University Press, 2004.

———. *Secret Cures of Slaves: People, Plants, and Medicine in the Eighteenth-Century Atlantic World*. Stanford, CA: Stanford University Press, 2017.

Schumpeter, Elizabeth Boody. *English Overseas Trade Statistics 1697–1808*. New York: Oxford University Press, 1960.

Schwartz, Stuart B., ed. *Tropical Babylons: Sugar and the Making of the Atlantic World, 1450–1680*. Chapel Hill: University of North Carolina Press, 2004.

Schwartzkopf, Stacey, and Kathryn E. Sampeck, eds. *Substance and Seduction: Ingested Commodities in Early Modern Mesoamerica*. Austin: University of Texas Press, 2017.

Schwarz, Philip. *Twice Condemned: Slaves and the Criminal Laws of Virginia, 1705–1865*. Baton Rouge: Louisiana State University Press, 1988.

Seth, Suman. *Difference and Disease: Medicine, Race, and the Eighteenth-Century British Empire*. New York: Cambridge University Press, 2018.

Seybolt, Robert Francis. "Lithotomies Performed by Dr. Gardiner, of Boston, 1738 and 1741." *New England Journal of Medicine* 202, no. 3 (1930): 109.

Shammas, Carole. "How Self-Sufficient Was Early America?" *Journal of Interdisciplinary History* 13, no. 2 (1982): 247–72.

———. *The Pre-Industrial Consumer in England and America*. New York: Oxford University Press, 1990.

———. "The Revolutionary Impact of European Demand for Tropical Goods." In McCusker and Morgan, *Early Modern Atlantic Economy*, 163–85.

Shaw, James E., and Evelyn S. Welch. *Making and Marketing Medicine in Renaissance Florence*. New York: Rodopi, 2011.

Shepherd, Verene A., ed. *Slavery Without Sugar: Diversity in Caribbean Economy and Society Since the 17th Century*. Gainesville: University Press of Florida, 2002.

Shepherd, Verene A., and Kathleen E. A. Monteith. "Pen-Keepers and Coffee Farmers in a Sugar-Plantation Society." In Shepherd, *Slavery Without Sugar*, 82–101.

Sheridan, Richard B. "The British Credit Crisis of 1772 and the American Colonies." *Journal of Economic History* 20, no. 2 (1960): 161–86.

———. "The Doctor and the Buccaneer: Sir Hans Sloane's Case History of Sir Henry Morgan, Jamaica, 1688." *Journal of the History of Medicine and Allied Sciences* 41, no. 1 (1986): 76–87.

———. *Doctors and Slaves: A Medical and Demographic History of Slavery in the British West Indies, 1680–1834*. New York: Cambridge University Press, 1985.

———. "Samuel Martin, Innovating Sugar Planter of Antigua 1750–1776." *Agricultural History* 34, no. 3 (1960): 126–39.

———. "Simon Taylor, Sugar Tycoon of Jamaica, 1740–1813." *Agricultural History* 45, no. 4 (1971): 285–96.

———. *Sugar and Slavery: An Economic History of the British West Indies, 1623–1775*. Baltimore: Johns Hopkins University Press, 1973.

Shryock, Richard H. *Medicine and Society in America, 1660–1860*. Ithaca, NY: Great Seal Book, 1962.

Simmons, Anna. "Medicines, Monopolies and Mortars: The Chemical Laboratory and Pharmaceutical Trade at the Society of Apothecaries in the Eighteenth Century." *Ambix* 53, no. 3 (2006): 221–36.

———. "Stills, Status, Stocks and Science: The Laboratories at Apothecaries' Hall in the Nineteenth Century." *Ambix* 61, no. 2 (2014): 141–61.

Siraisi, Nancy G. *Medieval and Early Renaissance Medicine: An Introduction to Knowledge and Practice*. Chicago: University of Chicago Press, 1990.

Sivin, Nathan. "William Lewis (1708–1781) as a Chemist." *Chymia* 8 (1962): 63–88.

Slavery and Justice: Report of the Brown University Steering Committee on Slavery and Justice. Providence, RI: Brown University, 2006.

Smallwood, Stephanie. *Saltwater Slavery: A Middle Passage from Africa to American Diaspora*. Cambridge, MA: Harvard University Press, 2007.

Smith, Kate. "Amidst Things: New Histories of Commodities, Capital, and Consumption." *Historical Journal* 61, no. 3 (2018): 841–61.

Smith, Pamela H. *The Body of the Artisan: Art and Experience in the Scientific Revolution*. Chicago: University of Chicago Press, 2004.

———. "Laboratories." In Daston and Park, *Cambridge History of Science*, 3:290–305.

Smith, S. D. *Slavery, Family, and Gentry Capitalism in the British Atlantic: The World of the Lascelles, 1648–1834*. New York: Cambridge University Press, 2006.

Smoak, Katherine. "The Weight of Necessity: Counterfeit Coins in the British Atlantic World, Circa 1760–1800." *William & Mary Quarterly* 74, no. 3 (2017): 467–502.

Soll, Jacob. "From Note-Taking to Data Banks: Personal and Institutional Information Management in Early Modern Europe." *Intellectual History Review* 20, no. 3 (2010): 355–75.

Solow, Barbara L., ed. *Slavery and the Rise of the Atlantic System*. New York: Cambridge University Press, 1991.

Solow, Barbara L., and Stanley L. Engerman, eds. *British Capitalism and Caribbean Slavery: The Legacy of Eric Williams*. New York: Cambridge University Press, 1987.

Sonnedecker, Glenn, ed. *Kremers and Urdang's History of Pharmacy*. 4th ed. Philadelphia: J. B. Lippincott, 1976.

Sperling, John G. *The South Sea Company: An Historical Essay and Bibliographical Finding List*. Boston: Baker Library, Harvard Graduate School of Business Administration, 1962.

Spillers, Hortense. *Black, White, and in Color: Essays on American Literature and Culture*. Chicago: University of Chicago Press, 2003.

Stander, S. "Transatlantic Trade in Pharmaceuticals During the Industrial Revolution." *Bulletin of the History of Medicine* 43, no. 4 (1969): 326–43.

Stanley, Bill. *Historic Sites of Norwich from the Beginning*. Norwich, CT: self-published, 2009.

Stark, James H. *The Loyalists of Massachusetts and the Other Side of the American Revolution*. Boston: W. B. Clarke, 1907.

Starr, Paul. *The Social Transformation of American Medicine*. New York: Basic Books, 1982.

Steele, Ian K. *Atlantic Merchant-Apothecary: Letters of Joseph Cruttenden 1710–1717*. Toronto: University of Toronto Press, 1977.

———. *The English Atlantic 1675–1740: An Exploration of Communication and Community*. New York: Oxford University Press, 1986.

Stern, Philip J. *The Company-State: Corporate Sovereignty and the Early Modern Foundations of the British Empire in India*. New York: Oxford University Press, 2011.

Stone, Lawrence, and Jeanne C. Fawtier Stone. *An Open Elite? England, 1540–1880*. New York: Oxford University Press, 1984.

Styles, John. "Involuntary Consumers? Servants and Their Clothes in Eighteenth-Century England." *Textile History* 33, no. 1 (2002): 9–21.

Sumner, James. "Status, Scale and Secret Ingredients: The Retrospective Invention of London Porter." *History & Technology* 24, no. 3 (September 2008): 289–306.

Sweet, James. *Domingo Álvares, African Healing, and the Intellectual History of the Atlantic World*. Chapel Hill: University of North Carolina Press, 2011.

Szechi, Daniel, and Geoffrey Holmes. *The Age of Oligarchy: Pre-Industrial Britain 1722–1783*. New York: Longman, 1993.

Taylor, Alan. *American Colonies: The Settling of North America*. New York: Viking Press, 2001.

———. "'A Kind of Warr': The Contest for Land on the Northeastern Frontier, 1750–1820." *William & Mary Quarterly* 46, no. 1 (1989): 3–26.

———. *Liberty Men and Great Proprietors: The Revolutionary Settlement on the Maine Frontier, 1760–1820*. Chapel Hill: Published for the Omohundro Institute of Early American History & Culture, Williamsburg, VA, by the University of North Carolina Press, 1990.

Teigen, Philip M. "Taste and Quality in 15th- and 16th-Century Galenic Pharmacology." *Pharmacy in History* 29, no. 2 (1987): 60–68.

Thayer, Theodore. *Israel Pemberton, King of the Quakers*. Philadelphia: Historical Society of Pennsylvania, 1943.

Tolles, Frederick Barnes. *Meeting House and Counting House: The Quaker Merchants of Colonial Philadelphia, 1682–1763*. New York: W. W. Norton, 1963.

"Tombs on Plantations." *Journal of the Barbados Museum and Historical Society* 1, no. 1 (1933): 26.

Tomes, Nancy. *Remaking the American Patient: How Madison Avenue and Modern Medicine Turned Patients into Consumers*. Chapel Hill: University of North Carolina Press, 2016.

Topik, Steven. "Historicizing Commodity Chains: Five Hundred Years of the Global Coffee Commodity Chain." In Bair, *Frontiers of Commodity Chain Research*, 37–62.

Trivellato, Francesca. *The Familiarity of Strangers: The Sephardic Diaspora, Livorno, and Cross-Cultural Trade in the Early Modern Period*. New Haven, CT: Yale University Press, 2009.

Turner, Sasha. *Contested Bodies: Pregnancy, Childrearing, and Slavery in Jamaica*. Philadelphia: University of Pennsylvania Press, 2017.

Ulrich, Laurel Thatcher. "'The Living Mother of a Living Child': Midwifery and Mortality in Post-Revolutionary New England." *William & Mary Quarterly* 46, no. 1 (1989): 27–48.

———. *A Midwife's Tale: The Life of Martha Ballard, Based on Her Diary, 1785–1812*. New York: Alfred A. Knopf, 1990.

Valenze, Deborah. *The Social Life of Money in the English Past*. New York: Cambridge University Press, 2006.

Valeri, Mark. *Heavenly Merchandize: How Religion Shaped Commerce in Puritan America*. Princeton, NJ: Princeton University Press, 2010.

Van Horn, Jennifer. *The Power of Objects in Eighteenth-Century British America*. Chapel Hill: Published for the Omohundro Institute of Early American History & Culture, Williamsburg, VA, by the University of North Carolina Press, 2017.

Vaughan, Megan. *Curing Their Ills: Colonial Power and African Illness*. Stanford, CA: Stanford University Press, 1991.

Veluwenkamp, J. W., and Joost Veenstra. "Early Modern English Merchant Colonies: Contexts and Functions." In Zakharov, Harlaftis, and Katsiardi-Hering, *Merchant Colonies*, 11–30.

Vernon, James. *Distant Strangers: How Britain Became Modern*. Berkeley: University of California Press, 2014.

Voeks, Robert. "African Medicine and Magic in the Americas." *Geographical Review* 83, no. 1 (1993): 66–78.

———. *Sacred Leaves of Candomble: African Magic, Medicine, and Religion in Brazil*. Austin: University of Texas Press, 1997.

Walker, Geoffrey. *Spanish Politics and Imperial Trade, 1700–1789*. New York: Macmillan, 1979.

Walker, Timothy D. "Remedies from the Carreira da India: Asian Influences on Portuguese Medicine During the Age of Enlightenment." *Portuguese Studies Review* 9 (2001): 170–93.

———. "The Role and Practices of the Female Folk Healer in the Early Modern Portuguese Atlantic World." In Owens and Mangan, *Women of the Iberian Atlantic*, 148–73.

Wallis, Patrick. "Competition and Cooperation in the Early Modern Medical Economy." In Jenner and Wallis, *Medicine and the Market*, 47–68.

———. "Consumption, Retailing, and Medicine in Early-Modern London." *Economic History Review* 61, no. 1 (2008): 26–53.

———. "Exotic Drugs and English Medicine: England's Drug Trade, c. 1550–c. 1800." *Social History of Medicine* 25, no. 1 (2012): 20–46.

———. "Introduction: The Growth of the Early Modern Medical Economy." *Journal of Social History* 49, no. 3 (2016): 477–82.

Wallis, Patrick, and Teerapa Pirohakul. "Medical Revolutions? The Growth of Medicine in England, 1660–1800." *Journal of Social History* 49, no. 3 (2016): 510–31.

Wallis, Peter John, R. V. Wallis, J. G. L. Burnby, and T. D. Whittet. *Eighteenth Century Medics: Subscriptions, Licenses, Apprenticeships*. Newcastle: Project for Historical Biobibliography, 1988.

Warden, G. B. "The Medical Profession in Colonial Boston." In [Colonial Society], *Medicine in Colonial Massachusetts*, 145–57.

Warren, Wendy. *New England Bound: Slavery and Colonization in Early America*. New York: Liveright, 2016.

Warren, William. "The Oldest Pharmacy in London." *Chemist and Druggist* 62 (January 31, 1903): 187–93.

Watson, G. M. "Some Eighteenth-Century Trading Accounts." In Poynter, *Evolution of Pharmacy*, 45–77.

Watters, David R. "A Turlington Balsam Phail from Montserrat, West Indies: Genuine or Counterfeit?" *Historical Archaeology* 15, no. 1 (1981): 105–8.

Wear, Andrew. *Knowledge and Practice in English Medicine, 1550–1680*. New York: Cambridge University Press, 2000.

———. "Medical Practice in Late Seventeenth- and Early Eighteenth-Century England: Continuity and Union." In French and Wear, *Medical Revolution of the Seventeenth Century*, 294–320.

Weatherill, Lorna. *Consumer Behavior and Material Culture in Britain, 1660–1760*. New York: Routledge, 1988.

Weaver, Karol K. *Medical Revolutionaries: The Enslaved Healers of Eighteenth-Century Saint Domingue*. Urbana: University of Illinois Press, 2006.

Webster, Colin. "Heuristic Medicine: The Methodists and Metalepsis." *Isis* 106, no. 3 (September 2015): 657–68.

Weisser, Olivia. *Ill Composed: Sickness, Gender, and Belief in Early Modern England*. New Haven, CT: Yale University Press, 2015.

Wennerlind, Carl. *Casualties of Credit: The English Financial Revolution, 1620–1720*. Cambridge, MA: Harvard University Press, 2011.

Whittet, T. D., and Juanita G. L. Burnby. "The Firm of Corbyn and Stacey." *Pharmaceutical Journal* (January 9, 1982): 42–48.

Wilson, Renate. *Pious Traders in Medicine: A German Pharmaceutical Network in Eighteenth-Century North America*. University Park: Pennsylvania State University Press, 2000.

———. "Trading in Drugs Through Philadelphia in the Eighteenth Century: A Transatlantic Enterprise." *Social History of Medicine* 26, no. 3 (August 2013): 352–63.

Winterbottom, Anna. "Of the China Root: A Case Study of the Early Modern Circulation of *Materia Medica*." *Social History of Medicine* 28, no. 1 (2015): 22–44.

Wood, Betty, and T. R. Clayton. "Slave Birth, Death and Disease on Golden Grove Plantation, Jamaica, 1765–1810." *Slavery & Abolition* 6, no. 2 (1985): 99–121.

Wood, Betty, and Martin Lynn, eds. *Travel, Trade and Power in the Atlantic, 1765–1884*. New York: Cambridge University Press, 2003.

Wood, Gordon S. *The Radicalism of the American Revolution*. New York: Alfred A. Knopf, 1993.

Wood, Nicholas. "A Time of Transition: Thomas Mayleigh—Pharmacist or Physician?" *Pharmaceutical Historian* 40, no. 2 (June 2010): 30–39.

Worling, Peter M. "Pharmacy in the Early Modern World, 1617 to 1841 AD." In Anderson, *Making Medicines*, 57–76.

Wright, Conrad Edick, and Katheryn P. Viens, eds. *Entrepreneurs: The Boston Business Community, 1700–1850*. Boston: Massachusetts Historical Society, 1997.

Wulf, Andrea. *The Brother Gardeners: Botany, Empire and the Birth of an Obsession*. London: Windmill Books, 2009.

Young, James Harvey. *The Toadstool Millionaires: A Social History of Patent Medicines in America before Federal Regulation*. Princeton, NJ: Princeton University Press, 1961.

Zahedieh, Nuala. *The Capital and the Colonies: London and the Atlantic Economy, 1660–1700*. New York: Cambridge University Press, 2010.

———. "Making Mercantilism Work: London Merchants and Atlantic Trade in the Seventeenth Century." *Transactions of the Royal Historical Society* 9 (1999): 143–58.

Zakharov, Victor N., Gelina Harlaftis, and Olga Katsiardi-Hering, eds. *Merchant Colonies in the Early Modern Period*. London: Pickering & Chatto, 2012.

Zebroski, Bob. *A Brief History of Pharmacy: Humanity's Search for Wellness*. New York: Routledge, 2016.

INDEX

Page numbers in italics refer to figures.